Wind Power Integration into Power Systems

Wind Power Integration into Power Systems: Stability and Control Aspects

Editors

Lasantha Meegahapola
Siqi Bu

MDPI • Basel • Beijing • Wuhan • Barcelona • Belgrade • Manchester • Tokyo • Cluj • Tianjin

Editors
Lasantha Meegahapola
School of Engineering
RMIT University
Melbourne
Australia

Siqi Bu
Department of Electrical
Engineering
The Hong Kong Polytechnic
University (PolyU)
Kowloon, Hong Kong
China

Editorial Office
MDPI
St. Alban-Anlage 66
4052 Basel, Switzerland

This is a reprint of articles from the Special Issue published online in the open access journal *Energies* (ISSN 1996-1073) (available at: www.mdpi.com/journal/energies/special_issues/WPI_PS).

For citation purposes, cite each article independently as indicated on the article page online and as indicated below:

LastName, A.A.; LastName, B.B.; LastName, C.C. Article Title. *Journal Name* **Year**, *Volume Number*, Page Range.

ISBN 978-3-0365-1610-3 (Hbk)
ISBN 978-3-0365-1609-7 (PDF)

© 2021 by the authors. Articles in this book are Open Access and distributed under the Creative Commons Attribution (CC BY) license, which allows users to download, copy and build upon published articles, as long as the author and publisher are properly credited, which ensures maximum dissemination and a wider impact of our publications.

The book as a whole is distributed by MDPI under the terms and conditions of the Creative Commons license CC BY-NC-ND.

Contents

About the Editors vii

Preface to "Wind Power Integration into Power Systems: Stability and Control Aspects" . . . ix

Lasantha Meegahapola and Siqi Bu
Special Issue: "Wind Power Integration into Power Systems: Stability and Control Aspects"
Reprinted from: *Energies* 2021, *14*, 3680, doi:10.3390/en14123680 1

Jiaxin Wen, Siqi Bu, Bowen Zhou, Qiyu Chen and Dongsheng Yang
A Fast-Algorithmic Probabilistic Evaluation on Regional Rate of Change of Frequency (RoCoF) for Operational Planning of High Renewable Penetrated Power Systems
Reprinted from: *Energies* 2020, *13*, 2780, doi:10.3390/en13112780 5

Sijia Tu, Bingda Zhang and Xianglong Jin
Research on DFIG-ES System to Enhance the Fast-Frequency Response Capability of Wind Farms
Reprinted from: *Energies* 2019, *12*, 3581, doi:10.3390/en12183581 19

Ting-Hsuan Chien, Yu-Chuan Huang and Yuan-Yih Hsu
Neural Network-Based Supplementary Frequency Controller for a DFIG Wind Farm
Reprinted from: *Energies* 2020, *13*, 5320, doi:10.3390/en13205320 39

Yi-Wei Chen and Yuan-Yih Hsu
Flexible Kinetic Energy Release Controllers for a Wind Farm in an Islanding System
Reprinted from: *Energies* 2020, *13*, 6135, doi:10.3390/en13226135 55

Alija Mujcinagic, Mirza Kusljugic and Emir Nukic
Wind Inertial Response Based on the Center of Inertia Frequency of a Control Area
Reprinted from: *Energies* 2020, *13*, 6177, doi:10.3390/en13236177 71

Xiangwu Yan, Linlin Yang and Tiecheng Li
The LVRT Control Scheme for PMSG-Based Wind Turbine Generator Based on the Coordinated Control of Rotor Overspeed and Supercapacitor Energy Storage
Reprinted from: *Energies* 2021, *14*, 518, doi:10.3390/en14020518 89

Yingzong Jiao, Feng Li, Hui Dai and Heng Nian
Analysis and Mitigation of Sub-Synchronous Resonance for Doubly Fed Induction Generator under VSG Control
Reprinted from: *Energies* 2020, *13*, 1582, doi:10.3390/en13071582 111

Yafeng Hao, Jun Liang, Kewen Wang, Guanglu Wu, Tibin Joseph and Ruijuan Sun
Influence of Active Power Output and Control Parameters of Full-Converter Wind Farms on Sub-Synchronous Oscillation Characteristics in Weak Grids
Reprinted from: *Energies* 2020, *13*, 5225, doi:10.3390/en13195225 129

Jianqiang Luo, Siqi Bu and Jiebei Zhu
Transition from Electromechanical Dynamics to Quasi-Electromechanical Dynamics Caused by Participation of Full Converter-Based Wind Power Generation
Reprinted from: *Energies* 2020, *13*, 6270, doi:10.3390/en13236270 147

Bingchun Liu, Shijie Zhao, Xiaogang Yu, Lei Zhang and Qingshan Wang
A Novel Deep Learning Approach for Wind Power Forecasting Based on WD-LSTM Model
Reprinted from: *Energies* **2020**, *13*, 4964, doi:10.3390/en13184964 . . . . **167**

Pei Zhang, Chunping Li, Chunhua Peng and Jiangang Tian
Ultra-Short-Term Prediction of Wind Power Based on Error Following Forget Gate-Based Long Short-Term Memory
Reprinted from: *Energies* **2020**, *13*, 5400, doi:10.3390/en13205400 . . . . **185**

Yurong Wang, Ruolin Yang, Sixuan Xu and Yi Tang
Capacity Planning of Distributed Wind Power Based on a Variable-Structure Copula Involving Energy Storage Systems
Reprinted from: *Energies* **2020**, *13*, 3602, doi:10.3390/en13143602 . . . . **199**

Hongwei Li, Kaide Ren, Shuaibing Li and Haiying Dong
Adaptive Multi-Model Switching Predictive Active Power Control Scheme for Wind Generator System
Reprinted from: *Energies* **2020**, *13*, 1329, doi:10.3390/en13061329 . . . . **221**

Yuan Li, Zengjin Xu, Zuoxia Xing, Bowen Zhou, Haoqian Cui, Bowen Liu and Bo Hu
A Modified Reynolds-Averaged Navier–Stokes-Based Wind Turbine Wake Model Considering Correction Modules
Reprinted from: *Energies* **2020**, *13*, 4430, doi:10.3390/en13174430 . . . . **233**

About the Editors

Lasantha Meegahapola

Dr. Lasantha Meegahapola is currently an Associate Professor in Electrical and Biomedical Engineering, at the School of Engineering, RMIT University, Australia. He received a BSc. Eng. degree in Electrical Engineering (First Class, Honours) from the University of Moratuwa, Sri Lanka, in 2006, and a PhD degree from Queen's University of Belfast, U.K., in 2010. From 2011 to 2014, he was employed as a Lecturer at the University of Wollongong (UOW), and has continued as an Honorary Fellow. Dr Meegahapola's research interests are Renewable Power Generation, Power System Dynamics and Stability, and Microgrid Stability and Control. He is a Senior Member of IEEE (SMIEEE) and a Member of the IEEE Power Engineering Society (PES) and the IEEE Industry Applications Society (IAS). Dr. Meegahapola serves as an Associate Editor of the *IEEE Transactions on Industry Applications* (TIA)/*IEEE Industry Applications Magazine IEEE ACCESS* and *IET Renewable Power Generation* journals.

Siqi Bu

Dr. Siqi Bu received a Ph.D. degree from the electric power and energy research cluster, Queen's University Belfast, Belfast, U.K., where he continued his postdoctoral research work before entering industry. He worked with National Grid UK as an experienced UK National Transmission System Planner and Operator. He is an Associate Professor at the Hong Kong Polytechnic University, Kowloon, Hong Kong, and is also a Chartered Engineer with the UK Royal Engineering Council, London, U.K. His research interests are power system stability analysis and operational control, including wind/solar power generation, PEV, HVDC, FACTS, ESS and VSG. He is an Editor of *IEEE Access, CSEE Journal of Power and Energy Systems*, *IEEE Open Access Journal of Power and Energy, Protection and Control of Modern Power Systems* and *PEERJ Computer Science*, a Guest Editor of *IET Renewable Power Generation, Energies, IEEE Access, IET Generation, Transmission & Distribution, Shock and Vibration* and *Frontiers in Energy*.

Preface to "Wind Power Integration into Power Systems: Stability and Control Aspects"

Power network operators are rapidly incorporating wind power generation into their power grids to meet the widely accepted carbon neutrality targets and facilitate the transition from conventional fossil-fuel energy sources to clean and low-carbon renewable energy sources. Complex stability issues, such as frequency, voltage, and oscillatory instability, are frequently reported in the power grids of many countries and regions (e.g., Germany, Denmark, Ireland, and South Australia) due to the substantially increased wind power generation. Control techniques, such as virtual/emulated inertia and damping controls, could be developed to address these stability issues, and additional devices, such as energy storage systems, can also be deployed to mitigate the adverse impact of high wind power generation on various system stability problems. Moreover, other wind power integration aspects, such as capacity planning and the short- and long-term forecasting of wind power generation, also require careful attention to ensure grid security and reliability. This book includes fourteen novel research articles published in this *Energies* Special Issue on Wind Power Integration into Power Systems: Stability and Control Aspects, with topics ranging from stability and control to system capacity planning and forecasting.

Lasantha Meegahapola, Siqi Bu
Editors

Editorial

Special Issue: "Wind Power Integration into Power Systems: Stability and Control Aspects"

Lasantha Meegahapola [1,*] and Siqi Bu [2,*]

1 Electrical and Biomedical Engineering, School of Engineering, RMIT University, Melbourne 3001, Australia
2 Department of Electrical Engineering, The Hong Kong Polytechnic University, Hong Kong 999077, China
* Correspondence: lasantha.meegahapola@rmit.edu.au (L.M.); siqi.bu@polyu.edu.hk (S.B.)

Citation: Meegahapola, L.; Bu, S. Special Issue: "Wind Power Integration into Power Systems: Stability and Control Aspects". *Energies* **2021**, *14*, 3680. https://doi.org/10.3390/en14123680

Received: 26 March 2021
Accepted: 16 June 2021
Published: 21 June 2021

Publisher's Note: MDPI stays neutral with regard to jurisdictional claims in published maps and institutional affiliations.

Copyright: © 2021 by the authors. Licensee MDPI, Basel, Switzerland. This article is an open access article distributed under the terms and conditions of the Creative Commons Attribution (CC BY) license (https://creativecommons.org/licenses/by/4.0/).

Power network operators are rapidly incorporating wind power generation into their power grids to meet the widely accepted carbon neutrality targets and facilitate the transition from conventional fossil-fuel energy sources to the clean and low-carbon renewable energy sources. Complex stability issues, such as frequency, voltage, and oscillatory instability, are frequently reported in the power grid of many countries and regions (e.g., Germany, Denmark, Ireland, South Australia) with the substantially increased wind power generation. Control techniques, such as virtual/emulated inertia and damping controls, could be developed to address these stability issues, and additional devices, such as energy storage systems, can also be deployed to mitigate the adverse impact of high wind power generation on various system stability problems. This Special Issue includes 14 novel research articles mainly covering various stability analyses and associated control techniques of modern power systems as affected by high penetration of wind power generation.

Tu et al. [1] proposed a doubly fed induction generator (DFIG)–energy storage (ES) based hybrid system to improve the fast frequency response from wind farms. The ES system was designed to provide a similar inertial response as a synchronous generator of a similar rating as the wind farm. The authors have also proposed a coordinated virtual inertial response scheme for the DFIG-ES system to provide frequency response during frequency excursions. The proposed scheme was based on a fuzzy logic scheme, and it could improve the frequency nadir and also could alleviate the secondary frequency dip. Therefore, the proposed scheme is a very useful control scheme to improve the frequency response from wind farms.

Li et al. [2] proposed a multi-model predictive control algorithm based on a clustering approach to deal with the randomness and uncertainty of wind power generation. The authors developed the multi-model prediction model by first clustering the measured data and then applying the forgetting factor recursive least square method. The model predictive controller was developed to control the pitch-angle to vary the power output of the wind generator. The accuracy of the developed model was verified by using a DFIG-based wind farm in Western China by applying field-measured wind speed data. The proposed model predictive controller has shown a high prediction accuracy compared with the methods reported in the literature.

Jiao et al. [3] presented a sub-synchronous resonance (SSR) analysis and mitigation strategy for a DFIG operated as a virtual synchronous generator (VSG). Since the weak network phenomenon is more prominent in most of the present power grids, the VSG control method is used in power-electronic converter interfaced sources, such as the DFIG. Therefore, this paper presented an impedance-based analysis to characterise the sub-synchronous resonance (SSR) phenomenon for a DFIG-based wind farm controlled in VSG mode connected to the grid by a series-compensated line. According to the study, damping of reactive power plays a major role in mitigating the SSR phenomenon in DFIG wind generators controlled in VSG mode.

Wen et al. [4] presented a probabilistic assessment of the regional rate-of-change-of-frequency (RoCoF) for operational planning of high renewable penetrated power systems. Regional RoCoF is becoming an imperative factor in system planning and operation studies. Large-scale power-electronic converter interfaced generation sources are installed in regional areas of the power networks, which do not naturally respond to frequency excursions. This paper established an analytical sensitivity of regional RoCoF to the stochastic output of RES and subsequently, a linear sensitivity-based analytical method was proposed to calculate the regional RoCoF and the corresponding probabilistic distribution. The proposed method appears to be less time consuming than the existing methods.

Wang et al. [5] presented capacity planning of distributed wind power based on a variable-structure copula involving energy storage systems. Since some countries (e.g., China) require distributed wind power to be consumed within sub-transmission level, the distributed wind power capacity should be planned carefully while ensuring the entire distributed generation capacity is consumed within the network. Authors have developed a load and wind power prediction model based on the autoregressive moving average (ARMA) model, and subsequently, variable-structure copula models are established based on different time segment strategies to correlate the wind power and load. Finally, the capacity planning model was proposed based on the investment and operation cost, and environmental benefit and line loss cost. Subsequently, the model has been extended to a collaborative capacity planning model for distributed wind power and energy storage systems. The fidelity of the proposed model was validated using a modified IEEE-33 bus network.

Li et al. [6] presented a wind turbine wake model based on the modified Reynolds-averaged Navier–Stokes approach. The new model was proposed to improve the existing wake effect models' accuracy by proposing correction factors for the aerodynamic and turbulence models. The study has shown that the proposed model's velocity and turbulent fields are in close agreement with the data obtained from real wind turbines. In addition, the proposed mesh partition method has improved the computational efficiency, and hence the proposed model could be deployed to effectively assess the impact of the wake effect of wind turbines in power system studies.

Liu et al. [7] proposed a deep learning approach for wind power forecasting based on a wavelet decomposition (WD)–long short-term memory (LSTM) neural network model. Uncertainty and intermittency associated with power generation add complexity to system operation, and inaccurate forecasts increase the power network's risk of instability. Thus, to address this pressing issue, this paper proposed a hybrid prediction model based on the combination of WD and LSTM neural network. In this model, the nonstationary time series is decomposed into multidimensional components to reduce the original time series' volatility and make them more stable and predictable by WD. Subsequently, it has been used as the input to the LSTM to predict wind power generation. The results showed that the proposed model predicts wind power generation much more accurately than the existing prediction models used in China.

Hao et al. [8] studied the impact of active power outputs and control parameters of full-converter wind farms on the damping characteristics of sub-synchronous oscillation in weak power grids. Eigenvalue and participation factor analyses were performed to identify the dominant oscillation modes of the system and investigate the damping characteristics. The analysis demonstrated that when the phase-locked loop (PLL) proportional gain is high, the sub-synchronous oscillation damping has worsened with the increase in the active power output. On the contrary, when the PLL proportional gain is small, the sub-synchronous oscillation damping is improved with the increase in the active power output. By adjusting the control parameters in the PLL and DC link voltage controllers, system sub-synchronous oscillatory stability can be improved.

Chien et al. [9] designed an artificial neural network (ANN)-based real-time supplementary frequency controller for a DFIG wind farm, as the optimal controller gain that gives the highest frequency nadir or lowest peak frequency is a complicated nonlinear function and hence is not easy to be derived by conventional analytical methods, especially for an online environment. In this work, the load disturbance, wind penetration, and wind speed were used as the inputs and the desired controller gain was used as the output, and the ANN can be employed to yield the desired gain in a very efficient manner, even when the operating condition was not included in the training set. It was demonstrated in the paper that the proposed ANN-based frequency control could yield a better frequency response than the fixed-gain controller.

Zhang et al. [10] presented a model using modified LSTM to predict ultra-short-term wind power. The error following forget gate (EFFG)-based LSTM model was developed, which can update the output of the forget gate using the difference between the predicted value and the actual value, thereby reducing the impact of the prediction error at the previous moment on the prediction accuracy at this time and improving the rolling prediction accuracy of wind power. Study results revealed that the root mean square error of the wind power prediction model is less than 3%, while the accuracy rate and qualified rate are more than 90%. Hence, the EFFG-based LSTM model provides better performance than the support vector machine (SVM) and standard LSTM model.

Chen et al. [11] proposed two types of flexible kinetic energy release controllers for the DFIG to improve frequency nadir following a disturbance and avoid under-frequency load shedding. A deactivation function-based integral controller was firstly presented and a second flexible kinetic energy release controller was designed using a proportional-integral controller with the gains being adapted in real-time with the particle swarm optimisation algorithm. The design only releases a small amount of kinetic energy in the initial transient period and more kinetic energy would be released when the frequency dip exceeds a pre-set threshold. The paper concluded that the frequency nadir could be maintained around the under-frequency load shedding threshold of 59.6 Hz using the proposed controllers.

Mujcinagic et al. [12] presented a control scheme of the virtual inertia response of wind power plants based on the centre of inertia (COI) frequency of a control area for the inertia-insufficient power systems. The PSS/E user written wind inertial controller was developed using FORTRAN. The efficiency of the controller was tested and applied to the real interconnected power system of Southeast Europe. The performed simulations showed certain conceptual advantages of the proposed controller in comparison to traditional schemes that use the local frequency to trigger the wind inertial response.

Luo et al. [13] investigated the participation of full converter-based wind power generation (FCWG) in electromechanical dynamics and uncovered an unusual transition of the electromechanical oscillation mode (EOM). Modal analysis was employed to quantify the FCWG participation in electromechanical dynamics, with two new mode identification criteria proposed. The impact of different wind penetration levels and controller parameter settings on the participation of FCWG was studied. It was revealed that if an FCWG oscillation mode (FOM) has a similar oscillation frequency to the system EOMs, strong interactions between FCWG dynamics and electromechanical system dynamics of the external power systems might be induced, and an EOM can be dominated by FCWG dynamics instead, and hence becomes a quasi-EOM. Some key findings on the mechanism of this special phenomenon are finally summarised and discussed.

Yan et al. [14] presented a novel coordinated control scheme, i.e., overspeed-while-storing control for permanent magnet synchronous generator (PMSG)-based WTG to enhance its LVRT capability. The proposed control scheme regulated the rotor speed to reduce the input power of the machine-side converter (MSC) during slight voltage sags. When the severe voltage sag occurs, the coordinated control scheme sets the rotor speed at the upper-limit to decrease the input power of the MSC, while the surplus power is absorbed by the supercapacitor energy storage (SCES) to reduce its maximum capacity. Moreover, the specific capacity configuration scheme of SCES was detailed. The effectiveness of the overspeed-while-storing control in enhancing the LVRT capability was validated under different levels of voltage sags and different fault types in MATLAB/Simulink.

The papers mentioned above included in this Special Issue provide new and valuable insights, effectively representing ongoing research efforts and stimulating future research activities in the relevant field. As guest editors, we would like to thank all the authors and reviewers who contributed to this Special Issue.

Conflicts of Interest: The authors declare no conflict of interest.

References

1. Tu, S.; Zhang, B.; Jin, X. Research on DFIG-ES System to Enhance the Fast-Frequency Response Capability of Wind Farms. *Energies* **2019**, *12*, 3581. [CrossRef]
2. Li, H.; Ren, K.; Li, S.; Dong, H. Adaptive Multi-Model Switching Predictive Active Power Control Scheme for Wind Generator System. *Energies* **2020**, *13*, 1329. [CrossRef]
3. Jiao, Y.; Li, F.; Dai, H.; Nian, H. Analysis and Mitigation of Sub-Synchronous Resonance for Doubly Fed Induction Generator under VSG Control. *Energies* **2020**, *13*, 1582. [CrossRef]
4. Wen, J.; Bu, S.; Zhou, B.; Chen, Q.; Yang, D. A Fast-Algorithmic Probabilistic Evaluation on Regional Rate of Change of Frequency (RoCoF) for Operational Planning of High Renewable Penetrated Power Systems. *Energies* **2020**, *13*, 2780. [CrossRef]
5. Wang, Y.; Yang, R.; Xu, S.; Tang, Y. Capacity Planning of Distributed Wind Power Based on a Variable-Structure Copula Involving Energy Storage Systems. *Energies* **2020**, *13*, 3602. [CrossRef]
6. Li, Y.; Xu, Z.; Xing, Z.; Zhou, B.; Cui, H.; Liu, B.; Hu, B. A Modified Reynolds-Averaged Navierâ-Stokes-Based Wind Turbine Wake Model Considering Correction Modules. *Energies* **2020**, *13*, 4430. [CrossRef]
7. Liu, B.; Zhao, S.; Yu, X.; Zhang, L.; Wang, Q. A Novel Deep Learning Approach for Wind Power Forecasting Based on WD-LSTM Model. *Energies* **2020**, *13*, 4964. [CrossRef]
8. Hao, Y.; Liang, J.; Wang, K.; Wu, G.; Joseph, T.; Sun, R. Influence of Active Power Output and Control Parameters of Full-Converter Wind Farms on Sub-Synchronous Oscillation Characteristics in Weak Grids. *Energies* **2020**, *13*, 5225. [CrossRef]
9. Chien, T.; Huang, Y.; Hsu, Y. Neural Network-Based Supplementary Frequency Controller for a DFIG Wind Farm. *Energies* **2020**, *13*, 5320. [CrossRef]
10. Zhang, P.; Li, C.; Peng, C.; Tian, J. Ultra-Short-Term Prediction of Wind Power Based on Error Following Forget Gate-Based Long Short-Term Memory. *Energies* **2020**, *13*, 5400. [CrossRef]
11. Chen, Y.; Hsu, Y. Flexible Kinetic Energy Release Controllers for a Wind Farm in an Islanding System. *Energies* **2020**, *13*, 6135. [CrossRef]
12. Mujcinagic, A.; Kusljugic, M.; Nukic, E. Wind Inertial Response Based on the Center of Inertia Frequency of a Control Area. *Energies* **2020**, *13*, 6177. [CrossRef]
13. Luo, J.; Bu, S.; Zhu, J. Transition from Electromechanical Dynamics to Quasi-Electromechanical Dynamics Caused by Participation of Full Converter-Based Wind Power Generation. *Energies* **2020**, *13*, 6270. [CrossRef]
14. Yan, X.; Yang, L.; Li, T. The LVRT Control Scheme for PMSG-Based Wind Turbine Generator Based on the Coordinated Control of Rotor Overspeed and Supercapacitor Energy Storage. *Energies* **2021**, *14*, 518. [CrossRef]

Article

A Fast-Algorithmic Probabilistic Evaluation on Regional Rate of Change of Frequency (RoCoF) for Operational Planning of High Renewable Penetrated Power Systems

Jiaxin Wen [1], Siqi Bu [1],*, Bowen Zhou [2], Qiyu Chen [3] and Dongsheng Yang [2]

1. Department of Electrical Engineering, The Hong Kong Polytechnic University, Hong Kong, China; Jiaxin.wen@connect.polyu.hk
2. College of Information Science and Engineering, Northeastern University, Shenyang 110819, China; zhoubowen@ise.neu.edu.cn (B.Z.); yangdongsheng@mail.neu.edu.cn (D.Y.)
3. Power System Department, China Electric Power Research Institute, Haidian District, Beijing 100192, China; chen.qiyu2009@hotmail.com
* Correspondence: siqi.bu@polyu.edu.hk

Received: 5 May 2020; Accepted: 28 May 2020; Published: 1 June 2020

Abstract: The high rate of change of frequency (RoCoF) issue incurred by the integration of renewable energy sources (RESs) into a modern power system significantly threatens the grid security, and thus needs to be carefully examined in the operational planning. However, severe fluctuation of regional frequency responses concerned by system operators could be concealed by the conventional assessment based on aggregated system frequency response. Moreover, the occurrence probability of a high RoCoF issue is actually a very vital factor during the system planner's decision-making. Therefore, a fast-algorithmic evaluation method is proposed to determine the probabilistic distribution of regional RoCoF for the operational planning of a RES penetrated power system. First, an analytical sensitivity (AS) that quantifies the relationship between the regional RoCoF and the stochastic output of the RES is derived based on the generator and network information. Then a linear sensitivity-based analytical method (LSM) is established to calculate the regional RoCoF and the corresponding probabilistic distribution, which takes much less computational time when comparing with the scenario-based simulation (SBS) and involves much less complicated calculation procedure when comparing with the cumulant-based method (CBM). The effectiveness and efficiency of the proposed method are verified in a modified 16-machine 5-area IEEE benchmark system by numerical SBS and analytical CBM.

Keywords: renewable energy sources (RESs); regional RoCoF; model-based operational planning; linear sensitivity-based method (LSM); cumulant-based method (CBM)

1. Introduction

The integration of renewable energy sources (RESs) brings an increasing number of stochastic disturbances into power systems [1–4] and meanwhile considerably reduces the system inertia [5–7], which hence incurs higher rate of change of frequency (RoCoF) than ever before [8,9], and sometimes even serious incidents [10]. The recent London blackout on 9 August 2019 has drawn wide attention, and the official investigation report [10] indicates that a sudden reduction in the power output of the Hornsea offshore wind farm has worsened the RoCoF significantly, which further causes the enormous loss of both generations and demands. Hence, there is a pressing need to evaluate the impact of stochastic variation of RESs on the RoCoF in modern operational planning.

To accommodate the uncertainties brought by RESs, the safe operation of the system under the assumed "worst-case scenario" is guaranteed by reserving excessive conventional generation in real-time operation. However, the "worst-case scenario" where the uncertain disturbances of all the RESs reach maximum simultaneously rarely happens in a highly RES-penetrated power system because of spatiotemporal uncorrelation among the same or different types of the RESs in the network. For different types of RESs, wind power plants often reach the maximal output in the night while the photovoltaic plants only generate during the daytime. For the same type of RESs located in different places, the correlation of their stochastic output can be quite low. Both factors above significantly reduce the occurrence probability of the simultaneous maximal output of renewable energy plants. Thereby, a two-dimensional evaluation including both the severity and the occurrence probability of the event could be more beneficial for the system planner to make a decision, which may further increase the allowed penetration level of RESs. There are two common approaches to achieve the two-dimensional evaluation mentioned above [11–14]. (1) Monte Carlo simulation (MCS), which aims to compute the probabilistic distribution of the concerned indices by generating a large number of random variables and thus, simulation results. In [13], scenario-based simulation (SBS), similar to MCS, is proposed to calculate the maximal renewable energy penetration limits to maintain the frequency performance by considering numerous potential operational scenarios. The results from the SBS are accurate, but its calculation procedure is very time-consuming, which is normally regarded as a verification tool. (2) Analytical method, e.g., cumulant-based analytical method (CBM), calculates the distribution of the concerned indices based on the sensitivity and the series expansion. This method can comfortably accommodate arbitrary types of continuous or noncontinuous distribution and correlation of stochastic variables [14], which is proven to be the most efficient and accurate way to conduct probabilistic small-signal stability analysis in [15]. In [16], a probabilistic assessment framework on system RoCoF is proposed based on the CBM for the operational planning of a power system with RESs. However, the calculation procedure of the CBM is very complicated and not easy to implement.

The system frequency response (SFR), as an overall performance of the system frequency, is aggregated by frequency responses of the individual generator [17] and normally required to remain within a specific range set by the system operator [18]. While the heterogeneity of different regional frequency responses would be more obvious because of the increasing penetration level of distributed RESs and uneven distribution of inertia sources, which cannot be simply revealed by an integrated SFR [19–21]. Reference [13] reports that regional RoCoF violates the given limits, whereas the system RoCoF operates safely after the disturbance, which demonstrates the necessity of regional RoCoF assessment. Moreover, the RoCoF at the disturbance instant (i.e., $t = 0^+$) is usually observed to be the worst RoCoF without any assistance from the system fast-acting control [19,22–24]. Hence, regional RoCoF deserves a careful investigation in the operational planning stage to avoid the potential risk of RoCoF violation.

Taking all the points above into consideration, the paper proposes a novel fast-algorithmic evaluation to efficiently determine the probabilistic distribution of regional RoCoF, which demonstrates a clear superiority over the time-consuming SBS and the complicated CBM. The main contributions of the paper are listed below accordingly:

1. By combining the analytical sensitivity (AS) of RoCoF and the linear sensitivity-based method (LSM), AS-LSM is proposed to calculate the RoCoF. AS can adequately reflect the essential relationship between the variation of RESs and the RoCoF. Together with AS, LSM enables the evaluation of the RoCoF considering a complex multi-RES environment by using a superimposing technique, which considerably facilitates the understanding and implementation.

2. The proposed AS-LSM can facilitate the calculation for the probabilistic distribution of regional RoCoF. As a combination of numerical and analytical methods for probabilistic computation, the AS-LSM has a higher computing efficiency compared with SBS and a more straightforward calculation procedure compared with CBM.

3. The proposed AS-LSM could determine the probabilistic distribution of regional RoCoF influenced by the correlation of wind speed distribution more accurately than AS-CBM (i.e., CBM based on AS).

The rest of the paper is organized as follows. In Section 2, regional analytical sensitivity (AS) of RoCoF is derived based on the generator and network information. Based on the derived regional AS and the linear sensitivity-based method (LSM), regional RoCoF in a multi-RES penetrated power system is calculated by the proposed AS-LSM in Section 3. Case studies are conducted in Section 4 to verify the effectiveness and efficiency of the proposed method with consideration of different wind speed correlations. The conclusion is drawn in Section 5.

2. Analytical Sensitivity (AS) of Regional RoCoF w.r.t a Single Disturbance

For a single-machine system, the RoCoF is directly expressed as (1) according to [25]:

$$RoCoF = \frac{f(t)}{dt} = \frac{1}{2H}P(t) \tag{1}$$

where H is the inertia, $f(t)$ is the frequency deviation from the nominal frequency f_0, and $P(t)$ is the imposed active power disturbance.

2.1. Generator-Level Power Disturbance Propagation and Its Distribution Coefficient

At the moment of the disturbance occurring ($t = 0^+$), the system active power disturbance (P_{Dist}), which is incurred by the sudden change of RES in this paper, would propagate in the system. The active power disturbance component distributed to each generator bus $P_i(0^+)$ can be determined by the synchronizing power coefficients (P_{sik}) between the location of the RES and the individual generator [17]. The propagating procedure is illustrated by Figure 1.

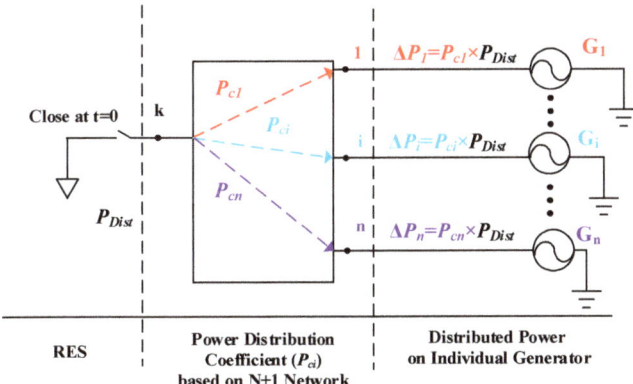

Figure 1. Active power disturbance propagation from renewable energy sources (RES) (i.e., bus k) to each generator bus.

First, the full network is reduced to the N+1 bus equivalent network, where N is the total number of the generators in the network, and "1" refers to the single RES. Second, the synchronizing power coefficients (P_{sik}) between RES, i.e., bus k and, the ith generator bus is calculated as (2) according to [17].

$$P_{sik} = V_i V_k (B_{ik} \cos \delta_{ik0} - G_{ik} \sin \delta_{ik0}) \tag{2}$$

where V_i and V_k are the voltage magnitude of bus i and bus k, respectively. B_{ik} and G_{ik} are the imaginary and real parts of the equivalent admittance between bus i and bus k separately. δ_{ik0} is the steady angle difference between bus i and bus k.

Then the distribution coefficient of power disturbance (P_c) is defined as a percentage in (3), which quantifies what percentage of active power disturbance from a single RES could arrive at the individual generator bus [17]. The active power disturbance component allocated to each generator bus, i.e., $P_i(0^+)$ w.r.t the stochastic output of RES (P_{Dist}) can be computed by (4).

$$P_{ci} = \frac{P_{sik}}{\sum_{i=1}^{N} P_{sik}} \tag{3}$$

$$P_i(0^+) = P_{ci} P_{Dist} \tag{4}$$

2.2. Regional Power Disturbance Propagation and Its Distribution Coefficient

Based on the above analysis, the active power disturbance component allocated to a region equals the sum of the active power disturbance distributed to the individual generator bus in this region.

$$P^j(0^+) = \sum_{i=1}^{G} P_i^j(0^+) \tag{5}$$

where, $P^j(0^+)$ is the located active power disturbance component in the jth area, P_i^j is the active power disturbance distributed to the ith generator bus in the jth area and G is the number of generators in the jth area. Substituting (4) into (5), the regional active power disturbance distributed from system active power disturbance source is obtained in (6), where the regional distribution coefficient of power disturbance is defined in (7).

$$P^j(0^+) = \sum_{i=1}^{G} P_{ci}^j P_{Dist} = P_c^j P_{Dist} \tag{6}$$

$$P_c^j = \frac{\sum_{i=1}^{G} P_{sik}}{\sum_{i=1}^{N} P_{sik}} \tag{7}$$

where P_c^j refers to the jth area P_c w.r.t the output of RES, and N is the total number of the generators in the network.

2.3. Analytical Sensitivity (AS) of Generator-Level RoCoF

The generator-level RoCoF is calculated as (8) by substituting (4) into (1), where the generator-level AS is defined in (9).

$$RoCoF_i = \frac{1}{2H_i} P_{ci} P_{Dist} = AS_i P_{Dist} \tag{8}$$

$$AS_i = \frac{1}{2H_i} \times \frac{P_{sik}}{\sum_{i=1}^{N} P_{sik}} \tag{9}$$

where $RoCoF_i$ is the RoCoF of the ith generator and the AS_i is the AS of $RoCoF_i$ w.r.t the output of the RES.

2.4. Analytical Sensitivity (AS) of Regional RoCoF

In a multi-machine system, an active power disturbance would cause various frequency responses of different generators in a power system. Traditionally, system frequency response, as an overall performance of all the frequency responses in the system, is aggregated based on the concept of the center of inertia (COI), where all generators are integrated into one equivalent generator with the sum of inertia under a base power capacity [17]. Hence, a similar method is applied here to calculate the regional center of inertia (RCOI), which is defined as follows.

First, the base power capacity of the system is selected, and the individual inertia constant under a base power capacity (H_i) is acquired in (10):

$$H_i = H_{i,o} \times \left(\frac{S_i}{S_{base}}\right) \quad (10)$$

where $H_{i,o}$ is the ith inertia constant w.r.t its rated power capacity S_i, and S_{base} is the base power capacity. Then the RCOI of the jth area (H^j) is defined in (11):

$$H^j = \sum_{i=1}^{G} H_i^j \quad (11)$$

where H_i^j is the inertia of the ith generator in the jth area, and G is the number of generators in the jth area. The superscript refers to the number of the area.

In [17], COI frequency is defined as $f_{COI} = \sum_{i=1}^{N} f_i H_i / \sum_{i=1}^{N} H_i$, where f_i is the frequency response of the ith generator, and N is the number of generators. It can be revealed that the COI frequency is the weighted average of the frequency response of each generator, and the weighted coefficient is the percentage of the inertia of individual generator over the system inertia. A similar approach is employed to calculate the RCOI frequency for the jth area (f_{RCOI}^j), as defined by (12):

$$f_{RCOI}^j = \frac{\sum_{i=1}^{G} H_i^j f_i^j}{\sum_{i=1}^{G} H_i^j} \quad (12)$$

where f_i^j is the frequency response of the ith generator in the jth area, and G is the number of the generators in the jth area.

By using the concept of a regional equivalent generator, the regional RoCoF ($RoCoF^j$) is derived in (13) by substituting (6) and (11) into (1), where the regional analytical sensitivity (AS^j) is defined in (14).

$$RoCoF^j = \frac{1}{2H^j} P^j(0^+) = \frac{1}{2\sum_{k=1}^{G} H_k^j} P_c^j \, P_{Dist} = AS^j P_{Dist} \quad (13)$$

$$AS^j = \frac{1}{2\sum_{k=1}^{G} H_k^j} \times \frac{\sum_{i=1}^{G} P_{sik}}{\sum_{i=1}^{N} P_{sik}} \quad (14)$$

where $RoCoF^j$ is the RoCoF of the jth area, the AS^j is the AS of the $RoCoF^j$, and G is the number of the generators in the jth area.

3. Probabilistic Distribution of Regional RoCoF in a Multi-RES Penetrated Power System

3.1. Regional Active Power Disturbance Integration

From the above analysis, the propagation and distribution of system active power disturbance from the RES depend on the "electrical distance" between the RES and each generator bus at t = 0$^+$ demonstrated by (4). In a multi-RES penetrated power system, it is reasonable to assume that the active power disturbance distributed to a generator bus equals the sum of the active power disturbance allocated to the same bus from different RESs, which is expressed by (15). As mentioned above the active power disturbance distributed to a region equals the sum of the active power allocated to the individual generator bus in a coherent area, the regional active power disturbance component in a multi-RESs penetrated system can be depicted in (16).

$$P_i(0^+) = \sum_{l=1}^{M} P_{Distil} = \sum_{l=1}^{M} \frac{P_{sil}}{\sum_{i=1}^{N} P_{sil}} P_{Distl} = \sum_{l=1}^{M} P_{cil} P_{Distl} \tag{15}$$

$$P^j(0^+) = \sum_{i=1}^{G} P_i(0^+) = \sum_{i=1}^{G} \sum_{l=1}^{M} P_{cil} P_{Distl} \tag{16}$$

where P_{Distil} and P_{cil} are the distributing active power and distribution coefficient of the ith generator bus from the lth RES, respectively. P_{Distl} is the active power disturbance of the lth RES. M is the number of the RES in the system, and G is the number of generators in the jth area.

The propagating procedure of the active power disturbance from multiple RESs is illustrated in Figure 2. Assume there are M RESs and N generators in the system. First, each RES spreads the active power disturbance to individual generator bus through the reduced N + 1 network, where P_{Distij} is the active power allocated to the ith generator bus from the jth RES. Then, the total active power (P_i) distributed to a single generator bus equals the sum of the active power distributed to this bus from different RESs. In addition, the regional active power disturbance equals the sum of the active power disturbance distributed to the generator bus in the coherent region. For example, when the first i generators are in Area 1, the active power disturbance distributed to Area 1 (P^1) is equivalent to the sum of the active power distributed to the generator bus in Area 1, i.e., P_k, $k = 1 \cdots i$, as shown in Figure 2.

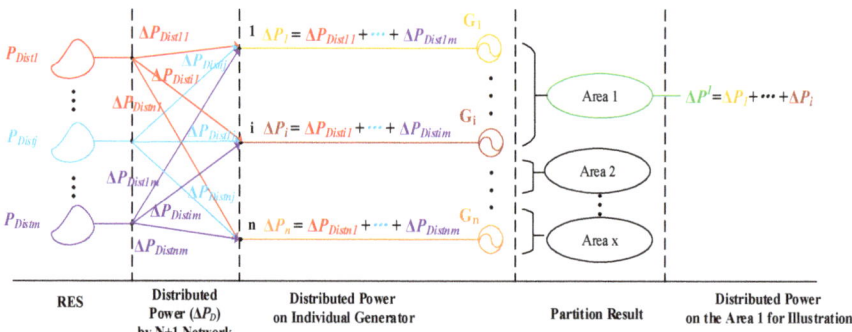

Figure 2. The propagating procedure of the active power disturbances in the multi-RES penetrated power system and the derivation of the regional active power disturbance.

3.2. Regional RoCoF Integration Based on Analytical Sensitivity and Linear Sensitivity-Based Method (AS-LSM)

A linear sensitivity-based method (LSM), which is capable of accommodating multiple stochastic variables (i.e., active power disturbance from RESs), is proposed here to compute the critical index (i.e., regional RoCoF) with a linear relationship (i.e., AS). Hence, the regional RoCoF based on AS-LSM is established in (17), and the full representation is given in (18).

$$RoCoF^j = \sum_{l=1}^{M} AS_l^j P_{Distl} \tag{17}$$

$$RoCoF^j = \frac{1}{2\sum_{k=1}^{G} H_k^j} \sum_{i=1}^{G} \sum_{l=1}^{M} \frac{P_{sil}}{\sum_{i=1}^{N} P_{sil}} P_{Distl} . \tag{18}$$

where G and N are the number of the generator in the *j*th area and the system, respectively, and M is the number of RES in the system. AS_l^j stands for AS of $RoCoF^j$ w.r.t the output of the *l*th RES.

The system-level RoCoF is a particular case of regional RoCoF when G = N and the (18) degrades to (19). Furthermore, when there is only one disturbance in the system, the (19) further degrades to (1).

$$RoCoF = \frac{1}{2\sum_{k=1}^{N} H_k} \sum_{l=1}^{M} \Delta P_{Distl} \tag{19}$$

3.3. Calculation Procedure of Probabilistic Distribution of Regional RoCoF

The flow chart of the calculation procedure of probabilistic distribution of regional RoCoF by AS-LSM is illustrated in Figure 3 and described as follows: (1) The information of the RES is obtained including type, number, capacity, steady output, probabilistic distribution of natural source, and the correlation coefficient matrix, based on which active power variation sample series is generated; (2) the information of the generator and the network is acquired; (3) on the basis of the above data, an analytical coherency identification method, e.g., slow coherency identification [26], is implemented to divide the system into several areas and the interested region is selected; (4) the concerned regional AS w.r.t the output of individual RES is calculated according to (14), and (5) AS-LSM is employed to determine the regional RoCoF based on the stochastic output of individual RES and the related regional AS by (17). This step repeats to get the probabilistic distribution of the regional RoCoF, and the number of the iterations depends on the number of generated sample series in step 2.

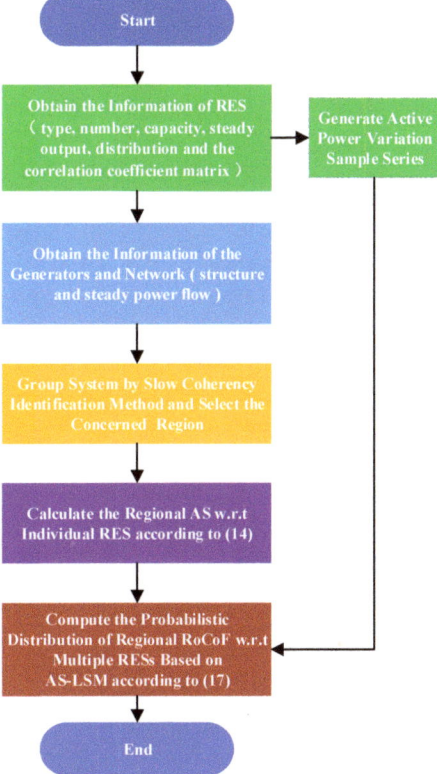

Figure 3. Flowchart of the calculation procedure for probabilistic distribution of regional rate of change of frequency (RoCoF) by analytical sensitivity-linear sensitivity-based method (AS-LSM).

4. Case Study

The effectiveness of the proposed AS-LSM is verified by SBS with 5000 times simulation, and AS-CBM is also applied to examine the probabilistic distribution of regional RoCoF for the first time because of its proven good performance on the probabilistic computation of system RoCoF [16]. The benchmark system is selected as a modified IEEE 16-machine 68-bus system with three wind farms (WFs) connected to bus 29, 32, and 41 respectively in Figure 4 partitioned by slow coherency identification method [26]. The probabilistic distributions of Region 4 and Region 5 are selected as the focus of the paper since they are the areas that contain more than one single generator. There are two scenarios studied in this section, i.e., with and without the correlations of wind speed.

The base capacity of the system is 100MVA. The operational state of the system decreases to 50% of the original level (system load, generation, and corresponding inertia). The capacity of each wind plant is 6 p.u, and the steady output is 2 p.u. The penetration of wind energy is defined by the ratio of the capacity of the WFs over the system load in [27], which is 19.7% in this section.

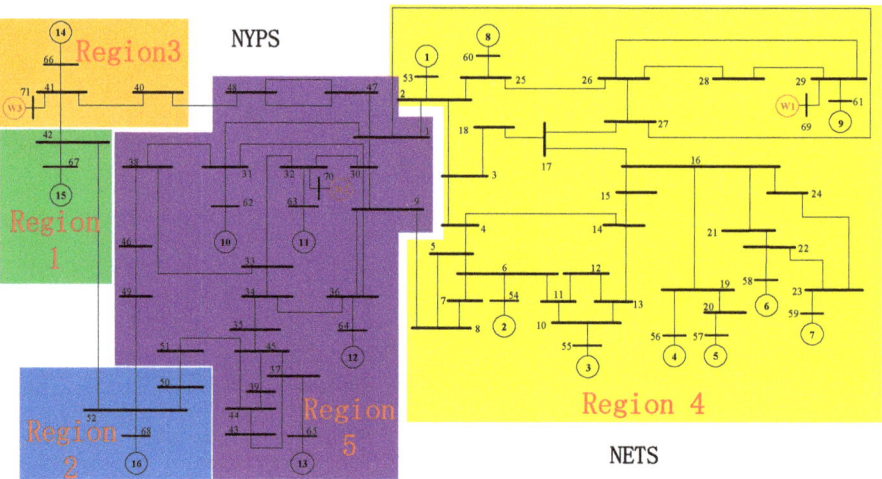

Figure 4. Line diagram of a modified IEEE 16-machine 5-area benchmark system with three wind farms.

Based on the calculation procedure in Figure 3, after information acquisition and the system partition, the regional ASs w.r.t the output of individual WF are calculated according to (14), and the results are represented in Table 1. For system-level analysis, the system is equivalent to one generator without considering the "electric distance," which is also proven by (14) when n equals g, and hence, the system AS w.r.t the output of different WFs are the same. However, there is a large difference among the AS of regional RoCoF w.r.t the output of different WFs due to the comprehensive influences from both "electric distance" and regional inertia. In details, the AS of Region 4 RoCoF w.r.t WF2 and 3 is small (i.e., 0.049267 and 0.004431), whereas the sensitivity w.r.t WF1 is relatively large (i.e., 0.294051), which is caused by different "electric distance." Furthermore, the maximal and minimal ASs of all regional RoCoFs w.r.t WF1 are 0.294051 and 0.000983 respectively, and the difference stems from the various regional inertia.

Table 1. The AS of system/regional RoCoF w.r.t the output of individual wind farms (WFs).

	WF1	WF2	WF3
System	0.050502	0.050502	0.050502
Region 1	0.000983	0.003597	0.019503
Region 2	0.003680	0.017777	0.001502
Region 3	0.006693	0.013478	0.299779
Region 4	0.294051	0.049267	0.004431
Region 5	0.019988	0.112672	0.003533

4.1. Scenario One (Uncorrelated Wind Speed)

The correlation between two wind power sources is closely related to their geographical distance, based on which correlation coefficient matrix $[\rho_{ij}]_{m \times m}$ for m grid-connected wind power sources is established [28], and the wind speed distribution in [29] is applied. In this scenario, the distances among each two WFs are assumed to be larger than 1200 km, which means there is no correlation among each WF, and hence the correlation matrix is a unit matrix.

Based on the AS in Table 1, AS-LSM and AS-CBM are employed to calculate the probabilistic distribution of the system/regional RoCoF, which are examined by SBS in Figure 5. The probabilistic density functions (PDFs) of the system, Region 4, and Region 5 RoCoF are exhibited in Figure 5a–c, respectively. The operational limit of the RoCoF is concerned by the system operator, which is set ±0.4 Hz/s for demonstration [16], and the detailed comparisons are given in Tables 2 and 3.

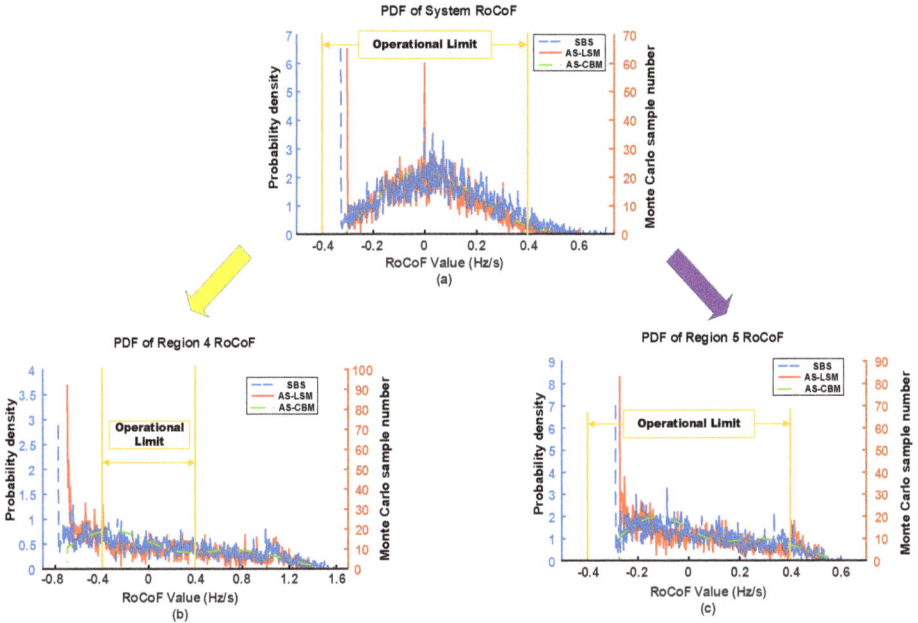

Figure 5. Probabilistic density functions (PDFs) of system/regional RoCoF by scenario-based simulation (SBS), AS-LSM, and analytical sensitivity-cumulant-based method (AS-CBM); (**a**) system, (**b**) Region 4, and (**c**) Region 5.

Table 2. Probabilistic distribution of the system, region 4, and region 5 RoCoF using SBS, AS-LSM, and AS-CBM within operational limits (uncorrelated wind speed).

No	SBS	AS-LSM	AS-CBM
System	96.9200%	98.0800%	98.0307%
Region 4	41.3200%	40.8400%	43.5933%
Region 5	92.8400%	90.7000%	92.9851%

Table 3. The absolute error of probabilities for the system, region 4, and region 5 RoCoF by AS-LSM and AS-CBM within operational limits (uncorrelated wind speed).

No	AS-LSM	AS-CBM
System	1.1600%	1.1107%
Region 4	0.4800%	2.2733%
Region 5	2.1400%	0.1451%

The AS-LSM and AS-CBM perform well in computing the probabilistic distribution of the RoCoF in the system, Region 4, and Region 5 intuitively, as displayed in Figure 5. Furthermore, it is also discovered that the shapes of the probabilistic distributions of the system RoCoF and the regional RoCoFs are different, but both methods could approach the trend, which is verified by the detailed result in both Tables 2 and 3. The absolute errors of the probabilistic results by both AS-LSM and AS-CBM, as presented in Table 3, reveal that the probabilistic distributions of system RoCoF calculated by both methods are relatively stable compared with that of regional RoCoFs. For example, the probabilistic result of Region 4 RoCoF can be estimated more accurately by AS-LSM than that by AS-CBM with less deviation (0.48% vs. 2.2733%). While the AS-CBM has a better performance than AS-LSM in calculating the probabilistic distribution of Region 5 RoCoF (0.1451% vs. 2.14%).

The computational time of each method are compared in Table 4. Both AS-LSM and AS-CBM are more than 1000 times faster than SBS, while the AS-LSM is a little faster because of the simple calculation procedure, which avoids a large amount of computation on Gram-Charlier expansion.

Table 4. The computational time of SBS, AS-LSM, and AS-CBM.

	SBS	AS-LSM	AS-CBM
Computational Time	2691.72s	1.95s	2.57s

4.2. Scenario Two (Correlated Wind Speed)

In this scenario, the correlation coefficient between WF2 and WF3 is set to be 0.8 (highly correlated) as (20).

$$[\rho]_{3\times 3} = \begin{bmatrix} 1 & 0 & 0 \\ 0 & 1 & 0.8 \\ 0 & 0.8 & 1 \end{bmatrix} \quad (20)$$

The PDFs of RoCoF on the system and Region 4 are given in Figures 6 and 7, respectively, for illustration, while the PDF of RoCoF associated with Region 5 is not given due to similar outcomes. The detailed probabilistic results and errors are also listed in Tables 5 and 6, respectively.

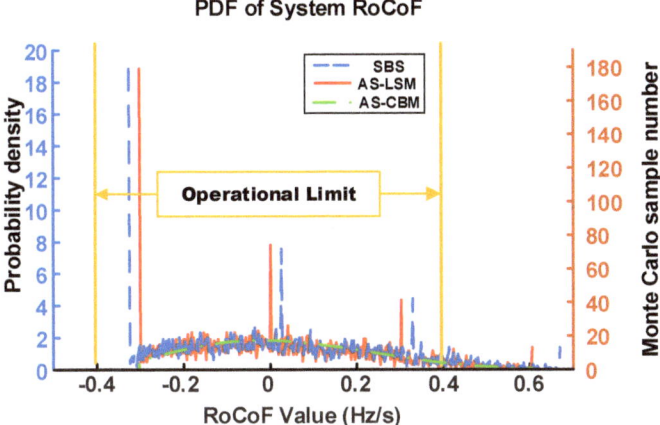

Figure 6. PDF of system RoCoF by SBS, AS-LSM, and AS-CBM.

Figure 6 illustrates the probabilistic distribution of system RoCoF carried out by SBS, AS-LSM, and AS-CBM. Compared with the real probabilistic distribution of system RoCoF in Figure 5a, there are a few noticeable "impulses" (i.e., occurrence probability) at a few points on the horizontal ordinate (i.e., RoCoF value, including maximal/minimal system RoCoF), which increases the probability of the "worst-case scenario" and deserves careful consideration in operational planning. The most apparent "impulse" in Figure 6 is the probability at the lowest RoCoF value, which is larger than the probability of the steady state (0 Hz/s). On the other hand, the total probability is 1, and this leads to a few decreases in the probabilities of other RoCoF values, which presents a smooth curve in Figure 6. Both methods evaluate the system RoCoF well according to the detailed probabilistic results in Tables 5 and 6.

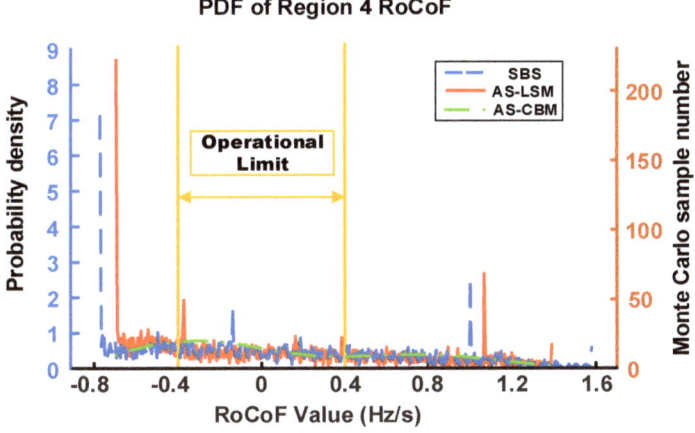

Figure 7. PDF of Region 4 RoCoF by SBS, AS-LSM, and AS-CBM.

Table 5. Probabilistic distribution of the system, region 4, and region 5 RoCoF using SBS, AS-LSM, and AS-CBM within operational limits (correlated wind speed).

No	SBS	AS-LSM	AS-CBM
System	93.6600%	94.9800%	95.2664%
Region 4	42.7000%	41.8000%	43.9902%
Region 5	91.4600%	90.9200%	92.7150%

Table 6. The absolute error of probabilities for the system, region 4, and region 5 RoCoF by AS-LSM and AS-CBM within operational limits (correlated wind speed).

No	AS-LSM	AS-CBM
System	1.3200%	1.6064%
Region 4	0.9000%	1.2902%
Region 5	0.5400%	1.2550%

As indicated in Figure 7, the "impulses" still occur in the probabilistic distribution of regional RoCoF, and the curve is much smoother compared with that in uncorrelated wind speed situations. The probabilistic distribution of regional RoCoF obtained by SBS is not bell-shaped, which could be depicted by both methods effectively, while the AS-LSM performs better than AS-CBM owing to less deviation, i.e., 0.9% vs. 1.2902% in Region 4 RoCoF and 0.54% vs. 1.255% in Region 5 RoCoF as given in Table 6.

5. Conclusions

The regional RoCoF is an important indicator for the safe operation of the power system, which needs to be carefully considered in operational planning. This paper proposes a fast-algorithmic assessment for the probabilistic distribution of regional RoCoF, which is more advantageous as it needs less time compared with SBS and provides a more straightforward calculating procedure than CBM. SBS validates the probabilistic results of both AS-LSM and AS-CBM with and without the consideration of wind speed correlation. Some important findings are summarized as follows:

(1) The probabilistic distributions of system RoCoF and regional RoCoF are different, i.e., bell-shaped vs. non-bell-shaped, which should be assessed separately. Both AS-LSM and AS-CBM can achieve the goal while AS-LSM has a better overall performance.

(2) When the wind speed correlation is considered, some evident "impulses" occur for the probabilistic distribution of both system and regional RoCoF as indicated by SBS. This phenomenon could be correctly reflected by both AS-LSM and AS-CBM, while AS-LSM performs better, which also demonstrates the flexibility and robustness of the proposed AS-LSM.

(3) The proposed AS-LSM converts a multi-disturbance problem into the superposition of a single-disturbance problem, which provides a more straightforward and convenient solution for their industrial implementation.

Author Contributions: J.W. completed the theory development and case study and wrote the paper. S.B. supervised the research throughout, coordinated the project, and revised the paper. B.Z., Q.C., and D.Y. provided guidance during the paper revision. All authors have read and agreed to the published version of the manuscript.

Funding: This research was funded by National Natural Science Foundation of China grant number (51807171), Guangdong Science and Technology Department grant number (2019A1515011226), Hong Kong Research Grant Council grant number (25203917), (15200418) and (15219619), and Department of Electrical Engineering, The Hong Kong Polytechnic University grant number (1-ZE68). The APC was funded by TPS Scheme of Hong Kong Polytechnic University.

Acknowledgments: The authors would like to acknowledge the support from National Natural Science Foundation of China for the Research Project (51807171), Guangdong Science and Technology Department for the Research Project (2019A1515011226), Hong Kong Research Grant Council for the Research Projects (25203917), (15200418) and (15219619), and Department of Electrical Engineering, The Hong Kong Polytechnic University for the Start-up Fund Research Project (1-ZE68).

Conflicts of Interest: The authors declare no conflict of interest.

Abbreviations

AS	Analytical sensitivity
CBM	Cumulant-based method
COI	Center of inertia
LSM	Linear sensitivity-based analytical method

MCS	Monte Carlo Simulation
PDF	Probabilistic density function
RCOI	Regional center of inertia
RES	Renewable energy source
RoCoF	Rate of change of frequency
SBS	Scenario-based simulation
SFR	System frequency response
WF	Wind farm
AS-CBM	CBM based on AS
AS-LSM	LSM based on AS

References

1. Cai, G.; Kong, L. Techno-economic analysis of wind curtailment/hydrogen production/fuel cell vehicle system with high wind penetration in China. *CSEE J. Power Energy Syst.* **2017**, *3*, 44–52. [CrossRef]
2. Li, G.; Li, G.; Zhou, M. Model and application of renewable energy accommodation capacity calculation considering utilization level of inter-provincial tie-line. *Prot. Control Mod. Power Syst.* **2019**, *4*, 1. [CrossRef]
3. Xu, D.; Wu, Q.; Zhou, B.; Li, C.; Bai, L.; Huang, S. Distributed multi-energy operation of coupled electricity, heating and natural gas networks. *IEEE Trans. Sustain. Energy* **2019**, *1*. [CrossRef]
4. Zhou, B.; Xu, D.; Chan, K.W.; Li, C.; Cao, Y.; Bu, S. A two-stage framework for multiobjective energy management in distribution networks with a high penetration of wind energy. *Energy* **2017**, *135*, 754–766. [CrossRef]
5. Xu, H.; Su, J.; Liu, N.; Shi, Y. A grid-supporting photovoltaic system implemented by a VSG with energy storage. *Energies* **2018**, *11*, 3152. [CrossRef]
6. Mcelroy, M.B.; Chen, X. Wind and solar power in the united states: Status and prospects. *CSEE J. Power Energy Syst.* **2017**, *3*, 1–6. [CrossRef]
7. Wu, D.; Javadi, M.; Jiang, J.N. A preliminary study of impact of reduced system inertia in a low-carbon power system. *J. Mod. Power Syst. Clean Energy* **2015**, *3*, 82–92. [CrossRef]
8. Tielens, P.; Van Hertem, D. The relevance of inertia in power systems. *Renew. Sustain. Energy Rev.* **2016**, *55*, 999–1009. [CrossRef]
9. Arkhangelski, J.; Roncero-Sánchez, P.; Abdou-Tankari, M.; Vázquez, J.; Lefebvre, G. Control and restrictions of a hybrid renewable energy system connected to the grid: A battery and supercapacitor storage case. *Energies* **2019**, *12*, 2776. [CrossRef]
10. Ofgem. *Technical Report on the Event of 9 August 2019*; Ofgem: London, UK, 2019.
11. Silva, A.M.L.D.; Castro, A.M.D. Risk assessment in probabilistic load flow via monte carlo simulation and cross-entropy method. *IEEE Trans. Power Syst.* **2019**, *34*, 1193–1202. [CrossRef]
12. Berizzi, A.; Bovo, C.; Delfanti, M.; Merlo, M.; Pasquadibisceglie, M.S. A monte carlo approach for TTC evaluation. *IEEE Trans. Power Syst.* **2007**, *22*, 735–743. [CrossRef]
13. Ahmadyar, A.S.; Riaz, S.; Verbic, G.; Chapman, A.; Hill, D.J. A framework for assessing renewable integration limits with respect to frequency performance. *IEEE Trans. Power Syst.* **2018**, *33*, 4444–4453. [CrossRef]
14. Bu, S.Q.; Du, W.; Wang, H.F.; Chen, Z.; Xiao, L.Y.; Li, H.F. Probabilistic analysis of small-signal stability of large-scale power systems as affected by penetration of wind generation. *IEEE Trans. Power Syst.* **2012**, *27*, 762–770. [CrossRef]
15. Preece, R.; Huang, K.; Milanović, J.V. Probabilistic small-disturbance stability assessment of uncertain power systems using efficient estimation methods. *IEEE Trans. Power Syst.* **2014**, *29*, 2509–2517. [CrossRef]
16. Bu, S.; Wen, J.; Li, F. A generic framework for analytical probabilistic assessment of frequency stability in modern power system operational planning. *IEEE Trans. Power Syst.* **2019**, *34*, 3973–3976. [CrossRef]
17. Anderson, P.M.; Fouad, A.A. *Power System Control and Stability*; Iowa State University Press: Ames, IA, USA, 1977.
18. Operator, A.E.M. *Integrating Renewable Energy-Wind Integration Studies Report*; AEMO: Melbourne, Australia, 2013.

19. Hong, Q.; Nedd, M.; Norris, S.; Abdulhadi, I.; Karimi, M.; Terzija, V.; Marshall, B.; Bell, K.; Booth, C. Fast frequency response for effective frequency control in power systems with low inertia. *J. Eng.* **2019**, *2019*, 1696–1702. [CrossRef]
20. NationalgridESO. *The Day in the Life of SOGL Series-Operational Planning Part 2*; NationalgridESO: Warwickshire, UK, 2018.
21. Huang, S.; Zhou, B.; Bu, S.; Li, C.; Zhang, C.; Wang, H.; Wang, T. Robust fixed-time sliding mode control for fractional-order nonlinear hydro-turbine governing system. *Renew. Energy* **2019**, *139*, 447–458. [CrossRef]
22. Markovic, U.; Chu, Z.; Aristidou, P.; Hug, G. Fast frequency control scheme through adaptive virtual inertia emulation. In Proceedings of the 2018 IEEE Innovative Smart Grid Technologies-Asia (ISGT Asia), Singapore, 22–25 May 2018; pp. 787–792.
23. Rubio, A.; Behrends, H.; Geißendörfer, S.; Maydell, V.K.; Agert, C. Determination of the required power response of inverters to provide fast frequency support in power systems with low synchronous inertia. *Energies* **2020**, *13*, 816. [CrossRef]
24. Choi, Y.W.; Kook, S.K.; Yu, R.G. Control strategy of BESS for providing both virtual inertia and primary frequency response in the Korean Power System. *Energies* **2019**, *12*, 4060. [CrossRef]
25. Gu, H.; Yan, R.; Saha, T.K. Minimum synchronous inertia requirement of renewable power systems. *IEEE Trans. Power Syst.* **2018**, *33*, 1533–1543. [CrossRef]
26. Chow, J.H. *Power Syetem Coherency and Model Reduction*; Spring: New York, NY, USA, 2013.
27. Ackermann, T. *Wind Power in Power Systems*, 2nd ed.; Hon Wiley & Sons: Chichester, UK, 2012.
28. Freris, L.; Infield, D. *Renewable Energy in Power Systems*; Wiley: New York, NY, USA, 2008.
29. Abouzahr, I.; Ramakumar, R. An approach to assess the performance of utility-interactive wind electric conversion systems. *IEEE Trans. Energy Convers.* **1991**, *6*, 627–638. [CrossRef]

© 2020 by the authors. Licensee MDPI, Basel, Switzerland. This article is an open access article distributed under the terms and conditions of the Creative Commons Attribution (CC BY) license (http://creativecommons.org/licenses/by/4.0/).

Article

Research on DFIG-ES System to Enhance the Fast-Frequency Response Capability of Wind Farms

Sijia Tu, Bingda Zhang * and Xianglong Jin

The Key Laboratory of Smart Grid of Ministry of Education, Tianjin University, Tianjin 300072, China; tusijia@tju.edu.cn (S.T.); 2017234016@tju.edu.cn (X.J.)
* Correspondence: bdzhang@tju.edu.cn; Tel.: +86-022-2740-4101

Received: 23 July 2019; Accepted: 18 September 2019; Published: 19 September 2019

Abstract: With the increasing penetration of wind power generation, the frequency regulation burden on conventional synchronous generators has become heavier, as the rotor speed of doubly-fed induction generator (DFIG) is decoupled with the system frequency. As the frequency regulation capability of wind farms is an urgent appeal, the inertia control of DFIG has been studied by many researchers and the energy storage (ES) system has been installed in wind farms to respond to frequency deviation with doubly-fed induction generators (DFIGs). In view of the high allocation and maintenance cost of the ES system, the capacity allocation scheme of the ES system—especially for fast-frequency response—is proposed in this paper. The capacity allocation principle was to make the wind farm possess the same potential inertial energy as that of synchronous generators set with equal rated power. After the capacity of the ES system was defined, the coordinated control strategy of the DFIG-ES system with consideration of wind speed was proposed in order to improve the frequency nadir during fast-frequency response. The overall power reference of the DFIG-ES system was calculated on the basis of the frequency response characteristic of synchronous generators. In particular, once the power reference of DFIG was determined, a novel virtual inertia control method of DFIG was put forward to release rotational kinetic energy and produce power surge by means of continuously modifying the proportional coefficient of maximum power point tracking (MPPT) control. During the deceleration period, the power reference smoothly decreased with the rotor speed until it reached the MPPT curve, wherein the rotor speed could rapidly recover by virtue of wind power so that the secondary frequency drop could be avoided. Afterwards, a fuzzy logic controller (FLC) was designed to distribute output power between the DFIG and ES system according to the rotor speed of DFIG and *SoC* of ES; thus the scheme enabled the DFIG-ES system to respond to frequency deviation in most cases while preventing the secondary frequency drop and prolonging the service life of the DFIG-ES system. Finally, the test results, which were based on the simulation system on MATLAB/Simulink software, verified the effectiveness of the proposed control strategy by comparison with other control methods and verified the rationality of the designed fuzzy logic controller and proposed capacity allocation scheme of the ES system.

Keywords: DFIG; ES; virtual inertia control; capacity allocation; fuzzy logic controller

1. Introduction

With the rising emerging energy crisis and environmental problems, governments around the world are actively investing in the utilization and development of new renewable energy [1]. Wind energy and solar energy, as reliable renewable energy sources, have been exploited rapidly in recent years. However, wind power generation is challenging the safety and stability of the power system, as it results in the increasing penetration of wind power generation into the power grid.

In recent years, wind farms have been required to have frequency regulation capability [2]. The frequency regulation participated in by wind farms commonly involves two aspects: fast-frequency response and primary frequency regulation [3]. The fast-frequency response, which is also called short-term frequency response, is the intrinsically inertial response from synchronous generators to suppress the frequency fluctuation and raise the frequency nadir by means of releasing a large amount of kinetic energy stored in the rotating masses. The fast-frequency response lasts until the frequency drop to the nadir or rise to the peak. The primary frequency regulation is the process where the output active power of generators is adjusted to strike a power balance between generation side and load side so that the frequency is restored within the safe range. Considering the great demand of support power and long duration of the primary frequency regulation, the participation of wind farms in primary frequency regulation means the wind turbines have to operate in the load reduction mode before the frequency fluctuation so as to be able to provide enough reserve active power, which, however, reduces the profit of wind farms because of the waste of wind resources [4]. In spite of the increasing proportion of wind power generation in electricity generation, the primary frequency regulation is still the main task of conventional synchronous generators in view of the fluctuation of wind power output caused by the fluctuation of wind speed [5]. Therefore, the research on wind farms to enhance fast-frequency response capability is more important for the current grid structure.

At present, doubly-fed induction generators (DFIGs) are commonly applied in wind farms. Nevertheless, the rotor speed and the system frequency are decoupled because of the existence of the electronic converter [6], which means that the rotational kinetic energy of the rotor is completely "hidden". The contribution of DFIG to the frequency regulation is almost none, so the frequency stability will be jeopardized with the increasing proportion of wind power generation and the decrease of inertia of the entire system [7]. Consequently, international scholars have put forward a series of schemes to deal with this problem.

Compared with conventional synchronous generators, DFIGs have greater inertial energy due to their wider range of speed regulation. Reference [8] added an accessional differential control into the maximum power point tracking (MPPT) control link in order to lift instant output power by releasing the stored rotational kinetic energy of the rotor once the frequency drop event occurred. The power reference was modified as the frequency deviation was set as input variable in the MPPT control link. Although this scheme enables DFIG to utilize its inertial energy to realize fast-frequency response, DFIG would absorb much energy from the grid to recover its rotor speed, which would lead to the secondary frequency drop. Reference [9] proposed a scheme in which the instant output power of DFIG was maintained at the maximum level before the rotor speed exceeded the safe range once the frequency drop event was detected. Then, the power reference was decreased dramatically so that the rotor speed was able to recover. This scheme enables DFIG to support the frequency with its best effort and prevent over-deceleration of rotor speed in theory. However, because of the slow response of rotor speed control, in practice, the secondary frequency drop still happens when the rotor speed regulation range is too large, as is shown in the simulation results. Taking rotor speed and wind power penetration level into account, reference [10] made DFIG maintain the incremental power for a preset period and forced the rotor speed converge to a stable operating range with power reference decreasing. Because the incremental power varies with the penetration level according to the calculation process of the power reference, DFIG is unable to provide sufficient power to support frequency when the power system has low wind penetration, which should be easier to deal with for wind farms. Furthermore, DFIG is unable to conduct fast-frequency response under the condition of extreme wind speed (too high or too low), as DFIG does not possess surplus inertial energy under extremely low wind speed and cannot increase the output power under extremely high wind speed. Thus, wind farms have difficulties relying on DFIGs alone to respond to system frequency.

The energy storage (ES) system has been widely applied in electrical fields to suppress the fluctuation of output power and regulate frequency because of its stable performance, flexible control, and fast response [10]. Thus, it is suitable for the fast-frequency response in wind farms. In reference [11], in order to avoid the frequency secondary drop caused by rotor speed recovery, the ES was integrated at the AC bus of wind farm to respond to frequency fluctuation with DFIG where DFIG was the main part and ES was the auxiliary part. ES simply provided the energy required in the period of rotor speed recovery or supplemented the insufficient energy supplied by DFIG, so the enormous potential of ES was not made full use of in this scheme. To make the best use of both the capability of ES and the wind resources, reference [12] made only ES take charge of the fast-frequency response, whereas DFIG operated in the MPPT mode without responding to frequency fluctuation. However, there is no doubt that the service life of ES would reduce and the maintenance cost of ES would increase in this scheme. In addition, the capacity of the ES system was given as 10% of the DFIGs' rated power without any mathematical deduction. Herein, whether this scheme could increase the profit of wind farms or not is questionable. In reference [13], a hybrid control strategy is proposed, considering the de-loading (DL) state of wind turbine generator (WTG) and the state of charge (*SoC*) of ES. However, the power reference of the DFIG-ES system for fast-frequency response is calculated on the premise of the power load deviation forecast, which is hard to acquire precisely in practice. Reference [14] integrated an ES system at the DC bus of a DFIG to share the frequency regulation burden, and proposed coordinated control strategy to improve the nadir with consideration of the *SoC* of ES and the operating state of DFIG. Nevertheless, the DFIG-ES system is unable to provide extra active power and participate in frequency regulation under extremely high wind speed because of the capacity limitation of the converter in DFIG. Moreover, this scheme increases the number of ES systems and creates unnecessary maintenance and management cost when the ES system is embedded at the DC bus of DFIG. In summary, the research works mentioned above did not involve an effective coordinated control method for DFIG and ES to respond to frequency disturbance with their most effort under any wind speed.

To enhance the fast-frequency response capability of wind farms, this paper puts forward a coordinated control scheme of the DFIG-ES system with the consideration of different wind speed. The overall power reference of the DFIG-ES system was calculated on the basis of the frequency response characteristic of synchronous generators. Therefore, according to the biggest inertial energy that synchronous generators can provide, a capacity allocation scheme of ES for the fast-frequency response was proposed, and the capacity allocation principle was to make the wind farm possess the same potential inertial energy as that of synchronous generators set with equal rated power. The capacity allocation of ES is meaningful because it reduces the allocation and maintenance cost of ES compared with other schemes. In order to enable DFIG to respond to frequency deviation with ES, once the power reference of DFIG was determined, a novel virtual inertia control method was put forward to release rotational kinetic energy and produce power surge by means of continuously modifying the proportional coefficient of maximum power point tracking (MPPT) control. During the deceleration period, the power reference smoothly decreased with the rotor speed until it reached the MPPT curve, then the rotor speed could rapidly recover by virtue of wind power, so that secondary frequency drop could be avoided. After the deficiencies of DFIG and ES solely accomplishing fast-frequency response were respectively analyzed, a fuzzy logic controller was designed to distribute the output power between DFIG and ES according to the rotor speed of DFIG and *SoC* of ES, which ensured the effective cooperation between DFIG and ES. In the end, the test results, which were based on the simulation system on MATLAB/Simulink software, verified the effectiveness of the proposed control scheme by comparison with other control methods and verified the rationality of the designed fuzzy logic controller and proposed capacity allocation scheme of the ES system.

2. Modeling of the DFIG-ES System

The model structure of the DFIG-ES system introduced in this paper is shown as Figure 1.

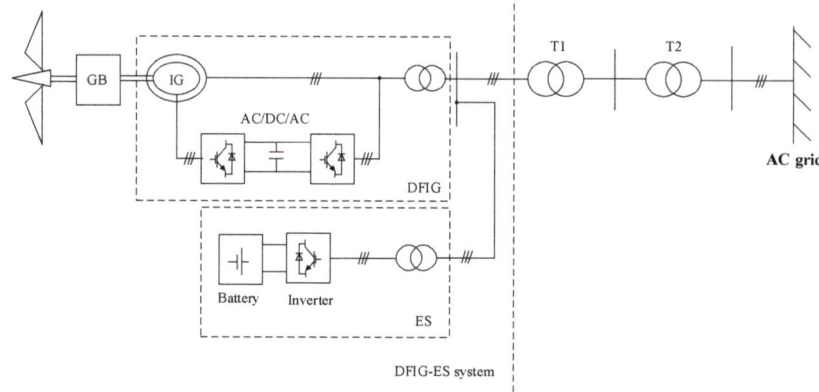

Figure 1. The structure diagram of the doubly-fed induction generator energy storage (DFIG-ES) system. GB is the gear box, IG is the induction generator, and T1 and T2 are transformer 1 and transformer 2, respectively.

2.1. Doubly-Fed Induction Generator Model

In the dq two-phase synchronous rotating coordinate system, the voltage, flux, power, and electromagnetic torque equations of the doubly-fed induction generator are shown as follows [15,16]:

$$\begin{cases} \Psi_{sd} = -L_s i_{sd} + L_m i_{rd} \\ \Psi_{sq} = -L_s i_{sq} + L_m i_{rq} \\ \Psi_{rd} = L_r i_{rd} - L_m i_{sd} \\ \Psi_{rq} = L_r i_{rq} - L_m i_{sq} \\ u_{sd} = p\Psi_{sd} - \omega_e \Psi_{sq} - R_s i_{sd} \\ u_{sq} = p\Psi_{sq} + \omega_e \Psi_{sd} - R_s i_{sq} \\ u_{rd} = p\Psi_{rd} - \omega_r \Psi_{rq} + R_r i_{rd} \\ u_{rq} = p\Psi_{rq} + \omega_r \Psi_{rd} + R_r i_{rq} \\ P_s = 1.5(u_{sd}i_{sd} + u_{sq}i_{sq}) \\ Q_s = 1.5(u_{sq}i_{sd} - u_{sd}i_{sq}) \\ T_e = 1.5p_n L_m(i_{sq}i_{rd} - i_{sd}i_{rq}) \end{cases} \quad (1)$$

where Ψ_{sd}, Ψ_{sq}, Ψ_{rd}, Ψ_{rq} are stator and rotor flux on the d and q axes, respectively; L_s, L_r, L_m are equivalent self-inductance and mutual inductance of the stator and rotor; i_{sd}, i_{sq}, i_{rd}, i_{rq} are stator and rotor current on the d and q axes; u_{sd}, u_{sq}, u_{rd}, u_{rq} are the stator and rotor voltage on the d and q axes; T_e is electromagnetic torque; p_n represents pole pairs of DFIG; ω_e is synchronized angular velocity; ω_r represents rotor angular velocity; P_s is the active power of DFIG; and Q_s is the reactive power of DFIG. Ignoring the change of stator resistance and stator flux, the d axis of the dq synchronous rotating coordinate system is oriented according to the space vector of grid voltage. The relationship between stator-side power and rotor current is shown as follows:

$$\begin{cases} i_{rd} = \frac{L_s}{L_m} \frac{2P_s}{3U_s} \\ i_{rq} = -\frac{L_s}{L_m} \frac{2Q_s}{3U_s} - \frac{U_s}{\omega_e L_m} \end{cases} \quad (2)$$

Once the stator voltage vector control is adopted, the active power of DFIG is controlled by the rotor current on the d axis, and the reactive power is controlled by the rotor current on the q axis. Thus, the active power and reactive power of DFIG are decoupled. The control diagram of the rotor-side converter is shown in the Figure 2 [8]. This control link contains double loops. The outer loop is a power loop, and the inner loop is a current loop. The rotor current reference, i^*_{rd} and i^*_{rq}, in the current loop are respectively determined by the maximum power point tracking control and the reactive power control in the outer loop. In the maximum power point tracking control, the relationship between active power reference value P^*_{opt} and rotor angular velocity ω_r is as follows:

$$P^*_{opt} = \begin{cases} k_{opt}\omega_1^3, \omega_0 < \omega_r < \omega_1 \\ \frac{P_{max} - k_{opt}\omega_1^3}{\omega_{max} - \omega_1}(\omega_r - \omega_{max}) + P_{max}, \omega_1 < \omega_r < \omega_{max} \\ P_{max}, \omega_{max} < \omega_r \end{cases} \quad (3)$$

where k_{opt} is the proportional coefficient of the maximum power point tracking control; ω_0 represents the lowest rotor angular velocity of DFIG; ω_1 is the rated rotor angular velocity of DFIG; ω_{max} is the maximum rotor angular velocity; and P_{max} is the maximum active power output of DFIG.

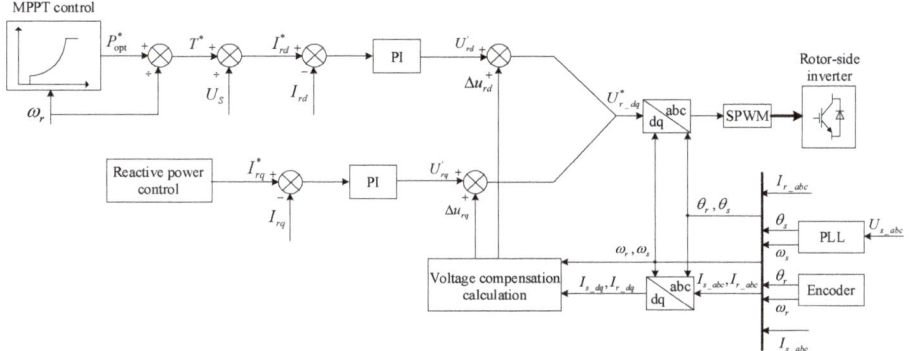

Figure 2. The control diagram of the rotor side converter of DFIG. MPPT: maximum power point tracking. PI: proportion and integral. SPWM: sinusoidal pulse width modulation. PLL: phase locked loop.

2.2. Energy Storage System Model

Different types of batteries can be chosen to make up an energy storage system, such as nickel-cadmium batteries, sulfur-sodium batteries, and lithium-ion batteries, among others. Among them, the energy storage technology of the lithium-ion batteries has developed rapidly over the years [17]. Thus, lithium-ion batteries are integrated into the ES system in this paper.

The ES affiliated to DFIG is controlled by an independent inverter, which features a fast response performance on the frequency regulation. Considering that the capacity of the ES is finite, the state of charge (SoC) of ES should be taken into consideration [18], which is calculated by

$$Q_{SoC}(t) = Q_{SoC}(0) - \frac{\int_0^t P_{ES}(\tau)d\tau}{E_{rated}} \times 100\% \quad (4)$$

where $Q_{SoC}(t)$ is the state of charge at time t; $Q_{SoC}(0)$ is the initial state of charge; $P_{ES}(\tau)$ is the output power (positive when ES is discharging) of ES at time τ; t is the operation time of ES; and E_{rated} is the rated capacity of ES.

Considering that the capacity of ES is a main factor influencing the frequency regulating capability of wind farms, especially in the case where DFIG is unable to respond to frequency deviation under extreme wind speed, a capacity allocation scheme of ES will be discussed later in the paper.

3. Capacity Allocation Scheme of ES

In reference to power system analysis, inertia time constant H is defined as the ratio of rotor inertial energy to rated capacity of generator at synchronous angular velocity [19]. Since the inertial energy of DFIG is not available naturally during a frequency fluctuation and the amount of virtual inertial energy provided by DFIG depends on the parameters in virtual inertial control link (discussed in the next section), the potential inertia time constant H_{SWE} of power systems containing wind farms and conventional power plants can be represented by

$$H_{SWE} = \frac{\sum_{i=1}^{n}(\frac{1}{2p_i^2}J_i\omega_e^2) + \sum_{j=1}^{m}E_{DFIG,j} + \sum_{k=1}^{l}E_{ES,k}}{S_{N\_all}} \quad (5)$$

where n, m, l is the number of synchronous generators of DFIG and ES systems, respectively; P_i is the pole pairs of synchronous generator i; J_i represents the rotational inertia of synchronous generator i; $E_{DFIG,j}$ is the rotational kinetic energy of DFIG j; $E_{ES,k}$ is the storage energy of ES k; and S_{N_all} is the total rated capacity of the generators in power system.

The DFIG operating in the MPPT mode, unlike the synchronous generator whose rotor speed is coupled with the system frequency, cannot exchange power with the grid when the system frequency fluctuates. Therefore, large-scale wind farms connected to the grid will inevitably lead to the reduction of system inertia, and the allocated capacity of ES will directly affect the inertia of the wind farm and even the whole power system. However, in view of the relatively high configuration and maintenance cost of ES, the larger the capacity of ES means the higher construction and maintenance cost of wind farms. Considering that synchronous generators contribute with the most electricity generation, adjusting the output energy of the wind farm to be approximately same as the inertial energy produced by conventional synchronous generators would reduce the negative impact of large-scale wind farms on frequency stability of the power system and be convenient for grid staff to dispatch the power grid [20]. Therefore, the capacity allocation scheme of ES installed in wind farms for fast-frequency response is proposed. The capacity allocation principle is to make the wind farm have the same potential inertial energy as that of synchronous generators set with equal rated power.

According to the power grid operating regulations in China [21], the rotor angular velocity of the synchronous generator is usually limited between 0.95 p.u and 1 p.u during frequency regulation. The maximum rotational kinetic energy $\Delta E_{S,MAX}$ released from the synchronous generator is

$$\Delta E_{S,Max} = \frac{1}{2}J(1 - 0.95^2)\omega_e^2 = 0.0488J\omega_e^2. \quad (6)$$

The rotational kinetic energy of the synchronous generator at rated angular velocity is

$$E_S = \frac{1}{2}J\omega_e^2 = \frac{1}{2}P_N T_J \quad (7)$$

where T_J is the inertia time constant of the synchronous generator; P_N is the rated output active power of the synchronous generator; $\Delta E_{S,MAX}$ can be regarded as the maximum inertial energy produced by the synchronous generator during fast-frequency response; The time during the frequency dropping from normal to the nadir is ΔT, and therein the virtual inertial energy released by ES ΔE_{ESS} is

$$\Delta E_{ESS} = \overline{P_{ESS}} * \Delta T = 0.0488 P_N T_J \quad (8)$$

where $\overline{P_{ESS}}$ is the average output power of ES during the fast-frequency response; and ΔT represents the time duration of the fast-frequency response. Then, $\overline{P_{ESS}}$ can also be expressed as

$$\overline{P_{ESS}} = 0.0488 P_N \frac{T_J}{\Delta T}. \quad (9)$$

Generally, the inertia time constant of the synchronous generator is about 4 to 18 s, and the fast-frequency response lasts about 7 to 15 s [22]. Thus, we set T_J as 11 s and ΔT as 11 s (average value), giving

$$\overline{P_{ESS}} = 0.0488 P_N. \tag{10}$$

Considering the huge instantaneous throughput of the ES system [23], the time required for the output power to reach the specified value, compared with the time duration of the fast-frequency response, can be negligible. In view of a certain output power fluctuation margin, the maximum output power of the ES is set as 5% of the rated power of the wind farm.

Deep charging and discharging will seriously reduce the service life of the ES system [24]. In this paper, the charging and discharging range of ES is maintained between 10% and 90% of the rated capacity. Normally, the initial *SoC* of the ES system is 50%, so the available storage energy used for fast-frequency response equals 40% of the rated capacity. As the time duration of fast-frequency response is supposed as 11 s, referring to Equation (10), the rated capacity of the ES E_{ESS} is calculated as follows:

$$E_{ESS} = \overline{P_{ESS}} * (11/3600)/0.4 = 0.00764 P_N. \tag{11}$$

Therefore, the rated capacity of the ES system affiliated to DFIGs equaled to 0.764% of the rated power of the wind farm.

4. Coordinated Control Strategy of DFIG-ES System with Consideration of Wind Speed

4.1. Division of Wind Speed Region

When DFIG is in normal operation, converter control and pitch control are comprehensively applied to adjust the operation state of wind turbine [10,25,26]. In the case of low wind speed, the control purpose is to make the operating point follow the maximum power point tracking curve and capture the maximum wind energy. With the increment of wind speed, the rotor speed reaches the limit and maintains at the maximum. In the case of high wind speed, the wind energy capturing efficiency is reduced by adjusting pitch angle so that the output power of DFIG is maintained at the maximum (rated power) [27]. The relationship among rotor speed, active power, pitch angle and wind speed is shown in Figure 3.

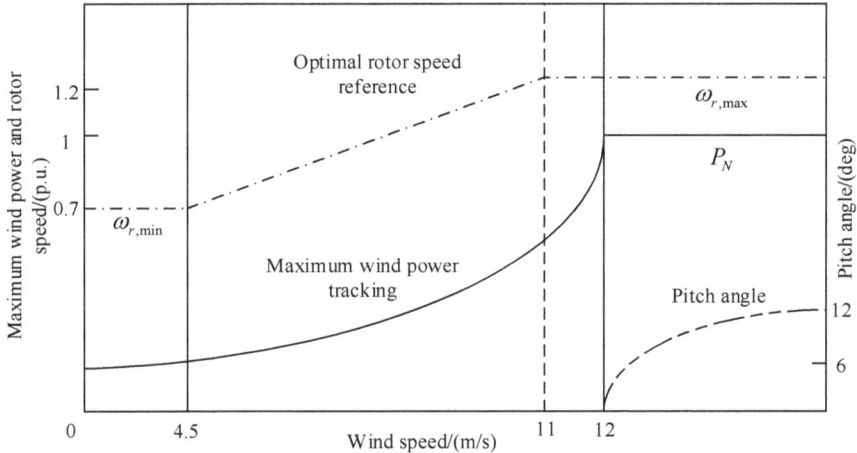

Figure 3. The operation state diagram of DFIG. $\omega_{r,min}$ is the minimum rotor angular velocity; $\omega_{r,max}$ is the maximum rotor angular velocity; and P_N is the rated active power of DFIG.

Because the operation state of DFIG is closely related to the wind speed, the wind speed was divided into three regions before the control strategy was discussed in this paper:

(1) Low speed region ($v_{wind} < 6$ m/s)—the rotor speed just maintains at ω_{min}, so the DFIG cannot respond to the frequency drop by releasing the rotational kinetic energy.

(2) Middle speed region (6 m/s ≤ v_{wind} ≤ 12 m/s)—the DFIG operates in MTTP mode to maximize the efficiency of capturing wind energy. The fast-frequency response can be achieved by releasing the rotational kinetic energy, except for the scenario where the output power is close to the maximum.

(3) High speed region (12 m/s < v_{wind})—although the rotational kinetic energy is huge, the output power of the DFIG is limited by the capacity of the converter and no excess energy can be generated, so the DFIG cannot undertake the task of fast-frequency response.

It can be seen that when the wind speed belongs to the low or high speed region, the DFIG cannot provide inertial energy and only relies on ES to complete fast-frequency response. Next, we will discuss the coordinated control strategy of the DFIG-ES system when the wind speed belongs to the middle speed region. It was noted that the control objective is very easy to achieve when frequency suddenly rises because DFIG can operate in load reduction mode and ES can operate in charging mode. Thus, in this paper, we only discuss frequency drop incident.

4.2. Virtual Inertia Control of DFIG-ES System

During the period of fast-frequency response, in order to make the wind farm emit the same amount of inertial energy as the synchronous generators set with equal rated power, the energy emitted by the DFIG-ES system can be represented as

$$\Delta E_{DFIG} + \Delta E_{ES} = \frac{1}{2p_n^2} J_{DFIG}[(\omega_{r0} + \Delta \omega_r)^2 - \omega_{r0}^2)] + \Delta E_{ES} \\ = \frac{1}{2p_n^2} J_{vir}[(\omega_e + \Delta \omega_e)^2 - \omega_e^2)] \quad (12)$$

where ΔE_{DFIG} is the energy emitted by DFIG; ΔE_{ES} is the energy emitted by ES; J_{DFIG} is the inertia constant of DFIG; J_{vir} represents the virtual inertia constant of the DFIG-ES system; $\Delta \omega_e$ is the change of synchronized angular speed; $\Delta \omega_r$ is the rotor speed change of DFIG; and ω_{r0} is the initial rotor angular velocity of DFIG.

Supposing that the energy emitted by DFIG accounts for ∂ ($0 \leq \partial \leq 1$) of the total energy emitted by the DFIG-ES system, the equivalent virtual inertia constant of the DFIG can be represented as

$$\begin{cases} J_{vir,DFIG} = \partial \frac{(2\omega_{r0}+\Delta\omega_r)\Delta\omega_r}{(2\omega_e+\Delta\omega_e)\Delta\omega_e} J_{DFIG} \approx \partial \frac{\omega_{r0}\Delta\omega_r}{\omega_e\Delta\omega_e} J_{DFIG} = \partial \lambda \frac{\omega_{r0}}{\omega_e} J_{DFIG} \\ \lambda = \frac{\Delta\omega_r}{\Delta\omega_e} \end{cases} \quad (13)$$

where λ is the rotor speed regulation coefficient, and $J_{vir,DFIG}$ is the equivalent virtual inertia constant of DFIG.

According to Equation (13), it can be seen that the equivalent virtual inertia constant of the DFIG is determined by not only the inertia constant of the DFIG, but also the initial angular velocity ω_{r0}, the rotor speed regulation coefficient λ, and the output power proportion of DFIG ∂.

On the basis of Equation (3) and Figure 2, in order to complete the fast-frequency response by releasing the kinetic energy, we can modify the proportional coefficient k_{opt} to adjust the operation state of DFIG. Supposing the frequency change is Δf, the proportional coefficient k_{opt} is modified to k_{opt}^* during the frequency regulation period, with the rotor angular velocity ω_0 decreasing to ω_1 to release rotational kinetic energy. The decrease range of rotor speed should not be too large (by limiting output power proportion of DFIG ∂ and active power reference) so as to ensure that the rotor will not absorb too much energy from the grid during the rotor speed recovery period. According to the maximum power point tracking curves under different wind speeds, if the rotor speed regulation is within the safe range, the secondary frequency drop would be avoided. Referring to Equations (13) and (3), supposing the range of rotor speed regulation is not too large, then we can find

$$\begin{cases} k_{opt}^* \omega_1^3 \approx k_{opt} \omega_0^3 \\ \omega_1 = \omega_0 + \Delta\omega_r = \omega_0 + \lambda^* \Delta\omega_e = \omega_0 + \partial \lambda 2\pi \Delta f \end{cases} \quad (14)$$

where λ^* is the modified rotor speed regulation coefficient with consideration of output power proportion ∂. Thus, the modified proportional coefficient k_{opt}^* can be represented as

$$k_{opt}^* = \frac{\omega_0^3}{(\omega_0 + \partial \lambda 2\pi \Delta f)^3} k_{opt}, \quad (15)$$

which follows the restrictions

$$\begin{cases} k_{opt}^* \leq \frac{P_{max}}{\omega_0^3} \\ \omega_{r,min} \leq \omega_1 = \omega_0 + \partial \lambda 2\pi \Delta f \leq \omega_{r,max} \end{cases}. \quad (16)$$

Therefore, if we input the frequency change Δf into the MPPT control link of DFIG, then the fast-frequency response and rotor speed recovery can be realized by modifying the proportional coefficient according to Equation (15). In particular, when the fast-frequency response is over, as the primary frequency regulation continues and the frequency deviation fades, the proportional coefficient k_{opt}^* changes smoothly to the original value and the operating point moves smoothly back to the original place. In the light of relatively long duration of primary frequency regulation, the energy required for rotor speed recovery depends mainly on the wind energy (supposing the wind speed remains), and secondly on the huge ramp rate of synchronous generators set. Therefore, as long as the speed regulation range is not particularly large, the active power shortage caused by the rotor speed recovery can be almost negligible.

The fast-frequency response progress of DFIG is shown in Figure 4. The DFIG originally operates at point A on the MPPT curve, assuming that the wind speed remains at 9 m/s throughout the period. When the frequency drop event is detected, the power reference is surged immediately and the operating point of DFIG moves rapidly to point P. Because the electromagnetic power of DFIG is larger than the mechanical power, the speed decreases rapidly. At the same time, the output power decreases until the operating point reaches the MPPT curve at point B, where the electromagnetic power is equal to the mechanical power. As is shown in Figure 4, the power shortage caused by rotor speed regulation is very small, and thus Equation (14) is reasonable. The frequency deviation decreases gradually as the primary frequency regulation processes, so the operating point moves smoothly from B to A. It is noteworthy that the energy required in rotor speed recovery period is mainly from the wind energy. Therefore, the secondary frequency drop can be avoided in this scheme, which will be proved in the study case.

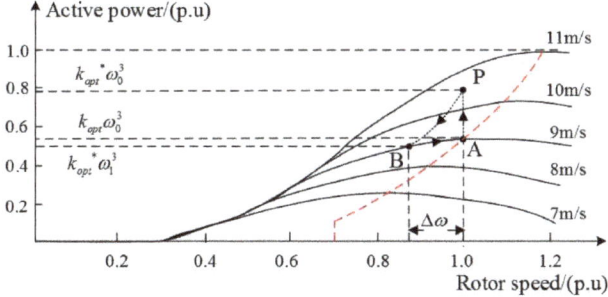

Figure 4. The fast-frequency response progress diagram of DFIG. $\Delta\omega$ is the rotor speed regulation during fast-frequency response. The red curve is the MPPT curve.

According to the Equations (7) and (12), the energy emitted by ES, which accounts for $1 - \partial$ ($0 \leq (1 - \partial) \leq 1$) of the output energy, can be expressed as

$$\Delta E_{ES} = \tfrac{1-\partial}{2} P_N T_J [(\omega_e + \Delta \omega_e)^2 - \omega_e^2)]$$
$$= 2(1-\partial)\pi^2 P_N T_J [1 - \tfrac{(f_N + \Delta f)^2}{f_N^2}] \qquad (17)$$

Considering that the output power released by ES is basically stable, the output power of ES ΔE_{ES} can be expressed as

$$\begin{cases} P_{ES} = 2(1-\partial)\pi^2 [1 - \tfrac{(f_N + \Delta f)^2}{f_N^2}] P_N \\ \left|\tfrac{df}{dt}\right| > threshhold \end{cases} . \qquad (18)$$

4.3. Determination of Output Power Proportion

Both DFIG and ES have to be taken into account to determine the output power proportion ∂. For DFIG, if its output power proportion is too large, it may face the risk of torsional vibration in the shaft system caused by too much rotor speed regulation [28], the risk of an unstable power grid caused by a rotor speed recovery range that is too large, and the risk of DFIG disconnection due to rotor speed declining below the threshold [29]. For ES, if its output power proportion is too large, it may face the risk of service life reduction as the result of frequent deep charging or discharging, which will cause excess maintenance cost. Therefore, the determination of output power proportion is a complicated problem where the appropriate mathematical expressions are difficult to establish.

The fuzzy logic control method shows its functionality when the control process cannot be described by a concrete mathematical model [29–31]. The principle diagram of the fuzzy logic control is shown in Figure 5.

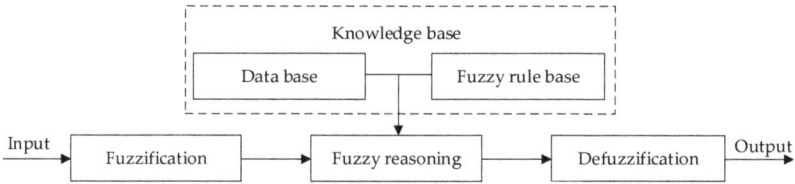

Figure 5. The principle diagram of fuzzy logic control.

Then the fuzzy logic controller (FLC) to determine the output power proportion can be designed as follows:

(1) The structure of fuzzy logic controller: The rotor speed ω_r of DFIG and *SoC* of ES are selected as input variables of FLC, and the output power proportion of DFIG ∂ is selected as the output variable of FLC. The corresponding FLC structure is shown in Figure 6.

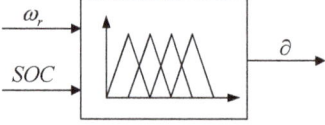

Figure 6. The structure diagram of fuzzy logic controller (FLC).

(2) The fuzzy state and the shape of membership function of fuzzy subset: Determining the fuzzy state means determining the quantity of linguistic values of variables. If the quantity of linguistic values is large then the control rules are precise, which will lead to the complexity of the controller.

If the quantity of linguistic values is small then the control rules are easy to implement, but too few linguistic values will result in a coarse controller [32]. Therefore, both simplicity and precision should be considered. Seven fuzzy domains for input variables are set: VS (very small), MS (middle small), S (small), M (middle), L (large), ML (middle large), VL (very large). Five fuzzy domains for output variable are set: VS (very small), S (small), M (middle), L (large), VL (very large). The common shapes of membership function are triangle and trapezoid. Compared with the SoC and ∂, the rotor speed is a more precise variable with clearer discrimination, so a triangle is chosen as the shape of the membership function of the rotor speed ω_r. The domain of ω_r is [0.7, 1.2] p.u. The trapezoidal is chosen as the shape of membership function of SoC. The domain of SoC is [0.1, 0.5] p.u. When $SoC > 0.5$ p.u, the value of SoC is set as 0.5 p.u. The trapezoidal is chosen as the shape of membership function of ∂. The domain of ∂ is [0, 1] p.u. The membership functions used in FLC are shown in Figure 7.

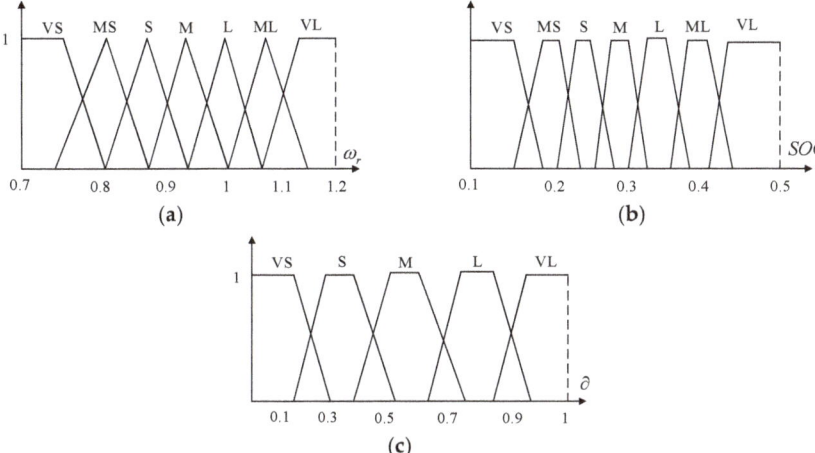

Figure 7. The membership functions used in FLC. (a) The membership function of ω_r. (b) The membership function of SoC. (c) The membership function of ∂.

(3) The fuzzy rules: The rules of the fuzzy logic controller are determined according to the engineering experience accumulated by the engineering experts during their long-term work [29]. The fuzzy rules designed in this paper are mainly based on the following engineering experience:

a. If the rotor speed of the DFIG is too low or too high, the DFIG cannot release excess energy by reducing the rotor speed. If the SoC of the ES is at a high level, the fast-frequency response is completed by the ES alone.

b. If the SoC of the ES is lower than 10%, the ES cannot emit energy to respond to frequency drop. If the rotor speed of the DFIG is relatively high, the fast-frequency response is completed by the DFIG alone.

c. If the rotor speed of the DFIG is relatively high and the SoC of the ES is higher than 10%, the frequency response is completed by DFIG and ES altogether, and the appropriate output power proportion of each device is determined according to their stored virtual inertial energy.

The fuzzy logic control rules are shown in Table 1.

Table 1. The rules of FLC.

∂ ω_r SOC	VS	MS	S	M	L	ML	VL
VS	VS	VS	M	--	--	--	--
MS	VS	VS	L	VL	--	--	--
S	VS	VS	M	VL	VL	VL	--
M	VS	VS	M	L	VL	VL	VL
L	VS	VS	S	M	VL	VL	VL
ML	VS	VS	S	M	L	VL	VL
VL	VS	VS	VS	S	S	VS	VS

(4) Defuzzification: Maximum of mean method, maximum average method, and center of gravity method are commonly used for defuzzification [29]. Considering the error mined by the center of gravity method is the least among these methods, the center of gravity method is applied for defuzzification. The equation is as follows:

$$\partial_{eout}{}^* = \frac{\int_a^b u_c(\partial_{eout})\partial_{eout}d\partial_{eout}}{\int_a^b u_c(\partial_{eout})d\partial_{eout}} \quad (19)$$

where ∂_{eout}^* is the value of output power proportion; ∂_{eout} is the non-fuzzy value of output power proportion; (a, b) is the range of ∂_{eout}; and $u_c(\partial_{eout})$ is the membership function of ∂_{eout}.

The coordinated control strategy of the DFIG-ES system for fast-frequency response is shown in Figure 8 (if the wind belongs to the high or low speed regions, the ∂ is set as 0).

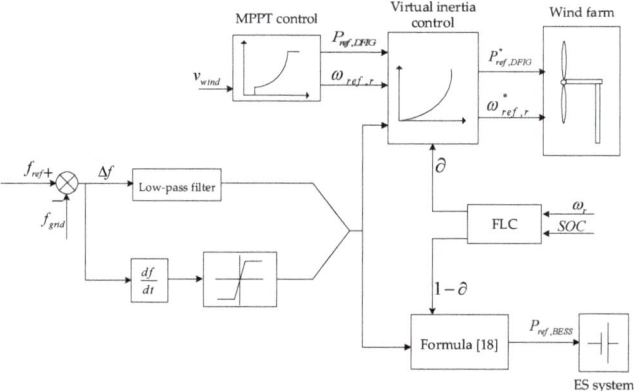

Figure 8. The control diagram of DFIG-ES system. f_{ref} is the reference of grid frequency; f_{grid} is the detected grid frequency; $P_{ref,BESS}$ is the active power reference of ES system; $P_{ref,DFIG}$ is the active power reference of DFIG; $P_{ref,DFIG}^*$ is the modified active power reference of DFIG; $\omega_{ref,r}$ is the reference of rotor speed; $\omega_{ref,r}^*$ is the modified reference of rotor speed; v_{wind} is the detected wind speed; and ω_r is the detected rotor speed.

5. Case Study

5.1. Introduction to the Simulation System

In order to verify the effectiveness of the proposed control strategy and the rationality of the designed fuzzy logic controller and proposed capacity allocation scheme of the ES system, a simulation system, whose structure is shown in Figure 9, was built on MATLAB/Simulink software. Because the proposed control strategy referring to single DFIG and single ES is universal, an equivalent DFIG was selected to represent the wind farm. The simulation system consisted of an ES system following the capacity allocation scheme, as well as three synchronous generators representing three power plants, transformers, loads, and equivalent lines. The battery ES model used in this paper was from MATLAB/Simulink original library, and this model was an integration of serial and parallel combinations of multiple single cells. The detailed parameters of each component in the simulation system were obtained from [12,33] and are shown in Table 2.

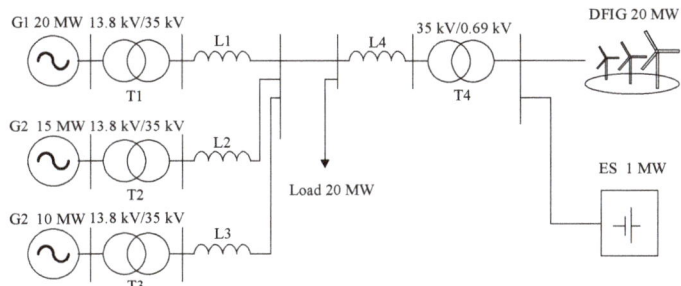

Figure 9. Structure diagram of simulation system.

Table 2. Parameters of each component in the simulation system. (a) Parameters of lines. (b) Parameters of DFIG. (c) Parameters of synchronous generator.

Line	R (Ω/Km)	X (Ω/Km)	B (S/Km)	Length (Km)
L1	0.065	0.315	3.65E-6	10
L2	0.065	0.315	3.65E-6	10
L3	0.065	0.315	3.65E-6	10
L4	0.065	0.315	3.65E-6	15

(a)

Rs	Rr	Xls	Xlr	Xlm	H
0.058 p.u.	0.105 p.u.	0.087 p.u.	0.118 p.u.	3.78 p.u.	5.75 p.u.

(b)

X_d	X_d'	X_d''	X_q	X_q'	X_q''	H
1.8 p.u.	0.1 p.u.	0.25 p.u.	1.7 p.u.	0.55 p.u.	0.25 p.u.	5 s
T_{do}'	T_{do}''	T_{qo}'	T_{qo}''	R_a	K_D	
8 s	0.03 s	0.4 s	0.05 s	0.025 p.u.	1	

(c)

As mentioned above, for the frequency rise event, the frequency response can be easily achieved by increasing the rotor speed or adjusting the pitch angle to reduce the active output of the wind farm. Moreover, the reduction of output power of two devices and power proportion between two devices can also be used in reference to the control strategy proposed in this paper. Therefore, simulation tests involved only the frequency drop incident. The simulation scenario was that the load power increased by 20 MW at 5 s.

5.2. Simulation Result

The coordinated control strategy of DFIG-ES, the inertia control of only DFIG [8], and no inertia control were simulated. The simulation results of these three schemes were compared to verify the effectiveness of the proposed method and its superiority over the traditional inertia control method of DFIG. Considering that the DFIG-ES system operating under low or high wind speed only relies on ES to achieve a fast-frequency response, this paper chose middle and high wind speed as simulation scenarios. Furthermore, in order to verify the rationality of the fuzzy logic controller proposed in this paper, the scenarios where the DFIG-ES system with different *SoC* under middle wind speed were simulated.

5.2.1. Simulation Result under Middle Wind Speed

In the simulation, the wind speed was set as 9 m/s and the initial *SoC* of ES was set as 50%. The simulation result under middle wind speed is shown in Figure 10. Figure 10a shows the system frequency curve. The output active power of DFIG and rotor speed curves are shown in Figure 10b,c. The output active power of ES is shown in Figure 10d.

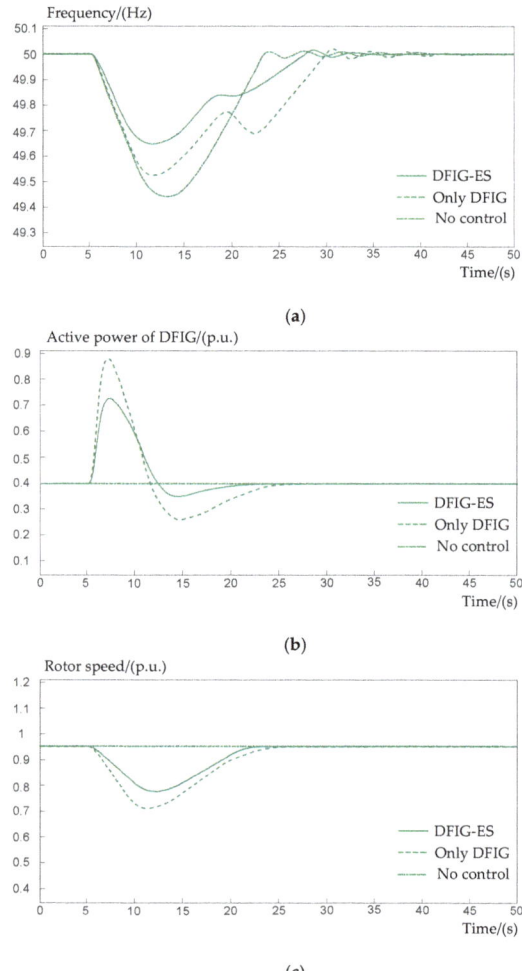

Figure 10. *Cont.*

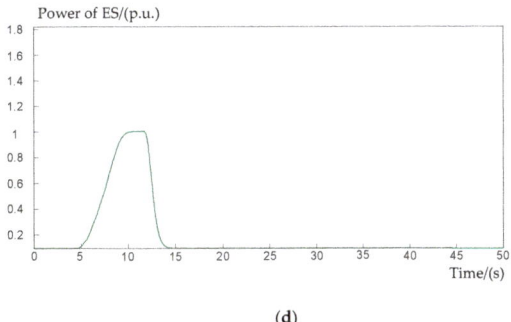

(d)

Figure 10. Simulation result under middle wind speed. (a) System frequency. (b) Active power of DFIG. (c) Rotor speed of DFIG. (d) Active power of ES.

5.2.2. Simulation Result under High Wind Speed

Under the conventional inertia control method, the output active power of DFIG reached the maximum power in this scenario, and thus the frequency response effect under inertia control was the same as that under no inertia control. Therefore, in this part, the coordinated control method and no control method were taken as comparison. The wind speed was set as 13 m/s and the initial SoC of ES was set as 50%. The frequency curves are shown in Figure 11. Figure 11a shows the system frequency curve. The output active power of the ES is shown in Figure 11b.

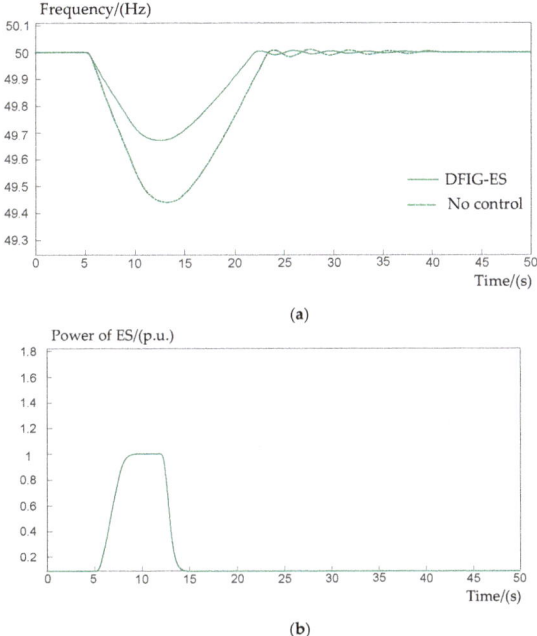

Figure 11. Simulation result under high wind speed. (a) System frequency. (b) Active power of ES.

5.2.3. Simulation Result under Different SoC

When the wind speed was 9 m/s, the initial SoC of ES was set as 30%, 40%, and 50%, respectively. The frequency curves are shown in Figure 12a. The output active power curves of DFIG, and the rotor speed curves and output active power curves of ES are respectively shown in Figure 12b–d.

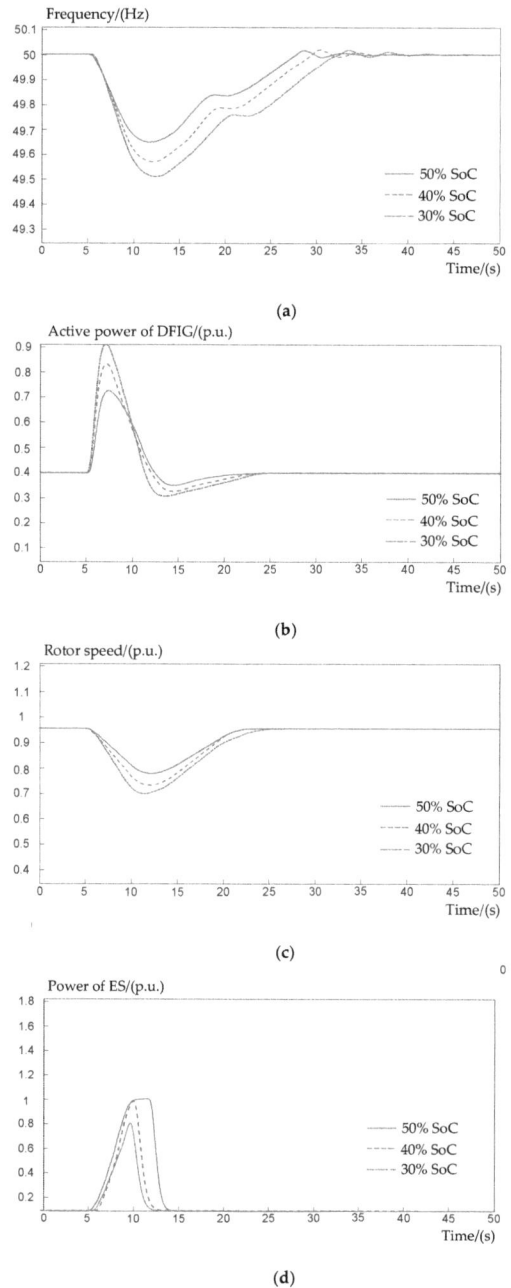

Figure 12. Simulation result under different state of charge (*SoC*). (**a**) System frequency. (**b**) Active power of DFIG. (**c**) Rotor speed of DFIG. (**d**) Active power of ES.

5.2.4. Simulation Result When *SoC* Is below 10%

When the wind speed was 9 m/s and *SoC* was below 10% (wherein the ES system cannot work), in the simulation scenario the load power increased by 10 MW at 5 s. The frequency curves are shown in Figure 13a. The output active power curves of DFIG and the rotor speed curves are respectively shown in Figure 13b,c.

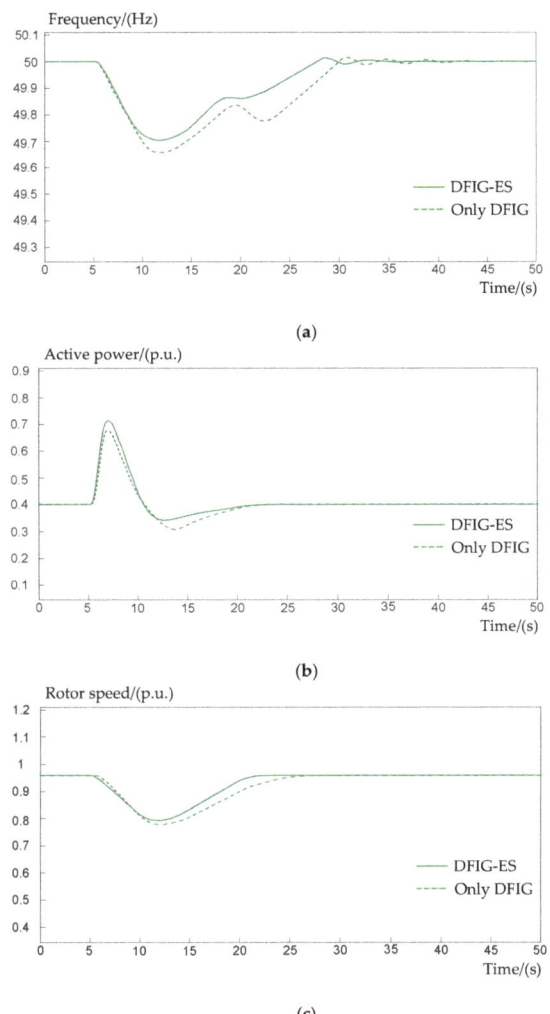

Figure 13. Simulation result when *SoC* was below 10%. (**a**) System frequency. (**b**) Active power of DFIG. (**c**) Rotor speed of DFIG.

6. Discussion

Under middle wind speed, the cooperative frequency support of DFIG and ES was the objective of the proposed control strategy. The nadir of frequency under the coordinated control method was 49.65 Hz, which was higher than the other two nadirs. This means the coordinated control method effectively decreased the frequency drop range and was conducive to maintaining the frequency stability of the power grid. In Figure 10a, the frequency curve under the conventional inertia control of DFIG incurred a second drop due to the large active power shortage caused by rotor recovery, which jeopardized the safety of the power grid. Because the output active power of DFIG was determined by Equation (15) and the fuzzy logic controller, it decreased with the decline of rotor speed until the fast-frequency response finished. It can be seen that the regulation range of rotor speed under the coordinated control method was the smallest because the frequency regulation was also supported by ES. The DFIG operated only in MPPT mode when no inertia control was taken, so it could not respond to frequency fluctuation. The output power of the ES, determined by Equation (18), climbed to the peak gradually and dropped fast once the fast-frequency response finished. Compared with frequency regulation effects of the other two methods, the coordinated control method not only completed the fast-frequency response, but also raised the nadir of frequency and alleviated the frequency fluctuation without causing a secondary frequency drop.

Under high wind speed, DFIG cannot respond to the frequency fluctuation, but can rely on the ES to complete the fast-frequency response. The frequency nadir under the coordinated control method was 49.68 Hz, which was obviously higher than the nadir under the no control method. Because of the huge throughput and stable output of the ES, the nadir under the high wind speed was slightly higher than that under the middle wind speed. Furthermore, frequency deviation during the primary frequency regulation was alleviated because of the lack of the rotor speed recovery. The active power output of the ES shown in Figure 11b was similar to that in Figure 10d, while the output power stayed on the peak for longer duration. The simulation example validated that the wind farm operating under high wind speed relied on the ES with proposed capacity to complete the fast-frequency response. The frequency fluctuation was effectively restrained and the frequency nadir was improved by ES, which made up for the inability to participate in the frequency response of the DFIG.

Under different SoC, the wind speed was set as 9 m/s, and thus the initial rotor speed was fixed. The inertial energy possessed by the DFIG-ES system increased with the increment of initial SoC. Therefore, the frequency nadir was 49.65 Hz, higher than the other two, when the SoC was 50%. The output power proportion between DFIG and ES was determined by fuzzy logic controller. According to the fuzzy rules, once the initial rotor speed is fixed, the output power proportion of DFIG decreases with the increment of SoC. It can be seen that the range of rotor speed regulation of DFIG was the smallest, the frequency oscillation during rotor speed recovery process was the smallest, and the overall output energy of DFIG was the smallest when the SoC of ES was 50%. This means that the higher the SoC, the more priority was given to reducing the mechanical loss of the DFIG; the lower the SoC, the more priority was given to prolonging the service life of the ES. This simulation verifies that the fuzzy logic controller effectively determined the output power proportion between DFIG and ES under the middle wind speed. The maintenance cost of both DFIG and ES was comprehensively reduced under the premise of meeting the demand of fast-frequency regulation.

When the SoC was below 10%, the ES system could not assist the DFIG to support frequency regulation. Thus, the DFIG only relied on itself to complete the fast-frequency response through virtual inertia control. Seemingly, the output active power of the DFIG-ES system showed the faster response and smaller power shortage. Therefore, the frequency nadir was improved to 49.71 Hz and the rotor speed gained a quick recovery under the proposed control method. However, when the frequency nadir was 49.64 Hz, a small secondary frequency drop incurred under the conventional inertia control method. This simulation result proves the excellent performance of DFIG in cases where the ES is unable to work, and verifies the superiority of the proposed virtual inertia control over the conventional inertia control.

7. Conclusions

In this paper, a novel coordinated control strategy was presented for the DFIG-ES system to enhance the fast-frequency response capability of wind farms.

The overall power reference of the DFIG-ES system was calculated on the basis of the frequency response characteristic of synchronous generators. On basis of this, the capacity allocation scheme of the ES system was presented and its principle was to make the wind farm possess the same potential inertial energy as that of synchronous generators set with equal rated power. The simulation results demonstrate that this scheme ensures the ES system is economical but possesses enough reserved energy to complete a fast-frequency response in most cases. Meanwhile, the scheme saves allocation and maintenance costs to some extent, compared with previous ES capacity allocation schemes.

A virtual inertia control of the DFIG-ES system was proposed to produce active power surge to improve the frequency nadir, and a fuzzy logic controller was designed to distribute the active power between DFIG and ES. In particular, the proposed virtual inertia control of DFIG ensured the rotor speed recovery without causing secondary frequency drop. To demonstrate the effectiveness of the proposed control scheme, a test system containing synchronous generators and a DFIG-ES system was built on MATLAB/Simulink software. The simulation results indicated the proposed control strategy made full use of the potential of both DFIG and ES to enhance the fast-frequency response capability of wind farms. The DFIG and ES could cooperate dynamically for fast-frequency response and improve the stability of the power grid and avoid the secondary frequency drop accident. Compared with conventional inertia control of only DFIG, the proposed control strategy is more reliable and adaptable.

Author Contributions: Conceptualization, S.T.; methodology, S.T.; software, S.T., X.J.; validation, S.T., X.J.; formal analysis, S.T.; investigation, S.T.; resources, S.T., X.J.; data curation, S.T.; writing—original draft preparation, S.T.; writing—review and editing, S.T., B.Z.; visualization, S.T.; supervision, B.Z.; project administration, B.Z.; funding acquisition, B.Z.

Funding: This research was funded by the National Natural Science Foundation of China, grant number 51477114.

Conflicts of Interest: The authors declare no conflict of interest.

References

1. Lu, M.S.; Chang, C.L.; Lee, W.J.; Wang, L. Combining the Wind Power Generation System with Energy Storage Equipment. *IEEE Trans. Ind. Appl.* **2009**, *45*, 2109–2115. [CrossRef]
2. Wang, P.; Gao, Z.; Bertling, L. Operational Adequacy Studies of Power System with Wind Farms and Energy Storages. *IEEE Trans. Power Syst.* **2012**, *27*, 2377–2384. [CrossRef]
3. Hong, N.; Nakanish, Y. Optimal Scheduling of an Isolated Wind-Diesel-Energy Storage System Considering Fast Frequency Response and Forecast Error. *Energies* **2019**, *12*. [CrossRef]
4. Vidyanandan, V.; Senroy, N. Primary Frequency Regulation by Deloaded Wind Turbines Using Variable Droop. *IEEE Trans. Power Syst.* **2013**, *28*, 837–846. [CrossRef]
5. Gevorgian, V.; Zhang, Y.; Ela, E. Investigating the impacts of wind generation participation in interconnection frequency response. *IEEE Trans. Sustain. Energy* **2015**, *6*, 1004–1012. [CrossRef]
6. Li, C.; Zhan, P.; Wen, J.; Yao, M.; Li, N.; Lee, W.J. Offshore Wind Farm Integration and Frequency Support Control Utilizing Hybrid Multiterminal HVDC Transmission. *IEEE Trans. Ind. Appl.* **2014**, *50*, 2788–2797. [CrossRef]
7. Dreidy, M.; Mokhlis, H.; Mekhilef, S. Inertia Response and Frequency Control Techniques for Renewable Energy Sources. *IEEE Trans. Sustain. Energy* **2017**, *69*, 144–155. [CrossRef]
8. Li, H.; Zhang, X.; Wang, Y.; Zhu, X. Virtual Inertia Control of DFIG-based Wind Turbines Based on the Optimal Power Tracking. *Proc. Chin. Soc. Electr. Eng.* **2012**, *32*, 32–39.
9. Yang, D.; Kim, J.; Kang, Y. Temporary Frequency Support of a DFIG for High Wind Power Penetration. *IEEE Trans. Power Syst.* **2018**, *33*, 3428–3437. [CrossRef]
10. Kang, M.; Kim, K.; Muljadi, E. Frequency Control Support of a Doubly-Fed Induction Generator Based on the Torque Limit. *IEEE Trans. Power Syst.* **2016**, *31*, 4575–4583. [CrossRef]
11. Qu, L.; Qiao, W. Constant Power Control of DFIG Wind Turbines with Supercapacitor Energy Storage. *IEEE Trans. Ind. Appl.* **2011**, *47*, 359–367. [CrossRef]
12. Miao, L.; Wen, J.; Xie, H.; Yue, C.; Lee, W.-J. Coordinated Control Strategy of Wind Turbine Generator and Energy Storage Equipment for Frequency Support. *IEEE Trans. Ind. Appl.* **2015**, *51*, 2732–2742. [CrossRef]
13. Delille, G.; Francois, B.; Malarange, G. Dynamic Frequency Control Support: A Virtual Inertia Provided by Distributed Energy Storage to Isolated Power Systems. In Proceedings of the 2010 IEEE PES Conference on Innovative Smart Grid Technologies Europe, Gothenburg, Sweden, 11–13 October 2010.
14. Kim, J.; Muljadi, E.; Gevorgian, V. Dynamic Capabilities of an Energy Storage-Embedded DFIG System. *IEEE Trans. Ind. Appl.* **2019**, *55*, 4124–4134. [CrossRef]

15. Xu, L.; Cartwright, P. Direct Active and Reactive Power Control of DFIG for Wind Energy Generation. *IEEE Trans. Energy Convers.* **2006**, *21*, 750–758. [CrossRef]
16. Xu, L.; Wang, Y. Dynamic Modeling and Control of DFIG-based Wind Turbines under Unbalanced Network Conditions. *IEEE Trans. Power Syst.* **2007**, *22*, 314–323. [CrossRef]
17. Diaz-Gonzalez, F.; Sumper, A.; Gomis-Bellmunt, O.; Villafafila-Robles, R. A Review of Energy Storage Technologies for Wind Power Applications. *IEEE Trans. Sustain. Energy* **2012**, *16*, 2154–2171. [CrossRef]
18. Ribeiro, P.F.; Johnson, B.K.; Crow, M.L.; Arsoy, A.; Liu, Y. Energy storage systems for advanced power applications. *Energy Syst. Proc. IEEE* **2001**, *89*, 1744–1756. [CrossRef]
19. Li, S.; Deng, C.; Long, Z.; Zhou, Q.; Zheng, F. Calculation of Equivalent Virtual Inertial Time Constant of Wind Farm. *Autom. Electr. Power Syst.* **2016**, *40*, 22–29.
20. Zhao, Z.; Cheng, W.; Dai, C.; Zhang, X. Influence of System Inertia Time Constants on Transient Stability Level of Interconnected AC Power Grid. *Power Syst. Technol.* **2012**, *36*, 102–107.
21. National Electric Power Regulatory Standardization Technical Committee. *Technical Rule for Connecting Wind Farm to Power System*; China Standard Press: Beijing, China, 2011.
22. Liu, J.; Yao, W.; Wen, J.; Ai, X.; Luo, W. A Wind Farm Virtual Inertia Compensation Strategy Based on Energy Storage System. *Proc. Chin. Soc. Electr. Eng.* **2015**, *35*, 1596–1605.
23. Mercier, P.; Cherkaoui, R.; Oudalov, A. Optimizing a Battery Energy Storage System for Frequency Control Application in an Isolated Power System. *IEEE Trans. Power Syst.* **2009**, *24*, 1469–1477. [CrossRef]
24. Divya, K.C.; Ostergaard, J. Battery Energy Storage Technology for Power Systems-An Overview. *Electr. Power Syst. Res.* **2009**, *79*, 511–520. [CrossRef]
25. Ashoori-Zadeh, A.; Toulabi, M.; Bahrami, S. Modification of DFIG's Active Power Control Loop for Speed Control Enhancement and Inertial Frequency Response. *IEEE Trans. Sustain. Energy* **2017**, *8*, 1772–1782. [CrossRef]
26. Ko, H.; Yoon, G.; Kyung, N.; Hong, W. Modeling and Control of DFIG-based Variable-speed Wind-turbine. *Electr. Power Syst. Res.* **2008**, *78*, 1841–1849. [CrossRef]
27. Lee, J.; Muljadi, E.; Srensen, P. Releasable Kinetic Energy-based Inertial Control of a DFIG Wind Power Plant. *IEEE Trans. Sustain. Energy* **2016**, *7*, 279–288.
28. Fateh, F.; White, W.; Gruenbacher, D. Torsional Vibrations Mitigation in the Drivetrain of DFIG-Based Grid-Connected Wind Turbine. *IEEE Trans. Ind. Appl.* **2017**, *53*, 5760–5767. [CrossRef]
29. Berenji, H.; Khedkar, P. Learning and Tuning Fuzzy Logic Controllers through Reinforcements. *IEEE Trans. Neural Netw.* **1992**, *3*, 724–740. [CrossRef] [PubMed]
30. Zhang, S.; Mishra, Y.; Shahidehpour, M. Fuzzy-Logic Based Frequency Controller for Wind Farms Augmented with Energy Storage Systems. *IEEE Trans. Power Syst.* **2016**, *31*, 1595–1603. [CrossRef]
31. Sloderbeck, M.; Meka, R.; Faruque Md, O.; Langston, J.; Steurer, M. New Neural Network and Fuzzy Logic Controllers to Monitor Maximum Power for Wind Energy Conversion System. *Energy* **2016**, *106*, 137–146. [CrossRef]
32. Bevrani, H.; Daneshmand, P. Fuzzy Logic-Based Load-Frequency Control Concerning High Penetration of Wind Turbines. *IEEE. Syst. J.* **2012**, *6*, 173–180. [CrossRef]
33. Liu, J.; Wen, J.; Yao, W. Solution to Short-term Frequency Response of Wind Farms by Using Energy Storage Systems. *IEEE. IET. Renew. Power. Gener.* **2016**, *10*, 669–678. [CrossRef]

© 2019 by the authors. Licensee MDPI, Basel, Switzerland. This article is an open access article distributed under the terms and conditions of the Creative Commons Attribution (CC BY) license (http://creativecommons.org/licenses/by/4.0/).

Article

Neural Network-Based Supplementary Frequency Controller for a DFIG Wind Farm

Ting-Hsuan Chien, Yu-Chuan Huang and Yuan-Yih Hsu *

Department of Electrical Engineering, National Taiwan University, EE Building 2, No. 1, Sec. 4, Roosevelt Rd., Taipei City 106, Taiwan; r07921020@ntu.edu.tw (T.-H.C.); r07921115@ntu.edu.tw (Y.-C.H.)
* Correspondence: hsuyy@ntu.edu.tw

Received: 7 September 2020; Accepted: 6 October 2020; Published: 13 October 2020

Abstract: An artificial neural network (ANN)-based supplementary frequency controller is designed for a doubly fed induction generator (DFIG) wind farm in a local power system. Since the optimal controller gain that gives highest the frequency nadir or lowest peak frequency is a complicated nonlinear function of load disturbance and system variables, it is not easy to use analytical methods to derive the optimal gain. The optimal gain can be reached through an exhaustive search method. However, the exhaustive search method is not suitable for online applications, since it takes a long time to perform a great number of simulations. In this work, an ANN that uses load disturbance, wind penetration, and wind speed as the inputs and the desired controller gain as the output is proposed. Once trained by a proper set of training patterns, the ANN can be employed to yield the desired gain in a very efficient manner, even when the operating condition is not included in the training set. Therefore, the proposed ANN-based controller can be used for real-time frequency control. Results from MATLAB/SIMULINK simulations performed on a local power system in Taiwan reveal that the proposed ANN can yield a better frequency response than the fixed-gain controller.

Keywords: doubly fed induction generator (DFIG); wind generation; frequency control; artificial neural network (ANN)

1. Introduction

To increase the percentage of green energy, wind farms are being built in Taiwan. It is expected that the installed capacities of these wind farms will reach 4.2 GW by 2025 [1]. For a system with a high penetration of wind power, the frequency regulation of the local power system becomes very important, especially when it is disconnected from the main grid due to a fault, resulting in islanding operation. How to design a proper supplementary frequency controller for the wind farm, such that the frequency for a local power system in the islanding operation can be controlled satisfactorily, is of major concern in this work.

Numerous works have been devoted to the design of a supplementary frequency controller for a doubly fed induction generator (DFIG) in order to improve the system frequency under disturbance conditions [2–14]. Both a proportional (droop) controller [2–6] and proportional (droop)–derivative (inertial) (PD) controller [7–9] have been extensively studied.

In the design of proper gains for the droop controller and inertia controller, the objective is to improve the dynamic system frequency response and keep the DFIG speed within the allowable range after a disturbance. Since the dynamic system frequency response is a complicated nonlinear function of load disturbance and system variables such as wind speed and the percentage of wind power penetration, it is not easy to derive an analytical formula relating the system frequency and DFIG speed to the load disturbance and system variables and get the desired optimal solutions for the droop and inertia gains using analytical methods. In the literature, numerous works have been reported [9–19] to get the desired gains using a simulation-based method. The frequency and speed responses with and without droop control were compared in [9]. The effect of proportional gain on the frequency response was investigated in [15], with the derivative gain being fixed at 15. The influence of governor speed, droop gain, and inertia gain on frequency deviation was examined in [16]. The effect of DFIG penetration on the frequency deviation was also studied [16]. System frequency and maximum transient frequency deviation under different values of droop and inertia gains were investigated in [17]. In [18], the frequency nadir and rate of change of frequency for different wind power penetrations, different percentages of steam turbines, and combined cycle gas turbines were analyzed in order to reach the optimal droop and inertia gain settings. It was found that the optimal inertia gain was near zero for most cases, and the inertia term could be neglected. In [19], the root locus and system frequency response for proportional and inertia control were depicted under different wind speeds. The effect of increased wind penetration on the root locus and frequency response was also studied.

The optimal controller gain that gives the highest frequency nadir for a system under a particular condition of load disturbance, wind speed, and wind penetration can be reached by an exhaustive search method in which the dynamic frequency response curves following a disturbance are simulated for all possible gains, and the gain that gives the highest frequency nadir is selected as the optimal gain. However, the exhaustive search method is not suitable for online applications, since it takes a long time to perform a great number of simulations.

The main purpose of this work is to design an artificial neural network (ANN) [20,21]-based frequency controller that gives the desired droop gain in a very efficient manner. The inputs to the ANN are the load disturbance and system variables, such as the wind speed and percentage of wind power penetration, which have significant impacts on the system frequency response [17–19]. Computer simulations are first conducted to obtain the optimal droop gains that give the highest frequency nadir (FN) (in the case of load increase) or lowest peak frequency (in the case of load decrease) for the system under different values of load disturbances, wind power penetrations, and wind speeds. The compiled ANN outputs (optimal droop gains) and the corresponding ANN inputs (load disturbance, wind penetration, and wind speed) are employed as the ANN training patterns. Once the ANN is trained, it can be used to provide the desired droop gain in a very efficient manner without any time-consuming simulations. Therefore, the proposed ANN-based frequency controller can be used for online applications.

The main contributions of the paper are summarized as follows:

1. The effects of load disturbance, percentage of wind penetration, and wind speed on the optimal droop gain are investigated.
2. The designed ANN-based frequency controller can yield the desired droop gain in a very efficient manner. Thus, it is suitable for real-time applications.
3. The proposed ANN-based frequency controller can give a better frequency response than the fixed-gain controller. In addition, the ANN can yield controller gains that are very close to the optimal gains, even when the input variables such as wind speed, wind penetration, and load disturbance are not included in the training patterns of the ANN.

2. System Model

The system under study is a local power system in Taiwan with wind farms that are lumped together as an equivalent DFIG [6]. The six fossil-fired steam turbine generators in the local power system are lumped together as an equivalent synchronous generator (SG). Figure 1 depicts the nonlinear block diagram for the synchronous generator frequency control and DFIG supplementary frequency control. Details on the block diagrams for the governor, turbine, synchronous generator, and DFIG were described in [6].

In this work, an ANN-based controller as shown in Figure 1 is proposed to adapt the gain K_{PD} for the DFIG supplementary frequency controller based on the load disturbance ΔP_{Load}, wind power penetration, and wind speed V_W.

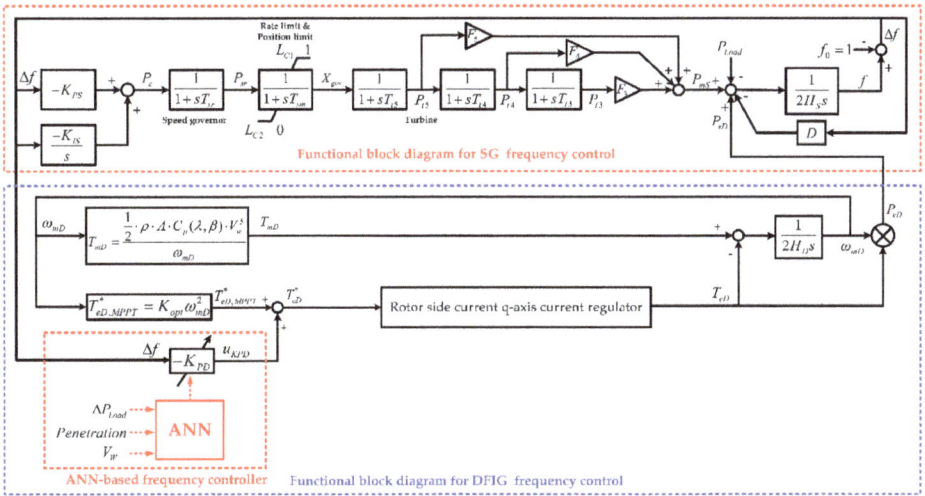

Figure 1. Block diagram for a frequency control system. SG: synchronous generator, ANN: artificial neural network, DFIG: doubly fed induction generator.

3. Effect of Load Disturbance, Wind Power Penetration, and Wind Speed on the Optimal Controller Gain

As shown in Figure 1, the system frequency f under disturbance conditions for a power system with a DFIG wind farm is governed by the swing equation of the synchronous machine:

$$2H_S \frac{df}{dt} = P_{mS} + P_{eD} - P_{Load} - D\Delta f \qquad (1)$$

where P_{mS} is the mechanical power of the synchronous machine, P_{eD} is the electrical power of the DFIG, P_{Load} is the load demand, and H_S and D are the equivalent per unit inertia constant for all synchronous generators in the power system and the load damping constant, respectively.

It is observed from Equation (1) that the system frequency deviates from its nominal value f_0 (1 pu or 60 Hz) when there is a disturbance ΔP_{Load} in the system load. In order to restore the system frequency to its nominal value, the mechanical power P_{mS} of the synchronous generators will be adjusted through the action of speed governors and turbines, and the DFIG electrical power output can also be modulated by the supplementary frequency controller denoted by $-K_{PD}\Delta f$ in Figure 1.

The main purpose of this work is to design a proper gain K_{PD} to meet the following objective function and constraints: [8,22,23]

$$\begin{array}{l} \text{Objective function:} \\ \quad \text{Maximize FN (in case of load increase) or} \\ \quad \text{Minimize peak frequency (in case of load decrease)} \end{array} \qquad (2)$$

$$\begin{array}{l} \text{Constraints:} \\ \text{(1) Frequency range: } 59.5 \text{ Hz} < f < 60.5 \text{ Hz.} \\ \text{(2) DFIG speed range: } 0.7 \text{ pu} \leq w_{mD} \leq 1.2 \text{ pu.} \\ \text{(3) Controller gain: } 16 \text{ pu} \leq K_{PD} \leq 50 \text{ pu (2–6\% droop).} \end{array} \qquad (3)$$

Note that only the frequency nadir or peak frequency is considered in the objective function, since the main purpose is to keep the system frequency within the allowable limit to avoid load shedding in an isolated power system. In addition, only the droop control is considered, since it was pointed out in [18] that the inertial constant is near zero and can be omitted.

It is observed from Figure 1 that the DFIG output power P_{eD} can be written as

$$P_{eD} = T_{eD}\omega_{mD} \quad (4)$$

where T_{eD} and ω_{mD} are the electrical torque and speed of the DFIG, respectively.

Note that the mechanical torque T_{mD} is given by

$$T_{mD} = \frac{\frac{1}{2}\rho A C_p(\lambda,\beta) V_W^3}{\omega_{mD}} \quad (5)$$

It is concluded from Equations (4) and (5) that the DFIG output power P_{eD} is a function of wind penetration and wind speed V_W. Therefore, the dynamic frequency response following a load disturbance will be affected by the load disturbance ΔP_{Load}, wind penetration, and wind speed V_W, and the optimal controller gain K_{PD} will be function of the three parameters of ΔP_{Load}, wind penetration, and V_W.

In the design of a fixed-gain supplementary frequency controller, the controller gain is usually determined based on a nominal operating condition, e.g., ΔP_{Load} = 30 MW, wind penetration = 29.4%, and V_W = 11 m/s.

Figure 2 depicts the dynamic response curves for the system subject to a load disturbance of ΔP_{Load} = 30 MW (wind penetration = 29.4% and V_W = 11 m/s). An observation of the frequency response in Figure 2a reveals that the controller gain K_{PD} = 32 gives the highest frequency nadir. Additionally shown in Figure 2a are the frequency response curves for the case with a smaller gain K_{PD} = 20 and for the case with a larger gain K_{PD} = 40. As shown in Figure 2c, the DFIG delivers less electrical power to the system and results in a lower frequency nadir when a smaller gain K_{PD} = 20 is employed. On the other hand, a larger gain of K_{PD} = 40 causes smaller frequency dips in the first few seconds following the disturbance and results in a lesser mechanical power increase for the synchronous generator and lower frequency nadir than the case of K_{PD} = 32. Therefore, an optimal gain of K_{PD} = 32 is selected for the base case of ΔP_{Load} = 30 MW, wind penetration = 29.4%, and V_W = 11 m/s.

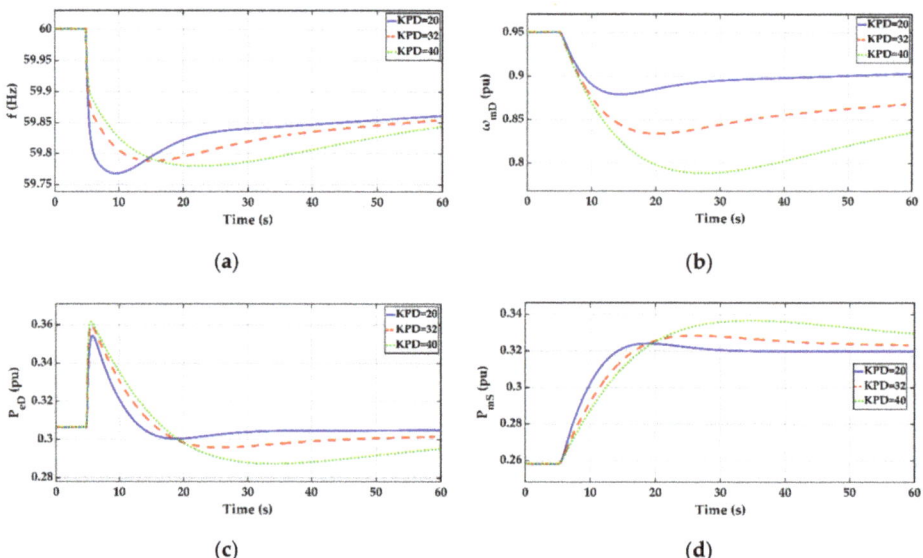

Figure 2. Dynamic response curves for a load disturbance of ΔP_{Load} = 30 MW (wind penetration = 29.4% and V_W = 11 m/s). (a) Frequency, (b) DFIG speed, (c) DFIG electrical power, and (d) SG mechanical power. K_{PD}: the optimal gain.

Since the optimal gain changes with the load disturbance ΔP_{Load}, wind penetration, and wind speed V_W, the effect of these parameters on the optimal gain is examined below.

3.1. Effect of Load Disturbance ΔP_{Load} on the Optimal Gain K_{PD}

Figure 3 depicts the dynamic response curves for the system subject to load increases of ΔP_{Load} = 62 MW and ΔP_{Load} = 63 MW, respectively.

It is observed from Figures 2 and 3a that the optimal gain K_{PD} decreases from 32 to 23 as the ΔP_{Load} is increased from 30 MW to 62 MW. It is also observed from Figures 2 and 3a that the frequency nadir decreases with the increasing load disturbance. Note that the frequency nadir reaches the lower limit of 59.5 Hz as the load disturbance is increased to 62 MW. If the load disturbance ΔP_{Load} is increased further to 63 MW, the frequency nadir is below 59.5 Hz when the gain K_{PD} remains at 23. When the gain is increased to 25, the DFIG speed drops to a value lower than 0.7, and the frequency nadir is still below 59.5 Hz. If the gain is decreased to 16, the DFIG speed will be higher than 0.7 but the frequency nadir lower than 59.5 Hz.

It is thus concluded from Figure 3 that it is impossible to find a proper gain within the allowable range (16 ≤ K_{PD} ≤ 50) when the load disturbance is 63 MW. Therefore, ΔP_{Load} = 62 MW is the upper limit for the load increase.

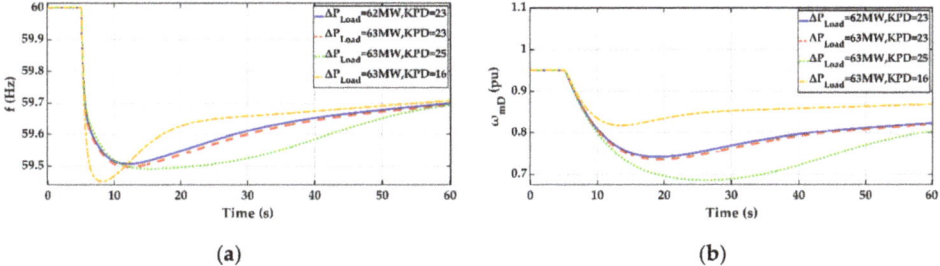

Figure 3. Dynamic response curves for the system subject to a load increase of ΔP_{Load} = 62 MW and 63 MW (wind penetration = 29.4% and V_W = 11 m/s). (**a**) Frequency and (**b**) DFIG speed.

Figure 4 depicts the optimal gain K_{PD} as a function of the load disturbance ΔP_{Load}. It is observed from Figure 4 that the optimal controller gain K_{PD} varies with the magnitude of the load disturbance ΔP_{Load}. This motivates the design of an ANN-based controller such that the controller gain can be adapted with the load disturbance, and the load disturbance must be selected as one of the inputs to the ANN-based controller.

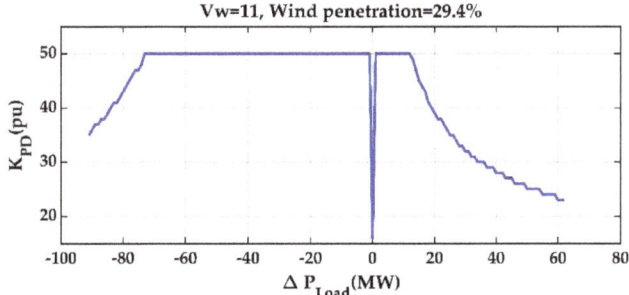

Figure 4. Optimal gain K_{PD} as a function of the load disturbance ΔP_{Load} (wind penetration = 29.4% and V_W = 11 m/s).

3.2. Effect of the Percentage of Wind Penetration on the Optimal Gain K_{PD}

Figure 5 depicts the optimal gain K_{PD} as a function of the percentage of wind penetration (ΔP_{Load} = 30 MW and V_W = 11 m/s).

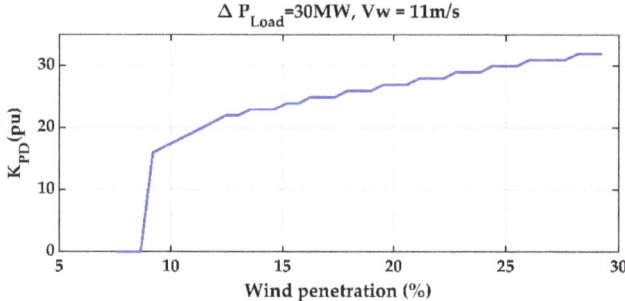

Figure 5. Optimal gain K_{PD} as a function of wind penetration (ΔP_{Load} = 30 MW and V_W = 11 m/s).

It is observed from Figure 5 that the optimal gain K_{PD} varies from 32 to 28 and 23 as the wind penetration changes from 29.4% to 22.05% and 14.7%, respectively. No feasible solution that satisfies the frequency and DFIG speed constraints can be found as the wind penetration is decreased to 7.35%. It is thus concluded that the wind penetration has a significant impact on the design of the optimal controller gain and must be employed as one of the inputs to the ANN-based controller.

3.3. Effect of the Percentage of Wind Speed on the Optimal Gain K_{PD}

Figure 6 depicts the optimal gain K_{PD} as a function of wind speed (ΔP_{Load} = 30 MW and wind penetration = 29.4%). It is observed from Figure 6 that the optimal gain K_{PD} varies from 34 to 32, 31, and 17 as the wind speed is decreased from 12 m/s to 11 m/s, 10 m/s, and 9 m/s, respectively. Therefore, the wind speed is selected as one of the inputs to the ANN-based controller, as it has a considerable effect on the optimal controller gain K_{PD}.

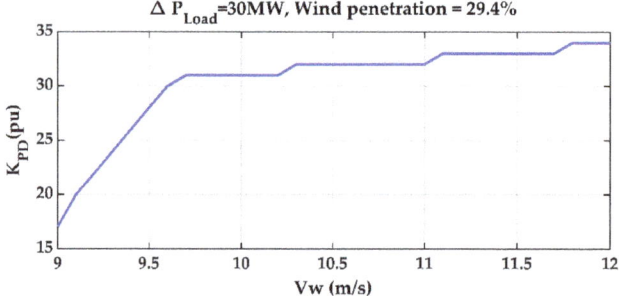

Figure 6. Optimal gain K_{PD} as a function of the wind speed V_W (ΔP_{Load} = 30 MW and wind penetration = 29.4%).

4. ANN-Based Frequency Controller

As shown in Figure 1, the gain for the supplementary frequency controller is adjusted by ANN based on the present load disturbance (ΔP_{Load}), wind penetration, and wind speed (V_W), which are provided as the inputs to the ANN. The output of the ANN is the desired gain K_{PD} for the DFIG supplementary frequency controller. A feedforward neural network with two hidden layers and ten nodes for each layer as shown in Figure 7 was used [21].

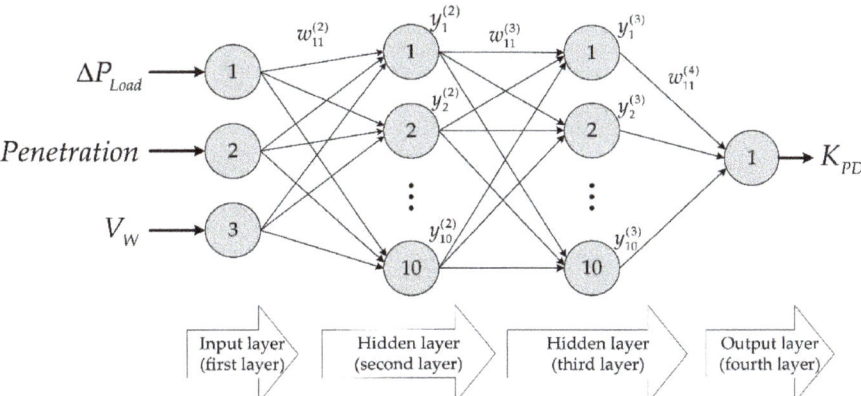

Figure 7. ANN for a frequency controller. ΔP_{Load}: load disturbance.

As shown in Figure 7, the output of the jth node in the nth layer, $y_j^{(n)}$, is a nonlinear function of the outputs from the nodes in the (n − 1)th layers, as described below:

$$y_j^{(n)} = g\left(\sum_{i=1}^{10} w_{ji}^{(n)} y_i^{(n-1)}\right) \quad (6)$$

where $w_{ji}^{(n)}$ is the connection weight between the ith node in the (n − 1)th layer, and the jth node in the nth layer and g is a nonlinear hyperbolic-tangent activation function.

The desired droop gain K_{PD} is obtained from the following equation:

$$K_{PD} = g\left(\sum_{i=1}^{10} w_{1i}^{(4)} y_i^{(3)}\right) \quad (7)$$

Before the ANN can be employed to yield the desired droop gain K_{PD}, the connection weights must be determined using a set of training patterns. In this paper, a total of 27,920 training patterns were used in the ANN training process to cover different combinations of ΔP_{Load}, wind penetration, and V_W.

The flow chart in Figure 8 was used to reach the desired ANN output (K_{PD}) for a particular combination of ANN inputs (ΔP_{Load}, wind penetration, and V_W).

The procedures to create training patterns are described as follows:

Step 1 Set the ΔP_{Load}, wind power penetration, and V_W that are considered in this work and the minimum value of the K_{PD}.
Step 2 Solve the dynamic frequency response of the system using the nonlinear model in Figure 1.
Step 3 If the dynamic response satisfies the requirements defined in Equation (3), record the K_{PD} and the frequency nadir.
Step 4 Find the K_{PD} that gives the highest frequency nadir under different scenarios and record the ΔP_{Load}, wind power penetration, V_W, and K_{PD}.

The created training patterns are depicted in Figure 9 for the cases under four different wind speeds: V_W = 9 m/s, V_W = 10 m/s, V_W = 11 m/s, and V_W = 12 m/s.

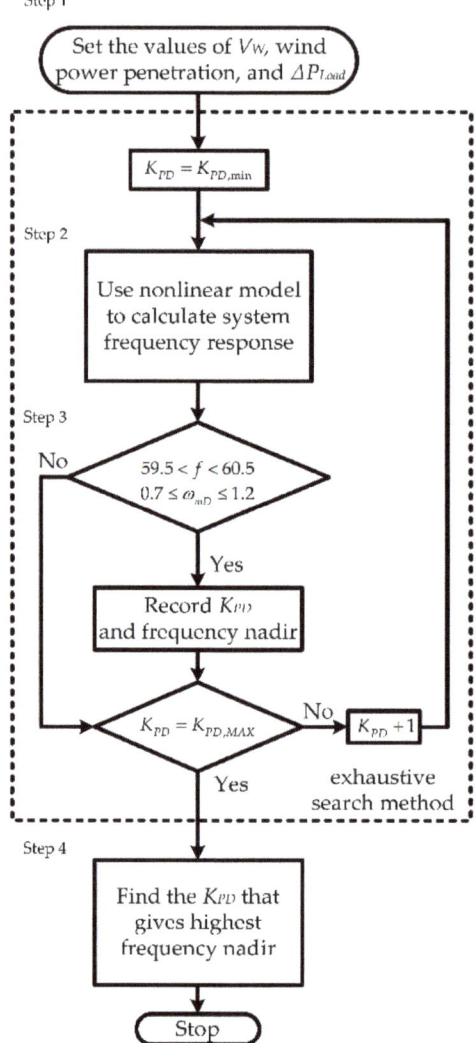

Figure 8. Flow chart to create training patterns.

Among the 27,920 training patterns as shown in Figure 9, 80% were used for training and 20% were used for testing. In the ANN training process, the connection weights between the inputs nodes, the nodes in the two hidden layers, and the output node are determined based on the criterion to make the droop gain from the ANN as close to the optimal gain in the training pattern as possible. In other words, the objective is to minimize the cost function, as described below:

$$E = \frac{1}{2}(K_{PD} - K_{PD}^*)^2 \tag{8}$$

where K_{PD} and K_{PD}^* are the droop gain from the ANN and the optimal droop gain in the training pattern. Detailed procedures to determine the connection weights from the cost function in Equation (8) can be found in [21].

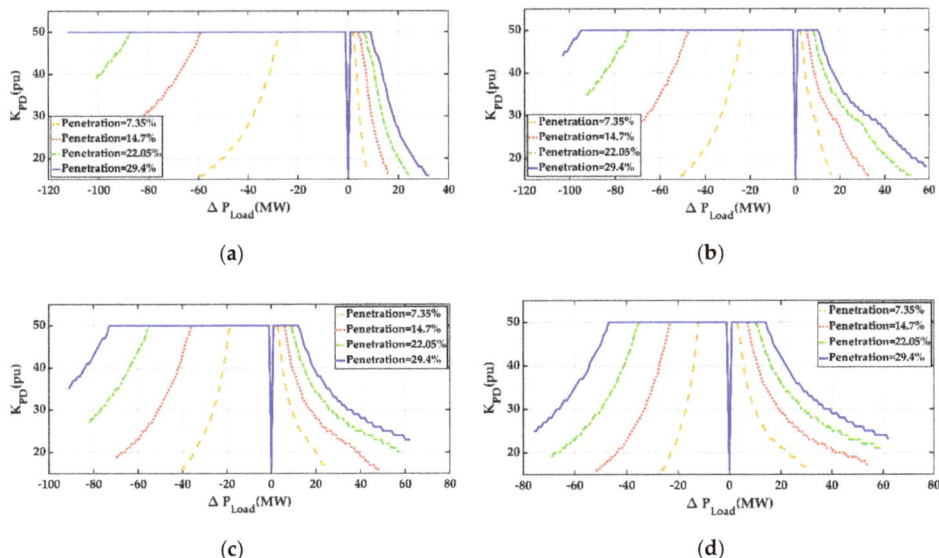

Figure 9. Training patterns under four different wind speeds: (**a**) V_W = 9 m/s, (**b**) V_W = 10 m/s, (**c**) V_W = 11 m/s, and (**d**) V_W = 12 m/s.

5. Case Studies

To demonstrate the effectiveness of the proposed ANN-based supplementary frequency controller, the local power system as shown by the block diagram in Figure 1 with the parameters in the Appendix A was simulated using MATLAB/SIMULINK. The results are described below.

5.1. Comparison of ANN-Based Controller and Fixed-Gain Controller under Different Load Disturbances

Figure 10 compares the droop gain K_{PD} from the ANN-based controller, optimal controller, and fixed-gain controller with K_{PD} = 32 under different load disturbances. It is observed from Figure 10 that the droop gain K_{PD} from the ANN-based controller is very close to those from the optimal controller. However, the ANN-based controller can be used for online applications, since the droop gain is reached in a very efficient manner. On the other hand, the optimal controller cannot be employed in real-time situations, since a great number of simulations are performed in order to reach the optimal droop gain by using the exhaustive search method.

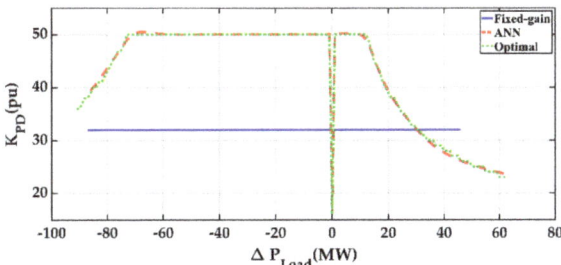

Figure 10. Droop gain K_{PD} from the ANN-based controller, optimal controller, and fixed-gain controller (wind penetration = 29.4% and V_W = 11 m/s) under different load disturbances.

Figure 11 depicts the frequency, DFIG speed, SG energy, and DFIG energy for a load disturbance of ΔP_{Load} = 62 MW. It is observed from Figure 11 that, in the case of a load increase of 62 MW, the DFIG speed fails to remain in the allowable range of 0.7 pu ≤ ω_{mD} ≤ 1.2 pu when the controller gain is fixed at 32. On the other hand, the satisfactory frequency and speed responses can be achieved by the ANN-based controller by adapting the gain to a lower value of 23.59.

The frequency, DFIG speed, SG energy, and DFIG energy for a load disturbance of ΔP_{Load} = 40 MW are depicted in Figure 12. As the observation of the response curves in Figure 12 indicates, the ANN-based controller gives better frequency and speed responses than the fixed-gain controller.

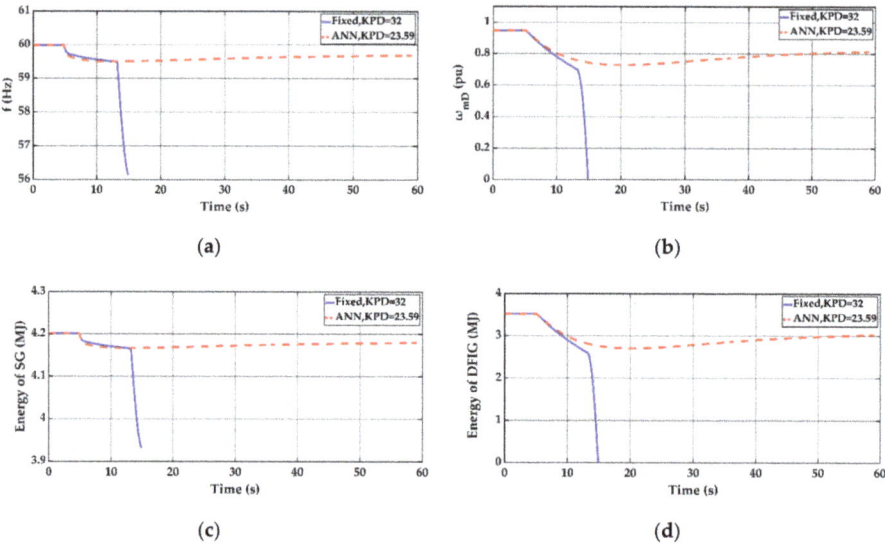

Figure 11. Dynamic response curves from the ANN-based controller and fixed-gain controller (ΔP_{Load} = 62 MW, wind penetration = 29.4%, and V_W = 11 m/s). (**a**) Frequency, (**b**) DFIG speed, (**c**) SG energy, and (**d**) DFIG energy.

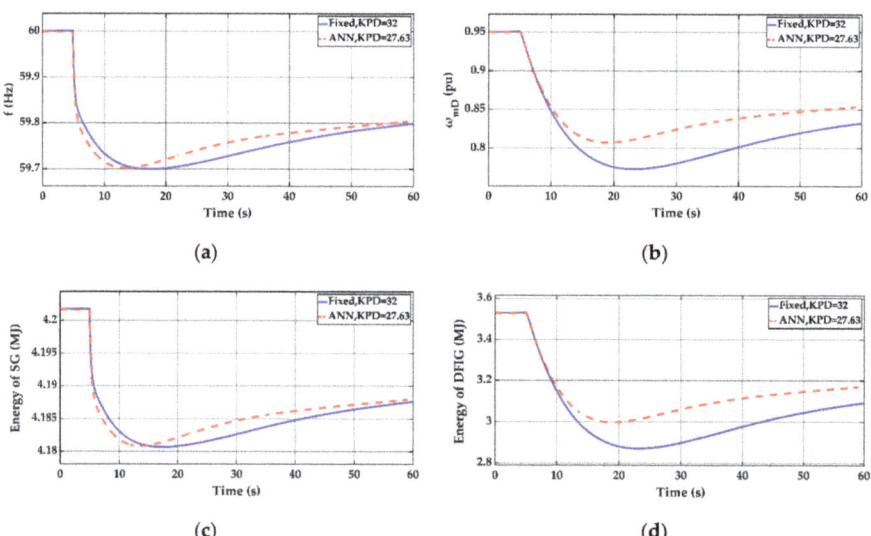

Figure 12. Dynamic response curves from the ANN-based controller and fixed-gain controller (ΔP_{Load} = 40 MW, wind penetration = 29.4%, and V_W = 11 m/s). (**a**) Frequency, (**b**) DFIG speed, (**c**) SG energy, and (**d**) DFIG energy.

5.2. Comparison of the ANN-Based Controller and Fixed-Gain Controller under Different Wind Power Penetrations

The droop gain and FN from the ANN-based controller and fixed-gain controller under different wind power penetrations are depicted in Figure 13. It is observed from Figure 13 that the ANN-based controller gives better frequency than the fixed-gain controller, since its droop gain is varied according to the percentage of the wind power penetration.

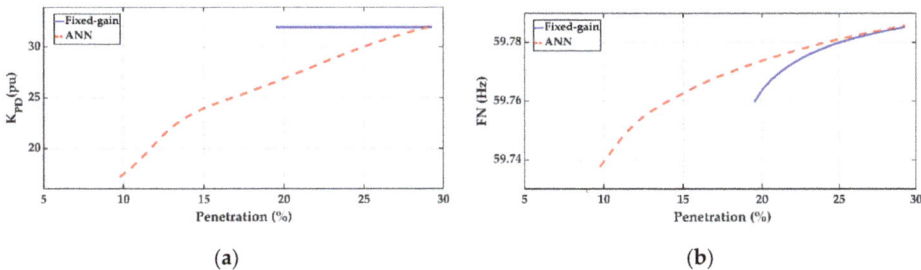

Figure 13. Droop gain K_{PD} and frequency nadir from the ANN-based controller and fixed-gain controller (ΔP_{Load} = 30 MW and V_W = 11 m/s) under different penetrations. (**a**) Droop gain K_{PD} and (**b**) frequency nadir (FN).

5.3. Comparison of the ANN-Based Controller and Fixed-Gain Controller under Different Wind Speeds

In order to examine the dynamic performance of the ANN-based controller under different wind speed conditions, the droop gain and FN under different wind speeds are shown in Figure 14. It is observed from Figure 14 that it is impossible to find a feasible solution that satisfies the frequency and speed constraints by using the fixed-gain controller when the wind speed is lower than 9.8 m/s. However, satisfactory frequency nadir can still be achieved by the ANN-based controller. It is also observed that the ANN-based controller gives better FN than the fixed-gain controller, since the droop gain is varied with the changing wind speed.

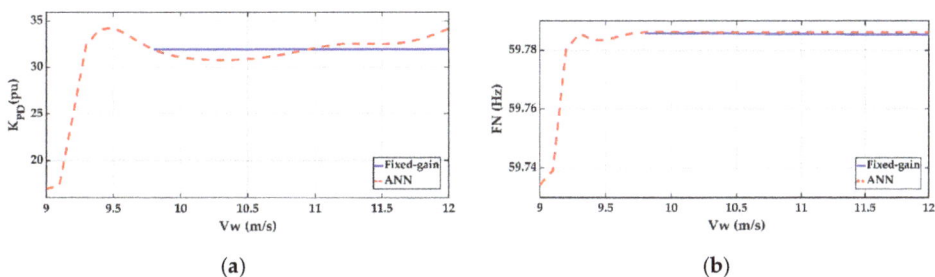

Figure 14. Droop gain K_{PD} and frequency nadir from the ANN-based controller and fixed-gain controller (ΔP_{Load} = 30 MW and wind penetration = 29.4%) under different wind speeds. (**a**) Droop gain K_{PD} and (**b**) frequency nadir (FN).

5.4. ANN Performance Test for Untrained Cases

A major feature of the ANN is that, once the ANN is trained using an appropriate set of training patterns, it can be used to generate the desired output (controller gain K_{PD}) directly even when the input variables (ΔP_{Load}, wind penetration, and wind speed) are not within the set of training patterns. Time-consuming simulations for the untrained cases can thus be avoided.

Figure 15 compares the droop gain K_{PD} from the ANN-based controller and the optimal controller for the case of V_W = 10.5 m/s, which was not included in the training patterns. It is observed from

Figure 15 that the ANN-based controller can yield droop gains that are very close to the optimal gain from the exhaustive search method even when the wind speed (V_W = 10.5 m/s) is different from the wind speeds in all training patterns for the ANN.

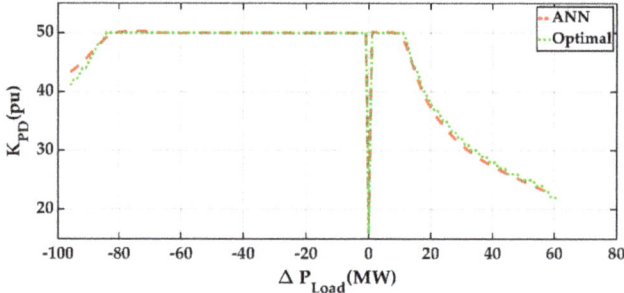

Figure 15. Droop gain K_{PD} from the ANN-based controller and optimal controller (V_W = 10.5 m/s and wind penetration = 29.4%).

5.5. Feasible Operating Regions for the ANN-Based Controller

The feasible operating regions for the ANN-based controller and fixed-gain controller are compared in Figure 16 for cases under four different wind speeds: V_W = 9 m/s, V_W = 10 m/s, V_W = 11 m/s, and V_W = 12 m/s. It is concluded from Figure 16 that the proposed ANN-based controller with variable gain provides a wider operating zone than the fixed-gain controller.

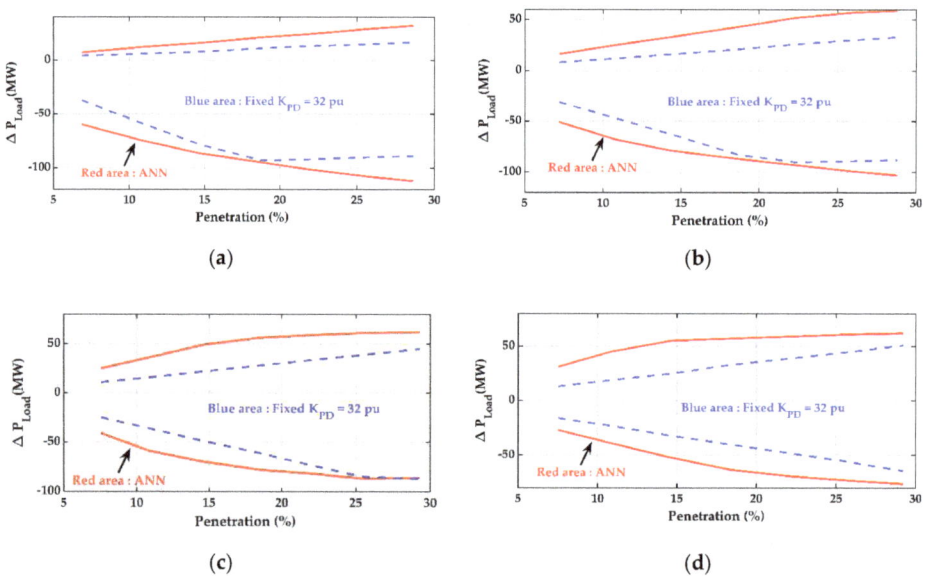

Figure 16. Comparison of the feasible operating regions for the ANN-based controller and fixed-gain controller. (**a**) V_W = 9 m/s, (**b**) V_W = 10 m/s, (**c**) V_W = 11 m/s, and (**d**) V_W = 12 m/s.

6. Conclusions

An ANN was designed to yield the droop gain K_{PD} for the supplementary frequency controller of a DFIG wind farm under different load disturbances, wind penetrations, and wind speeds. The effects of the load disturbances, wind penetrations, and wind speeds on the optimal gain were first studied.

It was found that the load disturbance, wind penetration, and wind speed had significant impacts on the optimal controller gain. Therefore, the three variables were employed as the inputs to the ANN, and the output of the ANN was the desired droop controller gain. The specific conclusions are as follows:

1. The droop gain K_{PD} decreases with the increasing magnitude of the load disturbances.
2. The droop gain should be increased when the wind power penetration is increased.
3. The droop gain increases with the increasing wind speed.
4. The ANN-based controller yields essentially the same droop gain as the optimal controller using the exhaustive search method. However, the ANN-based method is more efficient than the exhaustive search method, since time-consuming simulations can be avoided after the ANN is trained. Therefore, the ANN-based controller can be used in online applications, and the optimal controller using the exhaustive search method cannot be employed for real-time applications.
5. A major feature of the ANN-based controller is that it can be employed to provide the desired droop gain without the need to perform additional simulations, even when the load disturbance, wind penetration, and wind speed are not within the set of training patterns.
6. By using the ANN-based controller with different gains under different operating conditions, the feasible operating regions under different wind speeds and different wind penetrations can be expanded.
7. In practical applications, the load disturbance can be estimated from the rate of change of frequency ($\Delta P_{Load} = -2H_s \frac{df}{dt}$). The wind penetration is computed using the rated capacities of online units. The wind speed is assumed to be available at the local wind farm.

Author Contributions: Conceptualization, methodology, software, validation, investigation, data curation, and visualization, T.-H.C.; writing—review and editing, Y.-C.H.; and writing—original draft, supervision, and project administration, Y.-Y.H. All authors have read and agreed to the published version of the manuscript.

Funding: This research was funded by Ministry of Science and Technology of Taiwan grant number MOST 106-2221-E-002-147-MY3 and the APC was funded by Ministry of Science and Technology of Taiwan.

Acknowledgments: This work was supported by the Ministry of Science and Technology of Taiwan under Contract MOST 106-2221-E-002-147-MY3. Cheng-Hung Chuang and Tien-Kuei Lu provided valuable system data and comments.

Conflicts of Interest: The authors declare that they have no known competing financial interests or personal relationships that could have appeared to influence the work reported in this paper.

Nomenclature

D	load damping
H_S, H_D	equivalent inertia time constants of synchronous machine and DFIG
f_0, f	nominal frequency and system frequency
F_5, F_4, F_3	power fractions of the high, intermediate, and low-pressure turbines
K_{opt}	maximum power point tracking constant
K_{PD}	DFIG supplementary proportional controller gain
K_{PS}, K_{IS}	synchronous machine droop and integral controller gains
P_c, P_{sr}, X_{gov}	control signal, speed relay output signal, and steam valve position of the synchronous machine
P_{t5}, P_{t4}, P_{t3}	mechanical output power of the high, intermediate, and low-pressure turbines
P_{Load}	Load demand
P_{eD}	electromagnetic power of DFIG
P_{mS}	mechanical power of synchronous machine
T_{sr}, T_{sm}	speed relay and servo-motor time constants of the synchronous machine
T_{t5}, T_{t4}, T_{t3}	steam chest, reheater, and crossover time constants of the synchronous machine
T_{mD}, T_{eD}	mechanical torque and electromagnetic torque of DFIG
$T^*_{eD, MPPT}$	electromagnetic torque command of DFIG for MPPT operation

T^*_{eD}	DFIG torque command
ω_{mD}	DFIG speed
V_W	wind speed
C_P	wind turbine power coefficient
A	area swept by the wind turbine blades
ρ	air density
λ, β	wind turbine tip speed ratio and blade pitch angle
Δ	incremental quantity

Appendix A

Synchronous machine:
Rated power: 480 MVA.
Machine parameters: H_S = 3.3 s.

Speed governor and turbine:
Droop and integral controller gains: K_{PS} = 20 and K_{IS} = 0.1.
Speed relay and servo-motor time constants: T_{sr} = 0.1 s and T_{sm} = 0.3 s.
Steam chest, reheater and crossover time constants: T_{t5} = 0.68 s, T_{t4} = 5.3 s, and T_{t3} = 0.58 s.
Power fractions: F_5 = 0.241, F_4 = 0.399, and F_3 = 0.360.

DFIG:
Rated power: 200 MVA.
Machine parameters: H_D = 3.5 s.
Load parameters:
Load damping coefficient: D = 1.

References

1. Bureau of Energy, Ministry of Economic Affairs, Wind Power Generation Four Years Project. Taiwan, August 2017. Available online: https://www.moeaboe.gov.tw/ECW/populace/content/ContentDesc.aspx?menu_id=5493 (accessed on 12 October 2020).
2. Shahabi, M.; Haghifam, M.R.; Mohamadian, M.; Nabavi-Niaki, S.A. Microgrid dynamic performance improvement using a doubly fed induction wind generator. *IEEE Trans. Energy Convers.* **2009**, *24*, 137–145. [CrossRef]
3. Lee, J.; Jang, G.; Muljadi, E.; Blaabjerg, F.; Chen, Z.; Kang, Y.C. Stable short-term frequency support using adaptive gains for a DFIG-based wind power plant. *IEEE Trans. Energy Convers.* **2016**, *31*, 1068–1079. [CrossRef]
4. Arani, M.F.M.; Mohamed, Y.A.I. Analysis and impacts of implementing droop control in DFIG-based wind turbines on microgrid/weak-grid stability. *IEEE Trans. Power Syst.* **2015**, *30*, 385–396. [CrossRef]
5. Arani, M.F.M.; Mohamed, Y.A.I. Dynamic droop control for wind turbines participating in primary frequency regulation in microgrids. *IEEE Trans. Smart Grid* **2018**, *9*, 5742–5751. [CrossRef]
6. Yang, J.; Chen, Y.; Hsu, Y. Small-signal stability analysis and particle swarm optimisation self-tuning frequency control for an islanding system with DFIG wind farm. *IET Gener. Transm. Distrib.* **2019**, *13*, 563–574. [CrossRef]
7. Hwang, M.; Muljadi, E.; Park, J.; Sørensen, P.; Kang, Y.C. Dynamic droop-based inertial control of a doubly-fed induction generator. *IEEE Trans. Sustain. Energy* **2016**, *7*, 924–933. [CrossRef]
8. Vidyanandan, K.V.; Senroy, N. Primary frequency regulation by deloaded wind turbines using variable droop. *IEEE Trans. Power Syst.* **2013**, *28*, 837–846. [CrossRef]
9. Ramtharan, G.; Ekanayake, J.B.; Jenkins, N. Frequency support from doubly fed induction generator wind turbines. *IET Renew. Power Gener.* **2007**, *1*, 3–9. [CrossRef]
10. Li, Y.; Xu, Z.; Zhang, J.; Wong, K.P. Variable gain control scheme of DFIG-based wind farm for over-frequency support. *Renew. Energy* **2018**, *120*, 379–391. [CrossRef]
11. Pradhan, C.; Bhende, C.N.; Samanta, A.K. Adaptive virtual inertia-based frequency regulation in wind power systems. *Renew. Energy* **2018**, *115*, 558–574. [CrossRef]

12. Ochoa, D.; Martinez, S. Frequency dependent strategy for mitigating wind power fluctuations of a doubly-fed induction generator wind turbine based on virtual inertia control and blade pitch angle regulation. *Renew. Energy* **2018**, *128*, 108–124.
13. Camblong, H.; Vechiu, I.; Guillaud, X.; Etxeberria, A.; Kreckelbergh, S. Wind turbine controller comparison on an island grid in terms of frequency control and mechanical stress. *Renew. Energy* **2014**, *63*, 37–45. [CrossRef]
14. Papaefthymiou, S.V.; Lakiotis, V.G.; Margaris, I.D.; Papathanassiou, S.A. Dynamic analysis of island systems with wind-pumped-storage hybrid power stations. *Renew. Energy* **2015**, *74*, 544–554. [CrossRef]
15. Mauricio, J.M.; Marano, A.; Gomez-Exposito, A.; Ramos, J.L.M. Frequency regulation contribution through variable-speed wind energy conversion systems. *IEEE Trans. Power Syst.* **2009**, *24*, 173–180. [CrossRef]
16. Kayikci, M.; Milanovic, J.V. Dynamic contribution of DFIG-based wind plants to system frequency disturbances. *IEEE Trans. Power Syst.* **2009**, *24*, 859–867. [CrossRef]
17. Margaris, I.D.; Papathanassiou, S.A.; Hatziargyriou, N.D.; Hansen, A.D.; Sorensen, P. Frequency control in autonomous power systems with high wind power penetration. *IEEE Trans. Sustain. Energy* **2012**, *3*, 189–199.
18. Vyver, J.V.D.; Kooning, J.D.M.D.; Meersman, B.; Vandevelde, L.; Vandoorn, T.L. Droop control as an alternative inertial response strategy for the synthetic inertia on wind turbines. *IEEE Trans. Power Syst.* **2016**, *31*, 1129–1138. [CrossRef]
19. Wilches-Bernal, F.; Chow, J.H.; Sanchez-Gasca, J.J. A fundamental study of applying wind turbines for power system frequency control. *IEEE Trans. Power Syst.* **2016**, *31*, 1496–1505. [CrossRef]
20. Hafiz, F.; Abdennour, A. An adaptive neuro-fuzzy inertia controller for variable-speed wind turbines. *Renew. Energy* **2016**, *92*, 136–146. [CrossRef]
21. Haykin, S. *Neural Networks and Learning Machines*, 3rd ed.; Pearson: Upper Saddle River City, NJ, USA, 2008.
22. Taiwan Power Company. *Power System Operation Guide*; Taiwan Power Company: Taipei City, Taiwan, 2020.
23. Zhang, Z.; Sun, Y.; Lin, J.; Li, G. Coordinated frequency regulation by doubly fed induction generator-based wind power plants. *IET Renew. Power Gener.* **2012**, *6*, 38–47. [CrossRef]

© 2020 by the authors. Licensee MDPI, Basel, Switzerland. This article is an open access article distributed under the terms and conditions of the Creative Commons Attribution (CC BY) license (http://creativecommons.org/licenses/by/4.0/).

Article

Flexible Kinetic Energy Release Controllers for a Wind Farm in an Islanding System

Yi-Wei Chen and Yuan-Yih Hsu *

Department of Electrical Engineering, National Taiwan University, EE Building 2, No. 1, Sec. 4, Roosevelt Rd., Taipei 106, Taiwan; r06921069@ntu.edu.tw
* Correspondence: hsuyy@ntu.edu.tw

Received: 23 October 2020; Accepted: 19 November 2020; Published: 23 November 2020

Abstract: To improve frequency nadir following a disturbance and avoid under-frequency load shedding, two types of flexible kinetic energy release controllers for the doubly fed induction generator (DFIG) are proposed. The basic idea is to release only a small amount of kinetic energy stored at the DFIG in the initial transient period (1–3 s after the disturbance). When the frequency dip exceeds a preset threshold, the amount of kinetic energy released is increased to improve the frequency nadir. To achieve the goal of flexible kinetic energy release, a deactivation function based integral controller is first presented. To further improve the dynamic frequency response under parameter uncertainties and external disturbances, a second flexible kinetic energy release controller is designed using a proportional-integral controller, with the gains being adapted in real-time with the particle swarm optimization algorithm. Based on the MATLAB/SIMULINK simulation results for a local power system, it is concluded that the frequency nadir can be maintained around the under-frequency load shedding threshold of 59.6 Hz using the proposed controllers.

Keywords: doubly fed induction generator (DFIG); load frequency control (LFC); wind farm; particle swarm optimization; kinetic energy

1. Introduction

To improve the dynamic frequency response of a local power system with high penetration of wind power, a supplementary frequency controller (SFC) installed on the rotor side converter (RSC) of a doubly fed induction generator (DFIG) has been widely investigated in recent years. The main purpose of this work is to design a proper SFC such that the frequency nadir (FN) of the islanding system following a grid disconnection event can be maintained higher than the threshold of 59.6 Hz, which is equal to the sum of the low frequency load shedding limit of 59.5 Hz set by the local utility and 0.1 Hz safety margin. In other words, frequency nadir following a disturbance causing a power deficit is of major concern in this paper.

In the literature, a proportional (droop) controller which generates a control signal proportional to the frequency deviation has been proposed [1–10]. When the control signal is added to the RSC of the DFIG, the electrical power output of the DFIG can be modulated and the system frequency response can be improved. To further improve the dynamic frequency response, the gain of the proportional controller was varied based on the rate of change of frequency (ROCOF) [4], or DFIG rotor speed [5]. In [7], the gain was decreased linearly with time. In [8], the particle swarm optimization (PSO) technique was employed to adapt the proportional gain in real-time in order to have a good dynamic frequency response following a disturbance. Other self-tuning techniques such as artificial neural networks [11], model predictive controllers [12–15], and fuzzy set algorithm [16–18] have been proposed to provide the required supplementary frequency control signal.

A proportional (droop)-derivative (inertia) (PD) controller with the inertial control signal being proportional to the derivative of frequency deviation has also been widely studied [19–28]. The effect of the PD controller on the initial ROCOF [22,24] or frequency nadir [22] has been examined. It was pointed out in [7,22] that the initial ROCOF might be improved during the inertia period (around 1–2 s after the disturbance) by the inertia controller through injecting more DFIG electrical power to the power system. However, the inertia control might have a negative impact on frequency nadir since the DFIG injects more electrical power to reach a smaller frequency dip, causing the steam or gas turbine to deliver less mechanical power to the synchronous generator (SG). The increase in DFIG kinetic energy release and reduction in SG accumulated mechanical energy in the first few seconds might cause a lower frequency nadir in the subsequent primary frequency regulation period (around 2–50 s after the disturbance). Therefore, the optimal inertia controller gain to have the highest frequency nadir was found to be close to zero [22]. Therefore, the inertia control will not be considered in this work, since our goal is to reach the highest frequency nadir to avoid under-frequency load shedding.

In this paper, two flexible kinetic energy release controllers for a DFIG wind farm are proposed in order to improve the frequency nadir after a disturbance. The basic idea is to release less kinetic energy from the DFIG during the first few seconds after the disturbance when the system frequency is above a certain threshold. When the system frequency drops to a level below the threshold, the kinetic energy reserved in the DFIG during the first few seconds after the disturbance is gradually released to improve the frequency nadir.

In the first flexible kinetic energy release controller, an integral controller is added to the conventional droop controller when the system frequency is lower than 59.9 Hz. A deactivation function is proposed to gradually decrease the integral controller output to zero and force the DFIG to return to its steady-state maximum power point tracking (MPPT) operating mode when the system frequency eventually returns to its nominal value of 60 Hz.

The gains in the first flexible kinetic energy release controller are designed based on a particular operating point. To improve the dynamic frequency response for the system subject to variations in system parameters such as speed governor and steam turbine time constants or external disturbances such as wind speed variation, the second flexible kinetic energy release controller, in which the controller gains are adapted in real-time based on PSO algorithm, is proposed. The advantages and disadvantages of the proposed controllers and controllers referred to in the literature are summarized in Table 1.

Table 1. Summary of supplementary frequency controllers for doubly fed induction generator (DFIG).

Controller/Algorithm	Operating Conditions Dependent	Needs Training	Needs Rule Base	Needs Evaluation Function	Computational Burden
PID [1–10,19–28]	YES	NO	NO	NO	LOW
ANN [11]	NO	YES	NO	NO	LOW
MPC [12–15]	NO	NO	NO	YES	HIGH
Fuzzy [16–18]	NO	NO	YES	NO	LOW
Type I	YES	NO	NO	NO	LOW
Type II	NO	NO	NO	YES	HIGH

2. Test System Model

The test system under study is a local power system as illustrated in Figure 1.

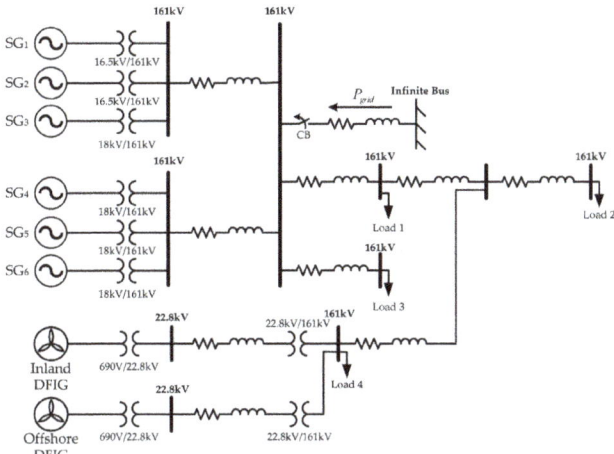

Figure 1. One-line diagram for the studied local power system.

The six synchronous generators in the local power system in Figure 1 are lumped together as an equivalent SG [29] and the inland and offshore wind farms are lumped together as an equivalent DFIG in this study [1]. When the grid is subject to a fault, the local power system is disconnected from the grid and is operated at an islanding operation mode. The functional block diagram for the equivalent SG and equivalent DFIG frequency control system is depicted in Figure 2.

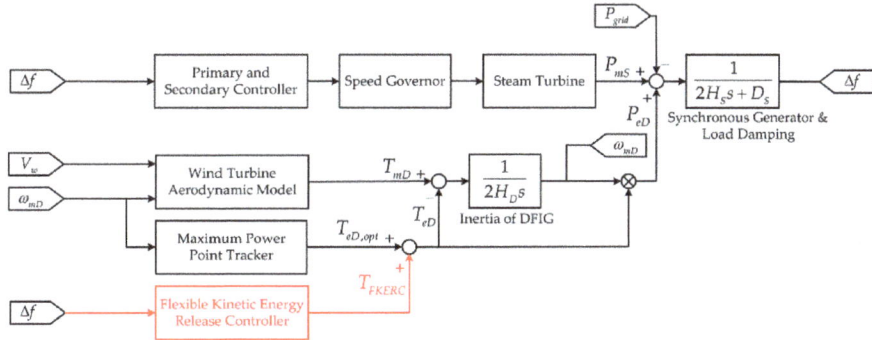

Figure 2. Frequency control functional block diagram for the islanding system.

To avoid considerable revenue losses, the approaches of power reserve such as pitch angle control and de-loaded operating strategy are not considered in this work. Therefore, the DFIG operates at the MPPT mode in a normal operation to achieve the maximum harvest of the wind power. The DFIG electromagnetic torque under the MPPT mode is expressed as:

$$T_{eD} = T_{eD,opt} \qquad (1)$$

where $T_{eD,opt}$ is the optimum electromagnetic torque for MPPT, which is a quadratic function of the DFIG rotor speed ω_{mD}. In the case of frequency drop due to grid disconnection, the DFIG kinetic energy is the only energy source from a wind farm that can be released to the system. To improve the dynamic response of system frequency, a supplementary frequency control signal from the flexible kinetic energy release controller T_{FKERC} is added to $T_{eD,opt}$ to obtain the desired electromagnetic torque command of the DFIG as shown below:

$$T_{eD} = T_{eD,opt} + T_{FKERC} \qquad (2)$$

In this work, the main purpose of the DFIG flexible kinetic energy release controller is to inject additional real power into the islanding system such that the system frequency nadir is as high as possible during the entire post-disturbance transient period. Due to the fast reaction of the RSC, the frequency response is dominated initially by DFIG. As mentioned earlier, the flexible kinetic energy release controller should be designed for the DFIG to release minimal kinetic energy in the initial transient period. Then, the reserved kinetic energy of the DFIG can be released afterwards to improve the frequency nadir. Details on the design of the two types of flexible kinetic energy release controllers proposed in this work are described in Sections 3 and 4.

3. Type I Flexible Kinetic Energy Release Controller Using the Deactivation Function Based Integral Controller

In this section, an innovative Type I flexible kinetic energy release controller, as depicted in Figure 3, is proposed to release the kinetic energy flexibly and effectively from a DFIG wind farm.

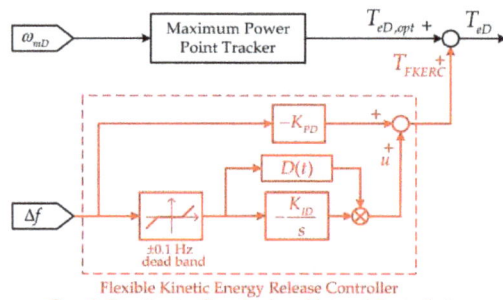

Figure 3. Deactivation function based integral controller.

As shown in Figure 3, the Type I flexible kinetic energy release control signal T_{FKERC} from the deactivation function based integral controller can be written as:

$$T_{FKERC} = \begin{cases} -K_{PD}\Delta f(t), & \text{if } f > 59.9 \text{ Hz} \\ -\left(K_{PD} + D(t)\frac{K_{ID}}{s}\right)\Delta f(t), & \text{otherwise} \end{cases} \qquad (3)$$

In order for the DFIG to deliver only a small amount of kinetic energy in the initial transient period, only the droop control is employed in Equation (3) when $f > 59.9$ Hz. When the system frequency is below 59.9 Hz, the deactivation function based integral controller is started. With the proposed Type I flexible kinetic energy release controller, an integral control signal u is gradually increased such that the goal to have a smaller control signal T_{FKERC} in the beginning and a larger control signal afterwards can be met. Therefore, the system frequency nadir can be improved through the action of an integral gain with a deferred and accumulated control output.

Since the integral controller output will not be zero after the transient period is over, the DFIG may work at an operating point which is different from that before the disturbance. For the DFIG to return to its MPPT mode operation when the system frequency approaches its nominal value, the integral controller output must be gradually decreased to zero. In the present work, a deactivation function $D(t)$, as depicted in Figure 4, is proposed to gradually decrease the integral control output to zero in a very smooth manner.

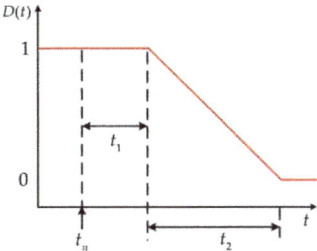

Figure 4. Deactivation function for the integral controller.

As shown in Figure 4, the integral controller output must be gradually decayed to zero in a period of decay time (t_2) after a delay time (t_1) from the instant of the lowest system frequency (the nadir time t_n). The dynamic frequency response curves for different delay times (t_1) and decay times (t_2) are shown in Figure 5a,b, respectively, for the case of $V_w = 11$ m/s and $P_{grid} = 30$ MW at $t = 10$ s.

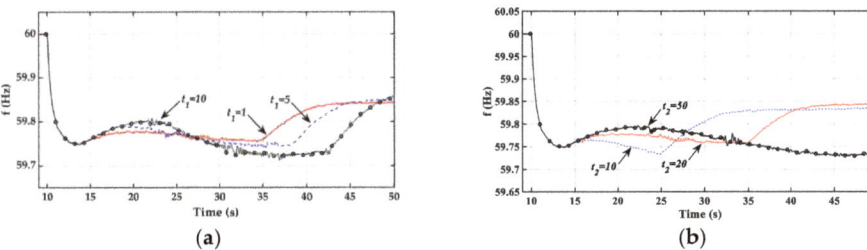

Figure 5. Effects of different delay times (t_1) and decay times (t_2) on system frequency. (**a**) Different delay, Table 1. (**b**) Different decay times (t_2).

It is observed from Figure 5a that, after a brief period of frequency rise, the system frequency will drop again to a new low value when the integral controller output is gradually decreased to zero. The decay time t_2 is chosen to be 20 s, and the PI controller gains are $K_{PD} = 20$ and $K_{ID} = 1.5$ in Figure 5a. As shown from the response curves in Figure 5a, the delay time t_1 should be kept as small as possible in order to decrease the subsequent frequency drop. This is due to the fact that a longer delay time will cause a greater drop in the DFIG kinetic energy and a lower system frequency following the removal of the integral controller. In the present work, a delay time of 1 s is chosen for t_1 to cover the time required to detect the system frequency nadir.

Figure 5b compares the frequency responses for different decay times t_2. The PI controller gains are chosen to be $K_{PD} = 20$ and $K_{ID} = 1.5$. It is observed that the system frequency after the decay of the integral controller output dips to a value even lower than the first nadir when a short decay time $t_2 = 10$ s is employed. This is due to the fact that the DFIG output power is decreased rapidly and the SG does not have enough time to increase its output power, causing a power deficit and subsequent frequency dip. On the other hand, the DFIG kinetic energy will be consumed too much, causing a later frequency drop if a long decay time such as $t_2 = 50$ s is used. In the present work, a moderate decay time $t_2 = 20$ s is employed.

Since it takes a long time to get a solution from MATLAB/SIMULINK circuit-level simulations, system-level simulations are conducted to approximately estimate the effects of integral controller gains K_{ID} on the system frequency nadir under different values of K_{PD} for the system subject to a disturbance of $P_{grid} = 45$ MW at $t = 10$ s, as depicted in Figure 6.

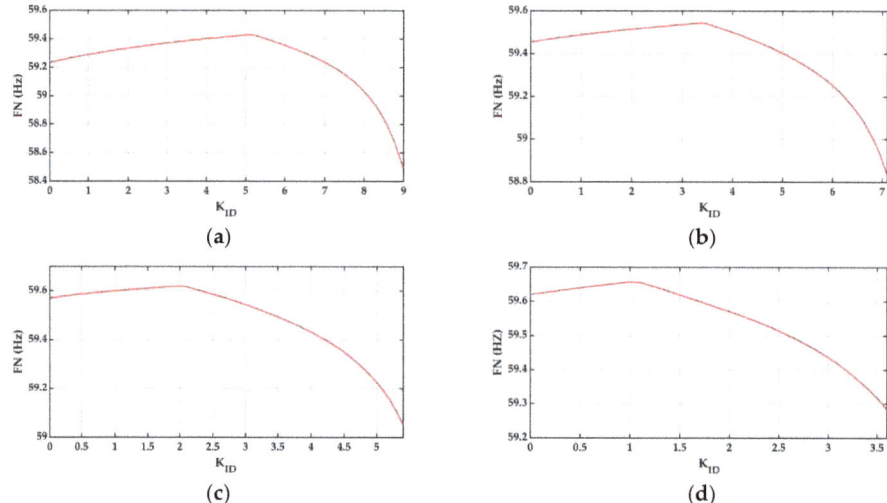

Figure 6. Effect of integral controller gain K_{ID} on frequency nadir under different values of K_{PD} (**a**) K_{PD} = 5, (**b**) K_{PD} = 10, (**c**) K_{PD} = 15, (**d**) K_{PD} = 20.

An observation of Figure 6 reveals that a higher frequency nadir can be achieved for the four different values of K_{PD} when a moderate gain of K_{ID} is employed. Table 1 lists the highest frequency nadir for various combinations of K_{PD} and K_{ID} under different values of P_{grid}.

It is observed from Table 2 that a combination of K_{PD} = 20 and K_{ID} = 1 gives the highest frequency nadir for the base case of P_{grid} = 45 MW. The preliminary gains of K_{PD} = 20 and K_{ID} = 1 from system-level simulations are further refined to be K_{PD} = 20 and K_{ID} = 1.1 using more time-consuming circuit-level simulations. In Section 5, the pair of gains K_{PD} = 20 and K_{ID} = 1.1 will be employed for the simulations.

Table 2. The highest frequency nadir for various combinations of K_{PD} and K_{ID} under different values of P_{grid}.

	K_{PD}	K_{ID}	FN
P_{grid} = 30 MW	5	5.5	59.63
	10	3.7	59.70
	15	2.3	59.75
	20	1.3	59.78
P_{grid} = 35 MW	5	5.4	59.56
	10	3.6	59.65
	15	2.2	59.71
	20	1.2	59.74
P_{grid} = 40 MW	5	5.3	59.50
	10	3.5	59.60
	15	2.1	59.67
	20	1.1	59.70
P_{grid} = 45 MW	5	5.1	59.43
	10	3.4	59.55
	15	2	59.62
	20	1	59.66

4. Type II Flexible Kinetic Energy Release Controller Using PSO

In the design of Type I flexible kinetic energy release controller, the controller gains are designed based on a certain set of system parameters and operating conditions in order to have a good frequency response under that particular operating point. However, the dynamic frequency response may become unsatisfactory when there is a change in system parameters or operating conditions. In order to have a good frequency response when the system is subject to variations in system parameters or operating conditions, the gains of the Type I flexible kinetic energy release controller may be adapted in real-time using the Type II PSO controller, as shown in Figure 7.

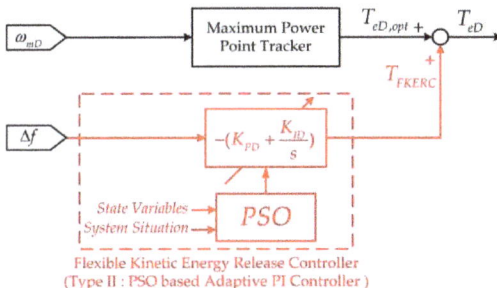

Figure 7. Particle swarm optimization (PSO) based adaptive Proportional-Integral (PI) controller.

The procedures followed by the proposed flexible kinetic energy release controller using PSO to adjust the controller gains have been described in [30,31]. As shown in Figure 8, in each iteration i, particle n has positions $K_n^i = [\ K_{PD,n}^i \quad K_{ID,n}^i\ ]^T$ and velocities $V_n^i = [\ V_{PD,n}^i \quad V_{ID,n}^i\ ]^T$. The positions and velocities are updated as follows:

$$K_n^{i+1} = K_n^i + V_n^{i+1} \qquad (4)$$

$$V_n^{i+1} = w_n^i \cdot V_n^i + r_1 \cdot \left(K_{n(pbest)} - K_n^i\right) + r_2 \cdot \left(K_{(gbest)} - K_n^i\right) \qquad (5)$$

where r_1 and r_2 are random numbers between 0 and 1, $K_{n(pbest)}$ is the best particle position, $K_{(gbest)}$ is the best global position, and w_n^i is a weighting factor expressed as:

$$w_n^i = w_{MAX} - \frac{(w_{MAX} - w_{min})}{N} i \qquad (6)$$

The number of particles N is chosen to be 12 and the total number of iterations is 15. The maximum and minimum values for the weight of the velocity vector are chosen to be $w_{MAX} = 0.5$ and $w_{min} = 0.1$, respectively. In this work, the evaluation function E in the PSO algorithm is defined as follows:

$$E = \text{MAX}\left|\Delta f_{pu}(t)\right| - \Delta f_{pu}^* \qquad (7)$$

where $\Delta f_{pu}(t) = f_{pu}(t) - 1$ pu and $\Delta f_{pu}^* = 1 - (59.6/60)$ pu. Note that the PSO algorithm is started only when the system frequency is lower than 59.6 Hz and the evaluation function is chosen to keep the system frequency as close to 59.6 Hz as possible. In this way, the retained DFIG kinetic energy can be released in critical conditions to avoid under-frequency load shedding.

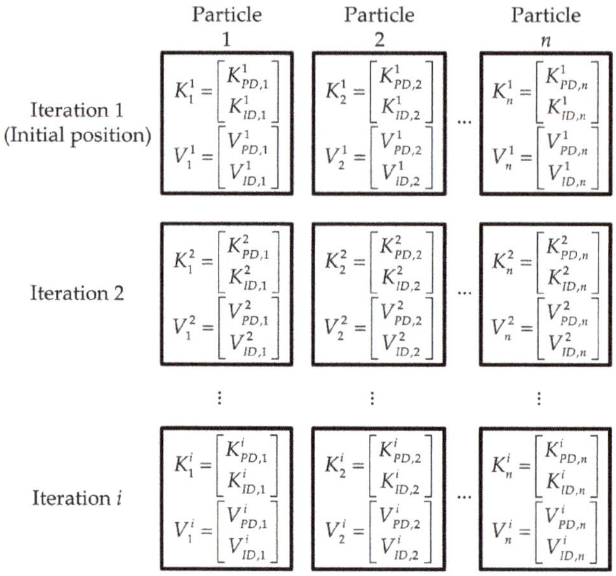

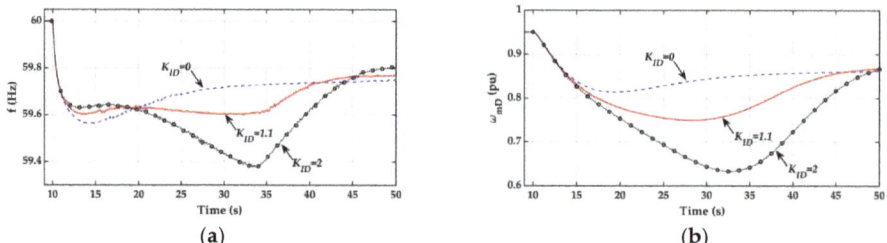

Figure 8. Schematic diagram of particle positions and velocities in every iteration.

5. Simulation Results

To demonstrate the effectiveness of the proposed flexible kinetic energy release controllers, the local power system in Figure 1 with the parameters in the Appendix A was simulated using MATLAB/SIMULINK. The dynamic responses from detailed circuit-level simulations are presented as follows.

5.1. Dynamic Performance of Type I Flexible Kinetic Energy Release Controller

To examine the dynamic performance of the Type I flexible kinetic energy release controller, dynamic response curves for the following three cases are depicted in Figure 9 under the condition of $K_{PD} = 20$, $P_{grid} = 45$ MW, and $V_w = 11$ m/s:

- Case 1: Droop control only ($K_{ID} = 0$).
- Case 2: Type I controller with $K_{ID} = 1.1$.
- Case 3: Type I controller with $K_{ID} = 2$.

Figure 9. Cont.

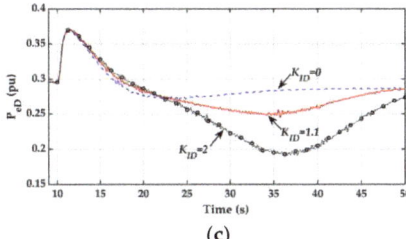
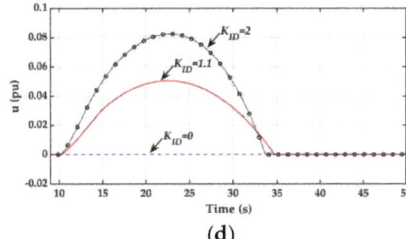

(c)   (d)

Figure 9. Comparison of dynamic response curves from different integral gains ($K_{PD} = 20$, $P_{grid} = 45$ MW, $V_w = 11$ m/s). (**a**) Frequency, (**b**) DFIG speed, (**c**) DFIG electrical power, (**d**) deactivation function based integral control signal.

The local power system was disconnected from the main grid at the instant of $t = 10$ s and remained in islanding operation afterwards. The power deficit of 45 MW immediately following the grid disconnection caused a frequency dip, which should be controlled by increasing the DFIG electrical power P_{eD} in order to present the frequency from dropping to a level lower than the under-frequency load shedding threshold of 59.6 Hz.

On comparing the response curves for the three cases in Figure 9, the following observations are in order:

1. The frequency response curves for the three cases are essentially the same during the inertia period (1–2 s after the disturbance or 10 s $\leq t \leq$ 11–12 s) since the frequency nadir is of major concern and inertia control is not considered in this work.
2. The frequency nadir for Case 1 with only a droop controller is lower than the prespecified threshold of 59.6 Hz.
3. With the proposed Type I flexible kinetic energy release controller, the frequency nadir is improved to a value higher than the threshold of 59.6 Hz. This is achieved through the addition of an integral controller with a gain of $K_{ID} = 1.1$. As shown in Figure 9c,d, both the DFIG electrical power P_{eD} and control signal are increased by the proposed Type I flexible kinetic energy release controller during the first few seconds in the primary frequency regulation period (12 s $\leq t \leq$ 22 s).
4. As evidenced by the response curve in Figure 9d, the control signal u from the proposed Type I flexible kinetic energy release controller is gradually increased when the system frequency drops to 59.9 Hz at $t = 10.5$ s. The deactivation function in Figure 4 begins to work at $t = 13$ s when the frequency nadir is detected. Due to the action of deactivation function, the rate of change of control signal u is gradually decreased and the control signal u is decreased to zero in a very smooth manner in order to avoid the second frequency nadir.
5. Although the DFIG output power P_{eD} can be further increased and the first frequency nadir at around $t = 13$ s can be improved further using a higher integral gain of $K_{ID} = 2$ for the Type I flexible kinetic energy release controller, the DFIG speed will drop significantly and the DFIG kinetic energy will be exhausted in the primary frequency regulation period. As a result, a very low second frequency nadir (59.4 Hz) will be observed at $t = 34$ s. A moderate gain of $K_{ID} = 1.1$ seems to be a good choice for the study system.

5.2. Dynamic Performance of Type II Flexible Kinetic Energy Release Controller

Detailed comparisons of the dynamic frequency responses for the system with Type I and Type II flexible kinetic energy release controllers, when the system is subject to parameter uncertainties and external disturbances, are described below.

5.2.1. Dynamic Response Curves under Uncertainties in System Parameters Servo-Motor Time Constant T_{sm} and Reheater Time Constant T_{t4} (P_{grid} = 45 MW, V_w = 11 m/s, T_{sm} = 0.375 s, T_{t4} = 6.625 s)

To investigate the dynamic performance of the Type II controller under parameter uncertainties, the servo-motor time constant T_{sm} and the reheater time constant T_{t4}, were changed from 0.3 and 5.3 s to 0.375 and 6.625 s, respectively. It is assumed that the Type II controller is unaware of the change in system parameters. The dynamic response curves from the Type I and Type II controllers are compared in Figure 10. Note that the initial gains for the Type II controller were set to be K_{PD} = 10 and K_{ID} = 0.

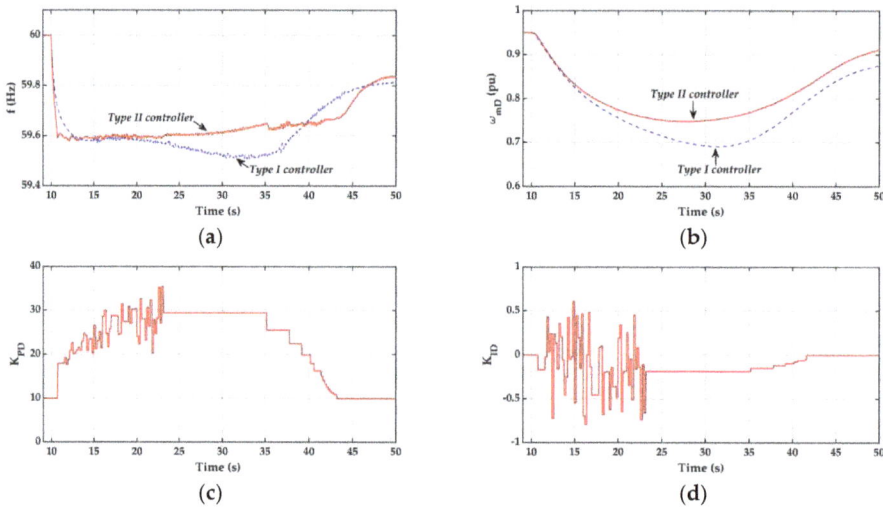

Figure 10. Comparison of dynamic response curves from the Type I and Type II controllers under change of system parameters T_{sm} and T_{t4} (P_{grid} = 45 MW, V_w = 11 m/s, T_{sm} = 0.375 s, T_{t4} = 6.625 s). (**a**) Frequency, (**b**) DFIG speed, (**c**) PSO proportional gain, (**d**) PSO integral gain.

It is observed from the frequency response curves in Figure 10a, for the case of T_{sm} = 0.375 s and T_{t4} = 6.625 s, that the system frequency nadir is lower than the threshold of 59.6 Hz when a Type I controller with the same controller gains of K_{PD} = 20, K_{ID} = 1.1, and deficit power P_{grid} = 45 MW as those used in Figure 9 was employed. Recall that the Type I controller with the gains K_{PD} = 20 and K_{ID} = 1.1 gave a satisfactory frequency response for the system with the original parameters (T_{sm} = 0.3 s and T_{t4} = 5.3 s), as shown in Figure 9a. Therefore, it is concluded that the frequency response from a Type I controller is sensitive to system parameter variations.

On the other hand, as evidenced by the response curve in Figure 10a, the frequency nadir remained around 59.6 Hz when the Type II controller was employed. As shown in Figure 10c,d, an improvement of the system frequency response was achieved by the Type II controller through the use of a pair of lower initial gains (K_{PD} = 10, K_{ID} = 0) immediately following the grid disconnection. As explained earlier, more kinetic energy retained in the DFIG as a result of lower initial gains during the first few seconds after disturbance enabled the DFIG to deliver more electrical power to the islanding system and to improve frequency response.

As shown in Figure 10c,d, the Type II controller was started when the frequency was lower than 59.6 Hz and higher gains for K_{PD} and K_{ID} from the Type II controller forced the DFIG to deliver more electrical power to the system. With the increase of the electrical power output from the DFIG, the system frequency was improved. In order to return to the MPPT operation mode, when the system frequency exceeded 59.65 Hz, the PSO gains were gradually decayed to the initial values (K_{PD} = 10, K_{ID} = 0).

5.2.2. Dynamic Response Curves under Change of Wind Speed

To examine the dynamic performance of the Type II controller under a change of wind speed as shown in Figure 11, Figure 12 depict the dynamic response curves for this case.

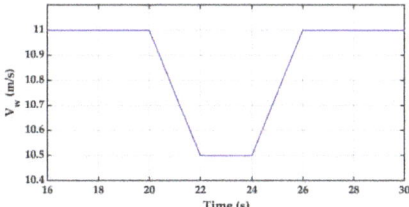

Figure 11. Variation of wind speed.

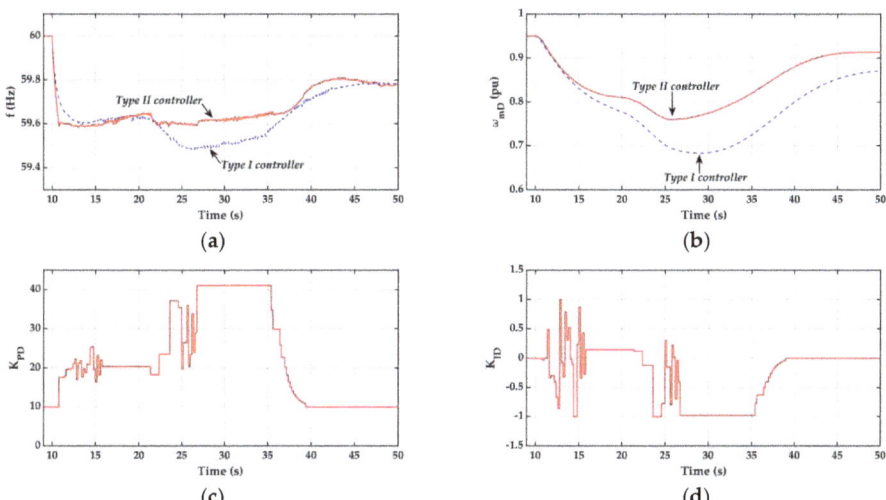

Figure 12. Comparison of dynamic response curves from the Type I and Type II controllers under change of wind speed (P_{grid} = 45 MW). (**a**) Frequency, (**b**) DFIG speed, (**c**) PSO proportional gain, (**d**) PSO integral gain.

Since the gains of the Type I controller had been selected based on a fixed wind speed of V_w = 11 m/s, the system frequency nadir failed to meet the requirement of 59.6Hz when the system was subject to a wind speed change, as evidenced by the frequency response curve in Figure 12a. However, the Type II controller can still maintain a satisfactory frequency response by adjusting the controller gains in real-time when the system is subject to a wind speed change, as shown in Figure 12c,d.

5.2.3. Dynamic Response Curves under Uncertainties in System Parameters and Fluctuation of Wind Speed in Different P_{grid}

In order to assess the effectiveness of the Type II controller, the dynamic responses for different values of P_{grid} under the combined effects of the SG time constant change, as described in Section 5.2.1, and the wind speed fluctuation, as shown in Figure 13, are depicted in Figure 14.

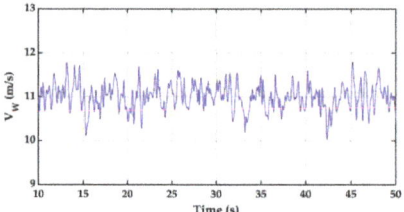

Figure 13. Fluctuation of wind speed.

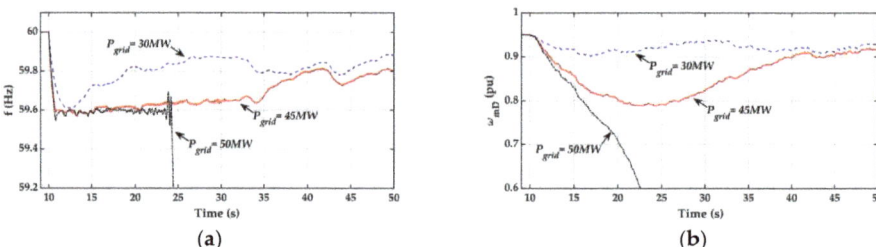

(a) (b)

Figure 14. Dynamic response curves of the Type II controller under parameters uncertainties and the fluctuation of wind speed at different deficit powers. (a) Frequency, (b) DFIG speed.

Based on the dynamic response curves in Figure 14, the following observations can be made.

1. When the deficit power P_{grid} is 30 MW, the PSO algorithm will not be initiated since the frequency response during the entire post-fault period is higher than the threshold of 59.6 Hz. Therefore, the controller gains for the Type II controller will remain at the initial values (K_{PD} = 10, K_{ID} = 0).
2. As the deficit power P_{grid} is increased to 45 MW, the controller gains for the Type II controller can be adjusted online to improve the frequency response, even though the PSO algorithm is unaware of the change in SG parameters and the fluctuations in wind speed.
3. When the system encounters a large deficit power, e.g., P_{grid} = 50 MW, the system frequency can be kept around 59.6 Hz in the first few seconds after disturbance using the proposed Type II controller. However, the DFIG stall and frequency collapse are observed afterwards since the large deficit power exceeds the upper limit of the DFIG stored kinetic energy, which can be released to the power system under disturbance conditions.

6. Conclusions

Two flexible kinetic energy release controllers have been designed for a DFIG to improve the frequency nadir of an islanding system, comprising an equivalent SG and an equivalent DFIG. Specific conclusions are summarized as follows:

1. The Type II flexible kinetic energy release controller with the controller gains being adapted in real-time, using the PSO technique, has been found to be able to offer a better dynamic frequency response than the Type I controller when the system is subject to external disturbances or parameter variations.
2. In this paper, the pitch angle is set to be zero to avoid considerable revenue losses. However, the de-loaded operation may be used to improve the system frequency at the price of revenue loss when the deficit power exceeds the DFIG kinetic energy limit.
3. Although the dynamic frequency response can be improved by the proposed Type I and Type II controllers, the delay time and decay time in the Type I controller must be properly designed in order to achieve good performance. In addition, it takes a long time for the PSO algorithm to reach the desired optimal gains of K_{PD} and K_{ID}. In order to reduce the computational burden,

it is necessary to limit the number of particles per iteration. However, the local minimum may be experienced as a result of an insufficient number of particles.

4. Future work will be devoted to the implementation and field test of the proposed controllers. Furthermore, the coordination between the pitch angle controller and the proposed controller will be investigated.

Author Contributions: Conceptualization, methodology, validation, and investigation, Y.-W.C.; writing—original draft, supervision, and project administration, Y.-Y.H. Both authors have read and agreed to the published version of the manuscript.

Funding: This research was funded by the Ministry of Science and Technology of Taiwan, grant number MOST 106-2221-E-002-147-MY3 and the APC was funded by the Ministry of Science and Technology of Taiwan.

Acknowledgments: This work was supported by the Ministry of Science and Technology of Taiwan, under contract MOST 106-2221-E-002-147-MY3. Cheng-Hung Chuang and Tien-Kuei Lu provided valuable system data and comments.

Conflicts of Interest: The authors declare that they have no known competing financial interests or personal relationships, which could have appeared to influence the work reported in this paper.

Nomenclature

D_S	load damping
$D(t)$	deactivation function
H_S, H_D	equivalent inertia time constants of synchronous machine and DFIG
f	system frequency
K_{PD}, K_{ID}	DFIG supplementary proportional and integral controller gain
P_{grid}	grid power
P_{eD}	electromagnetic power of DFIG
P_{mS}	mechanical power of synchronous machine
T_{mD}, T_{eD}	mechanical torque and electromagnetic torque of DFIG
$T_{eD,opt}$	electromagnetic torque command of DFIG for MPPT
T_{FKERC}	flexible kinetic energy release control signal
t_1, t_2, t_n	decay time, delay time, and nadir frequency time
ω_{mD}	DFIG speed
V_W	wind speed
Δ	incremental quantity

Appendix A

Speed governor and turbine:

Droop and integral controller gains: $K_{PS} = 20$ and $K_{IS} = 0.1$.
Speed relay and servo-motor time constants: $T_{sr} = 0.1$ s and $T_{sm} = 0.3$ s.
Steam chest, reheater, and crossover time constants: $T_{t5} = 0.68$ s, $T_{t4} = 5.3$ s, and $T_{t3} = 0.58$ s.
Power fractions: $F_5 = 0.241$, $F_4 = 0.399$, and $F_3 = 0.360$.

Synchronous machine:

Rated power: 480 MVA.
Rated voltage: 18 kV.
Rated frequency: 60 Hz.
Number of poles: 2 poles.
Machine parameters: $H_S = 3.3$ s, $D_S = 0$.

DFIG:

Rated power: 200 MVA.
Rated voltage: 690 V.
Proportional and integral gains of q-axis rotor current regulator: $K_{pq} = 0.0756$, $K_{iq} = 0.8318$.
Stator, rotor, and mutual inductances: $L_s = 3.1$ pu, $L_r = 3.08$ pu, and $L_m = 3$ pu.
Rotor resistance: $R_r = 0.01$ pu.
Number of poles: 4 poles.
Machine parameters: $H_D = 3.5$ s.

Loads:

Load 1 = 26 MW; Load 2 = 160 MW; Load 3 = 50 MW; Load 4 = 24 MW.

Infinite bus parameters:

Rated short-circuit power: 11,072.92 MVA.
Rated short-circuit current: 39.71 kA.
Rated voltage: 161 kV.
X/R ratio: 31.6912.

References

1. Shahabi, M.; Haghifam, M.R.; Mohamadian, M.; Nabavi-Niaki, S.A. Microgrid dynamic performance improvement using a doubly fed induction wind generator. *IEEE Trans. Energy Convers.* **2009**, *24*, 137–145. [CrossRef]
2. Mauricio, J.M.; Marano, A.; Gomez-Exposito, A.; Ramos, J.L.M. Frequency regulation contribution through variable-speed wind energy conversion systems. *IEEE Trans. Power Syst.* **2009**, *24*, 173–180.
3. Arani, M.F.M.; Mohamed, Y.A.I. Analysis and impacts of implementing droop control in DFIG-based wind turbines on microgrid/weak-grid stability. *IEEE Trans. Power Syst.* **2015**, *30*, 385–396.
4. Hwang, M.; Muljadi, E.; Park, J.; Sørensen, P.; Kang, Y.C. Dynamic droop–based inertial control of a doubly-fed induction generator. *IEEE Trans. Sustain. Energy* **2016**, *7*, 924–933.
5. Lee, J.; Jang, G.; Muljadi, E.; Blaabjerg, F.; Chen, Z.; Kang, Y.C. Stable short-term frequency support using adaptive gains for a DFIG-based wind power plant. *IEEE Trans. Energy Convers.* **2016**, *31*, 1068–1079.
6. Arani, M.F.M.; Mohamed, Y.A.I. Dynamic droop control for wind turbines participating in primary frequency regulation in microgrids. *IEEE Trans. Smart Grid* **2018**, *9*, 5742–5751.
7. Garmroodi, M.; Verbič, G.; Hill, D.J. Frequency support from wind turbine generators with a time-variable droop characteristic. *IEEE Trans. Sustain. Energy* **2018**, *9*, 676–684. [CrossRef]
8. Yang, J.; Chen, Y.; Hsu, Y. Small-signal stability analysis and particle swarm optimisation self-tuning frequency control for an islanding system with DFIG wind farm. *IET Gener. Transm. Distrib.* **2019**, *13*, 563–574.
9. Li, Y.; Xu, Z.; Zhang, J.; Wong, K.P. Variable gain control scheme of DFIG-based wind farm for over-frequency support. *Renew. Energy* **2018**, *120*, 379–391. [CrossRef]
10. Hwang, M.; Muljadi, E.; Jang, G.; Kang, Y.C. Disturbance-adaptive short-term frequency support of a DFIG associated with the variable gain based on the ROCOF and rotor speed. *IEEE Trans. Power Syst.* **2017**, *32*, 1873–1881. [CrossRef]
11. Hafiz, F.; Abdennour, A. An adaptive neuro-fuzzy inertia controller for variable-speed wind turbines. *Renew. Energy* **2016**, *92*, 136–146. [CrossRef]
12. Qudaih, Y.S.; Bernard, M.; Mitani, Y.; Mohamed, T.H. Model predictive based load frequency control design in the presence of DFIG wind turbine. In Proceedings of the 2011 2nd International Conference on Electric Power and Energy Conversion Systems (EPECS), Sharjah, UAE, 15–17 November 2011; pp. 1–5.
13. Mohamed, T.H.; Morel, J.; Bevrani, H.; Hiyama, T. Model predictive based load frequency control_design concerning wind turbines. *Int. J. Electr. Power Energy Syst.* **2012**, *43*, 859–867. [CrossRef]
14. Tielens, P.; Hertem, D.V. Receding horizon control of wind power to provide frequency regulation. *IEEE Trans. Power Syst.* **2017**, *32*, 2663–2672. [CrossRef]
15. Wang, H.; Yang, J.; Chen, Z.; Ge, W.; Ma, Y.; Xing, Z.; Yang, L. Model predictive control of PMSG-based wind turbines for frequency regulation in an isolated grid. *IEEE Trans. Ind. Appl.* **2018**, *54*, 3077–3089. [CrossRef]
16. Zhang, S.; Mishra, Y.; Shahidehpour, M. Fuzzy-logic based frequency controller for wind farms augmented with energy storage systems. *IEEE Trans. Power Syst.* **2016**, *31*, 1595–1603. [CrossRef]
17. Ghafouri, A.; Milimonfared, J.; Gharehpetian, G.B. Fuzzy-adaptive frequency control of power system including microgrids, wind farms, and conventional power plants. *IEEE Syst. J.* **2018**, *12*, 2772–2781. [CrossRef]
18. Abazari, A.; Monsef, H.; Wu, B. Load frequency control by de-loaded wind farm using the optimal fuzzy-based PID droop controller. *IET Renew. Power Gener.* **2019**, *13*, 180–190. [CrossRef]
19. Wu, Y.; Yang, W.; Hu, Y.; Dzung, P.Q. Frequency regulation at a wind farm using time-varying inertia and droop controls. *IEEE Trans. Ind. Appl.* **2019**, *55*, 213–224. [CrossRef]
20. Geng, H.; Xi, X.; Yang, G. Small-signal stability of power system integrated with ancillary-controlled large-scale DFIG-based wind farm. *IET Renew. Power Gener.* **2017**, *11*, 1191–1198. [CrossRef]

21. Lee, J.; Muljadi, E.; Srensen, P.; Kang, Y.C. Releasable kinetic energy-based inertial control of a DFIG wind power plant. *IEEE Trans. Sustain. Energy* **2016**, *7*, 279–288. [CrossRef]
22. Vyver, J.V.d.; Kooning, J.D.M.D.; Meersman, B.; Vandevelde, L.; Vandoorn, T.L. Droop control as an alternative inertial response strategy for the synthetic inertia on wind turbines. *IEEE Trans. Power Syst.* **2016**, *31*, 1129–1138. [CrossRef]
23. Vidyanandan, K.V.; Senroy, N. Primary frequency regulation by deloaded wind turbines using variable droop. *IEEE Trans. Power Syst.* **2013**, *28*, 837–846. [CrossRef]
24. Kayikci, M.; Milanovic, J.V. Dynamic contribution of DFIG-based wind plants to system frequency disturbances. *IEEE Trans. Power Syst.* **2009**, *24*, 859–867.
25. Ramtharan, G.; Ekanayake, J.B.; Jenkins, N. Frequency support from doubly fed induction generator wind turbines. *IET Renew. Power Gener.* **2007**, *1*, 3–9.
26. Morren, J.; Pierik, J.; De Haan, S.W.H. Inertial response of variable speed wind turbines. *Electr. Power Syst. Res.* **2006**, *76*, 980–987. [CrossRef]
27. Morren, J.; Haan, S.W.H.d.; Kling, W.L.; Ferreira, J.A. Wind turbines emulating inertia and supporting primary frequency control. *IEEE Trans. Power Syst.* **2006**, *21*, 433–434. [CrossRef]
28. Zhang, X.; Zhu, Z.; Fu, Y.; Li, L. Optimized virtual inertia of wind turbine for rotor angle stability in interconnected power systems. *Electr. Power Syst. Res.* **2020**, *180*, 106157. [CrossRef]
29. Kundur, P. *Power System Stability and Control*; McGraw-Hill Education: New York, NY, USA, 1994.
30. Eberhart, R.; Kennedy, J. A new optimizer using particle swarm theory. In *MHS'95. Proceedings of the Sixth International Symposium on Micro Machine and Human Science, Nagoya, Japan, 4–6 October 1995*; IEEE: New York, NY, USA; pp. 39–43.
31. Kennedy, J.; Eberhart, R. Particle swarm optimization. In Proceedings of the ICNN'95—International Conference on Neural Networks, Perth, WA, Australia, 27 November–1 December 1995; Volume 1944, pp. 1942–1948.

Publisher's Note: MDPI stays neutral with regard to jurisdictional claims in published maps and institutional affiliations.

© 2020 by the authors. Licensee MDPI, Basel, Switzerland. This article is an open access article distributed under the terms and conditions of the Creative Commons Attribution (CC BY) license (http://creativecommons.org/licenses/by/4.0/).

Article

Wind Inertial Response Based on the Center of Inertia Frequency of a Control Area

Alija Mujcinagic [1,*]**, Mirza Kusljugic** [2] **and Emir Nukic** [2]

1. State Electricity Regulatory Commission, 75000 Tuzla, Bosnia and Herzegovina
2. Faculty of Electrical Engineering, University of Tuzla, 75000 Tuzla, Bosnia and Herzegovina; mirza.kusljugic@untz.ba (M.K.); emir.nukic1@fet.ba (E.N.)
* Correspondence: amujcinagic@derk.ba; Tel.: +387-61-150-741

Received: 19 October 2020; Accepted: 23 November 2020; Published: 24 November 2020

Abstract: As a result of the increased integration of power converter-connected variable speed wind generators (VSWG), which do not provide rotational inertia, concerns about the frequency stability of interconnected power systems permanently arise. If the inertia of a power system is insufficient, wind power plants' participation in the inertial response should be required. A trendy solution for the frequency stability improvement in low inertia systems is based on utilizing so-called "synthetic" or "virtual" inertia from modern VSWG. This paper presents a control scheme for the virtual inertia response of wind power plants based on the center of inertia (COI) frequency of a control area. The PSS/E user written wind inertial controller based on COI frequency is developed using FORTRAN. The efficiency of the controller is tested and applied to the real interconnected power system of Southeast Europe. The performed simulations show certain conceptual advantages of the proposed controller in comparison to traditional schemes that use the local frequency to trigger the wind inertial response. The frequency response metrics, COI frequency calculation and graphical plots are obtained using Python.

Keywords: inertial response; low inertia; the center of inertia; frequency response metrics; wind integration; PSS/E; FORTRAN

1. Introduction

Over the last ten years, the structure and control of the electric power system (EPS) have changed dramatically due to the increased integration of variable power generation (from wind and photovoltaic solar) [1]. In 2019, 15.4 GW of new wind power plants (WPP) had been installed and connected to the European power system, so the total installed capacity of WPP in Europe reached 205 GW [2]. Southeast Europe (SEE) region is also experiencing a "wind boom". The total installed capacity of WPP in 2019 in the SEE reached 5 GW. The largest number of conventional power plants in the SEE region were built during the 1970s and 1980s. In the near future, due to carbon pricing (e.g., emissions trading system, carbon taxes or EU Carbon Border Adjustment mechanism) and the aging of conventional thermal power plants (TPP), it is expected that coal-fired TPPs will decrease production or will be decommissioned (e.g., the recent case of Romania). They will be mainly replaced by variable speed wind generators (VSWGs). Conventional generating units have an inherent property to support the grid frequency regulation since the generator's rotating mass provides kinetic energy to the grid (or absorbs it) in the case of a frequency deviation caused by an active power imbalance.

Since VSWGs are connected to the network via power back-to-back AC/DC/AC converters, there is no direct electrical coupling between the grid frequency deviation and VSWG active power generation [3], which means that they do not contribute to the total system inertia. As the amount of wind energy in EPS increases, the share of connected synchronous machines and the total system inertia will decrease during high wind power generation periods. A detailed assessment of the dynamic impact of wind generation on EPS frequency control and changing system frequency behavior trends following the largest generator's loss is studied in [4]. The assessment requires detailed modelling of an entire interconnection for different wind penetration and contingency scenarios [5]. Reduced total system inertia causes many challenges and research opportunities for frequency control of future EPS [6,7]. It has a direct impact on frequency stability, potentially leading to under-frequency load-shedding (UFLS) [8] or, in the worse cases, to system blackouts (e.g., the 2016 Australian blackout and the London blackout of 2019) [9,10]. From the EPS point of view, inertia is essential because inertia slows down the frequency change, giving balancing mechanisms more time to respond and stabilize frequency [11].

Modern power converter connected generation can provide inertia to the grid and support the frequency recovering process. The authors in [12,13] found that the virtual synchronous machine concept is a promising solution for improving the frequency response metrics in low-inertia EPS.

In the past few years, numerous strategies and control schemes have been developed to utilize WPP to inject additional active power during a frequency disturbance using synthetic or virtual inertia control [14–16]. A complete definition of synthetic inertia with a distinction from the general term of fast frequency response is well described in [17]. Synthetic inertia control is implemented to extract the kinetic energy stored in the wind turbine's rotating blades and in wind generators, and is used to improve the quality of frequency response (FR) after a disturbance. The world wind industry has started to integrate controllers on modern VSWG to provide a temporary inertial response during frequency deviation (e.g., General Electric WindINERTIA^{TM}).

Emulated inertia controllers are mainly based on two different approaches: releasing hidden inertia and reserve capacity in pitch [18]. However, both methods have focused on developing inertial controllers that track frequency change at the point of common coupling (PCC) of WPP. Utilizing frequency at PCC as the input signal for the wind inertial controller will result in different synthetic inertial responses depending on the wind power plant (WPP) locations.

This paper presents the WPP inertial response control scheme based on tracking the selected control area's center of inertia (COI) frequency. The control area is part of an interconnected system of the European Network of Transmission System Operators for Electricity (ENTSO-E). Usually, it coincides with a state's territory, administered by a transmission system operator. The rationale behind the proposed scheme is based on frequency response (FR) analysis and the hierarchical organization of operation and control of continental European EPS [19]. The role of wind power and the need for additional inertia in the European EPS until 2050 was quantified in [20]. In [21], insufficient system inertia in European EPS has to be compensated by the provision of synthetic inertia.

The COI frequency is computed based on synchronous machine rotor speeds and widely used to define the effect of primary and secondary frequency regulation. Furthermore, COI frequency is particularly useful to present the frequency group of coherent synchronous machines. Regarding practical implementation, some previous researches considered that the COI signal is not fully adequate to support local frequency controllers due to delays in the communication system [22]. Nevertheless, modern digital high-speed telecommunication infrastructure brings new research opportunities and boosts the COI signal possibilities. Wind inertial control based on COI frequency can provide a "firm frequency response" within the first few seconds after the disturbance, regardless of their locations.

This paper aims to open the door for further research of frequency control based on COI in future low inertia system. The main paper contributions may be summarized as follows:

- The frequency response (FR) analysis of real interconnected EPS of the Southeast Europe region is performed and evaluated. The novel FR quality indicator "concentration of inertia" is introduced and calculated.
- The proposed wind inertial response control scheme based on the center of inertia frequency is introduced. The PSS/E user written wind inertial controller is developed using FORTRAN. The efficiency of the controller is tested and applied to the real EPS of SEE Europe.

The paper is organized as follows. In Section 2, the theoretical background of the power system frequency response is described. ENTSO-E inertia specifics are presented. A novel frequency response indicator is introduced. Next, in Section 3, the concept of COI is described. In Section 4, WPP contribution to the grid frequency regulation is explained. A proposed wind inertia controller based on the frequency of COI (FCOI) is presented in detail. In Section 5, a model of the SEE power system developed in PSS/E is described. The efficiency of the proposed controller is tested. Results of dynamic simulations, performed using PSS/E and supported by Python's scripts, are presented. Finally, in Section 6, conclusions and some recommendations for future research are given.

2. Power System Frequency Response

The frequency of an interconnected electric power system is a fundamental quantity for estimation and control. The EPS must be operated within a safe frequency range. The frequency will remain at its standard value (e.g., 50 Hz in Europe) as long as active power from the generation and the load demand are balanced. The frequency quality parameters for the European Network of Transmission System Operators for Electricity (ENTSO-E) synchronous areas, continental Europe (CE), Great Britain (GB) and Nordic system (NO) are provided in Table 1 [23].

Table 1. Frequency Quality Defining Parameters of the Synchronous Areas (Continental Europe (CE), Great Britain (GB) and Nordic system (NO)).

Parameters	CE	GB	NO
Standard frequency range	±50 mHz	±200 mHz	±100 mHz
Max. instantaneous frequency deviation	800 mHz	800 mHz	1000 mHz
Max. steady state frequency deviation	200 mHz	500 mHz	500 mHz
Time to recover frequency	n/a	1 min	n/a
Time to restore frequency	15 min	10 min	5 min
Alert state trigger time	5 min	10 min	5 min

According to the ENTSO-E, the interconnected power system of Continental Europe has to survive any frequency deviations due to a significant sudden change in load or generation (active power imbalance). The frequency response is the traditional metric used to describe how an EPS has stabilized frequency after the active power imbalance. To comply with the parameters given in Table 1, the FR process is realized through several phases: an inertial response, governor response (slow primary response) as illustrated in Figure 1, automatic generation control (secondary control) and tertiary control. In the initial phase of the incident (disturbance), which occurs during a few seconds (0–3 s) after the frequency changes, the synchronous generators' rotor releases or absorbs part of its kinetic energy. The swing equation mathematically describes this process:

$$\frac{2H_i}{f_0}\frac{df_i}{dt} = P_{m.i} - P_{e,i} = \Delta P_i \qquad (1)$$

where, H_i (s) is the inertia constant of i-th turbine-generator, f_i (Hz) is the frequency of i-th generator, f_0 is the rated frequency, P_{mi} (p.u.) is the mechanical power of i-th turbine-generator, P_{ei} (p.u.) is the electrical power of i-th generator and ΔP_i (p.u.) is active power imbalance. Equation (1) shows an imbalance between the turbine generators' mechanical and electrical power results in a frequency derivative. The rate of change of system frequency (ROCOF) is mostly used to evaluate the EPS frequency response dynamic. The ROCOF is directly proportional to the amount of active power imbalance:

$$\text{ROCOF} = \frac{df}{dt} = -\frac{\Delta P f_0}{2H_{sys}} \qquad (2)$$

where, H_{sys} (s) is the total system inertia constant, ΔP (p.u.) power imbalance of the system and f_0 (Hz) system rated frequency. If the ROCOF becomes too high during a disturbance, it could lead to under frequency load shedding. The ROCOF is lower as the inertia of the power system is higher.

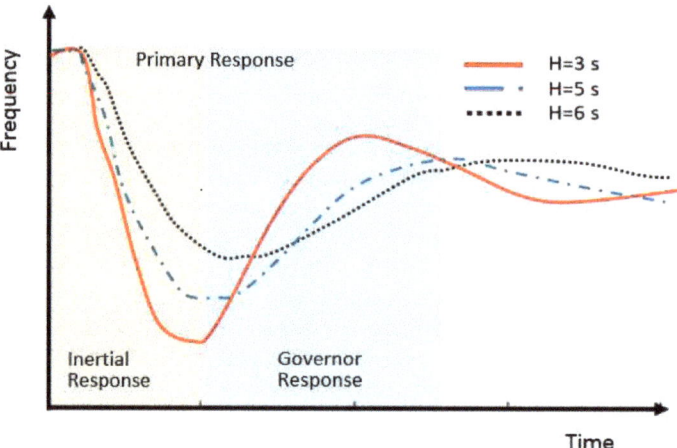

Figure 1. EPS frequency response time frame with different total system inertia.

Large conventional generators (like coal, gas and oil-fired thermal power plants, nuclear and hydropower plants), which use synchronous machines to convert the mechanical power from their turbines to electrical power, are the primary sources of inertia in today's EPS. Inertia constant H (s) describes the inertia of each generator:

$$H = \frac{1}{2}\frac{J\omega_n^2}{S_n} \text{ (s)} \tag{3}$$

where J (kg m^2) is the moment of inertia of generator and turbine, ω_n (rad/s) is the rotor angular speed and S_n (VA) is generator rated power. Inertia constant H falls typically in the wide range of 2–9 (s) [24]. The sum of all inertia of the spinning generators (and loads like motors) connected to the network is referred to as the total system inertia:

$$H_{sys} = \frac{\sum_{i=1}^{N} S_{ni} H_i}{S_{n,sys}} \text{ (s)} \tag{4}$$

where S_{ni} (MVA) is the rated power of the i-th generator, $S_{n,sys}$ is system rated power equal to the sum of S_{ni} and H_i (s) is the inertia constant of i-th turbine-generator. The number N in Equation (4) is the number of rotating synchronous generators connected to the system.

2.1. ENTSO-E Inertia Specifics

In Figure 2, the indicative inertia contribution of each European country (control area) at the time of the minimum total system inertia is presented. The contribution of SEE countries varies. Each country's contribution depends on the generation portfolio and the number of connected conventional generators. Due to the low share of wind and PV solar in their generation portfolio, some countries (e.g., Serbia and BIH) have substantial inertia contributions. Others, like Greece and Croatia, have slightly lower system inertia. Generally, EPS' stability and frequency control in the SEE region is still based on conventional power plants' dynamic characteristics and controllability.

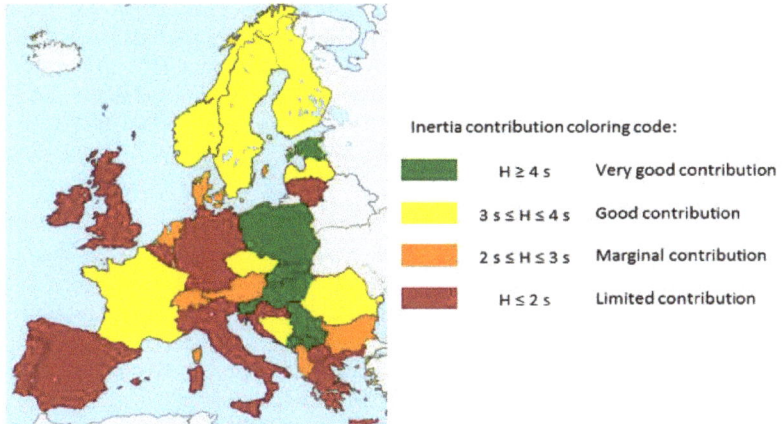

Figure 2. Indicating the contribution of each Transmission System Operator (TSO) to the total system inertia (European Network of Transmission System Operators for Electricity (ENTSO-E)—Ten Year Network Development Plan 2016).

2.2. Novel FR Indicator—The Concentration of Inertia

Modern interconnected EPS with a high share of converter interfaced production units will need a sufficient amount of inertia (kinetic energy). Calculation and estimation of each real interconnected EPS' available kinetic energy are crucial from the aspect of controllability and sharing inertial support among power systems. This paper introduced a novel FR metric called the "concentration" of inertia H_c (s). Instead of expressing inertia of a power system in seconds, it is often more convenient to calculate the kinetic energy stored in rotating masses of the system in megawatt seconds (MWs) [25]:

$$E_{k,sys} = \sum_{i=1}^{N} S_{ni} H_i \text{ (MWs)} \tag{5}$$

where S_{ni} is the rated apparent power of generator i [MVA] and H_i is the inertia constant of turbine-generator i (s).

This enhanced metrics (H_c) is determined using the ratio of the total kinetic energy and the total active power load (P_L) in the individual EPS:

$$H_c = \frac{E_{k,sys}}{P_L} \text{ (s)} \tag{6}$$

The proposed indicator—"concentration of inertia" characterizes each EPS and its contribution to the total system inertia. It is especially important that EPSs import electricity, and at the same time, have a significant share of wind in overall production. In the case of an imbalance that occurs in such EPS, a considerable part of the missing inertia will be provided from the interconnection. If the inertia (kinetic energy) of an EPS is low (small H_c), the participation of WPP in the inertial response should be required.

3. Center of Inertia Frequency Specifics

Real EPS consists of a large number of generators, and each of them, during the response to an active power disturbance, will operate at an individual frequency. Observing the entire system, there is an overall mean deceleration or acceleration of the fictitious center of inertia (COI) [26]. Individual generator frequency describes the power balance of the individual generator at a given location that is, due to electromechanical swings, oscillatory. In contrast, COI's frequency describes an aggregated active power balance of the entire system, which does not contain electromechanical oscillations. The amplitude and slope of individual electromechanical oscillations of synchronous generators with respect to the COI frequency in the inertial phase of FR will depend on their turbine-generator models' parameters and their electrical distance from the disturbance location. Based on the weighted average frequency (speed) of the individual generators, the frequency of the center of inertia can be calculated as:

$$f_{COI} = \frac{\sum_{i=1}^{N} H_i f_i}{\sum_{i=1}^{N} H_i} \tag{7}$$

The COI concept is the most common way to estimate the system frequency in transient stability analysis. The COI frequency could be calculated for a part of the interconnection, e.g., for a national EPS or a control area, which then contains inter-area electromechanical oscillations. Thus, the COI frequency is suited to study inter-area oscillations among coherent machine groups (clusters). In practice, instead of using COI, transmission system operators very often use the frequency estimated at a "pilot bus" of the system (typically a bus with a high short-circuit ratio). However, the "pilot bus" does not represent the system's average frequency, as it follows the dynamics of the closest synchronous generators.

4. WPP Inertial Controller Based on COI Frequency of a Control Area

Since variable speed wind generators (VSWG) are connected to the grid via power converters, their rotational speed is isolated from the system frequency and does not contribute to the total system inertia. A controller that emulates a synchronous generator's inertial response is often referred to as a synthetic, emulated, or virtual inertial response. Today, wind inertia controller activation primarily is based on the local frequency input signal. Using this type of controller, the active power output of VSWG is a function of the measured frequency and/or ROCOF. Estimating local frequency at PCC is commonly based on phase-locked loop (PLL) techniques. To emulate synthetic inertia, an initial additional active power ΔP_i must be controlled with the following equation [27]:

$$\Delta P_i = \frac{2 P_{max} H_{gen}}{f_n} \frac{df}{dt}\bigg|_{(t=0^+)} \tag{8}$$

where P_{max} is the generator's active power, H_{gen} is generator inertia constant and f_n the nominal frequency. In order to emulate the inertial response properly, synthetic (virtual) inertia must be very fast (0–3 s). The exact values of the inertial constant vary depending on the manufacturer of WT. However, for research purposes, the WT inertial constant as a function of its power P (W) can be estimated [28]:

$$H_{wt} \approx 1.87 P^{0.0597} \tag{9}$$

The total inertia of the wind turbine-generator system presents a sum of the inertia constant for the generator and the turbine. According to Equation (1), system frequency excursion in the inertial phase of disturbance is not the same in all system nodes. Utilizing the local frequency (at PCC) as the input signal for the wind inertial controller will result in different synthetic inertial responses depending on the wind power plant (WPP) locations.

Proposed Wind Inertial Control Scheme and Controller

The impact of emulated inertia on a real interconnected EPS frequency response is usually quantified and analyzed through time-domain simulations using commercial programs like Power Systems Simulation for Engineers (PSS/E) software. Unlike modelling classical thermal or hydro generating units, modelling of WPP is very specific. Wind turbine (WT) manufacturers develop their models, which are more or less complex, but substantially different. Modelling and transient stability simulations of EPS with many different WT commercial models could be hard work and quite frustrating. Generally, the idea is to create generic models that are parametrically adjustable to represent specific wind turbines available in the market. Developing a user-defined model and integrating it into a modular machine-model structure enables implementing a wide range of influencing factors in the power system dynamic simulations. PSS/E wind-related machine models with defined input-output dependencies can be integrated into the unique structure by developing a user-defined auxiliary signal model that controls the principle of operation of the entire model during the dynamic simulations.

The proposed wind inertia control scheme based on the control area's COI frequency is presented in Figure 3. The input signal $\Delta\omega_{COI}$ presents deviation of the frequency of the COI of a control area with N synchronous generators, computed according to:

$$\Delta\omega_{COI} = \frac{\sum_{i=1}^{Narea} H_i \Delta\omega_i}{\sum_{i=1}^{Narea} H_i} \quad (10)$$

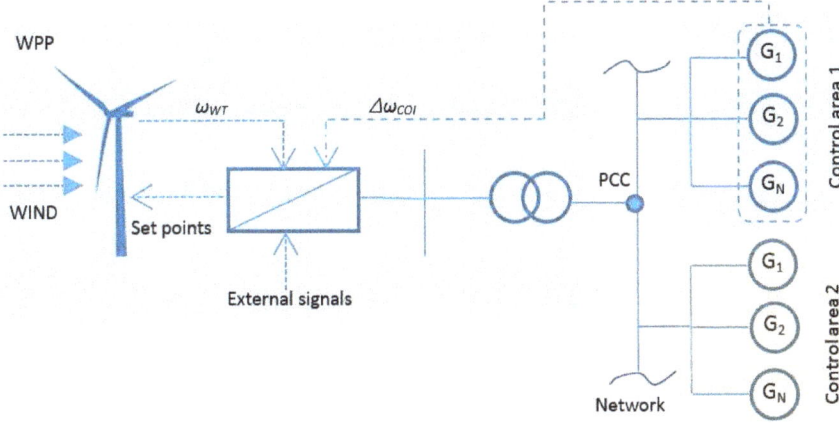

Figure 3. Frequency of center of inertia (FCOI) wind inertia control scheme.

Based on the proposed scheme, a novel generic model of wind inertial controller *FCPCAU1* is developed. The structure of *FCPCAU1* is presented in Figure 4. *FCPCAU1* is written in FORTRAN. PSS/E Environment Manager is used to compile code and create a dynamic linked library (DLL) [29]. The application of available wind generator and wind electrical PSS/E models combined with a user-defined wind auxiliary control model (*FCPCAU1*) implies the integration of the modified logic into the dynamic models' structure. This inertial controller model is integrated with the existing generic wind electrical model *REECAU1* and generic wind generator/convertor model *REGCAU1*. These models are contained in the PSS/E dynamics models library and they are widely used to represent Type 3, double fed induction generator (DFIG) or Type 4, fully fed generators.

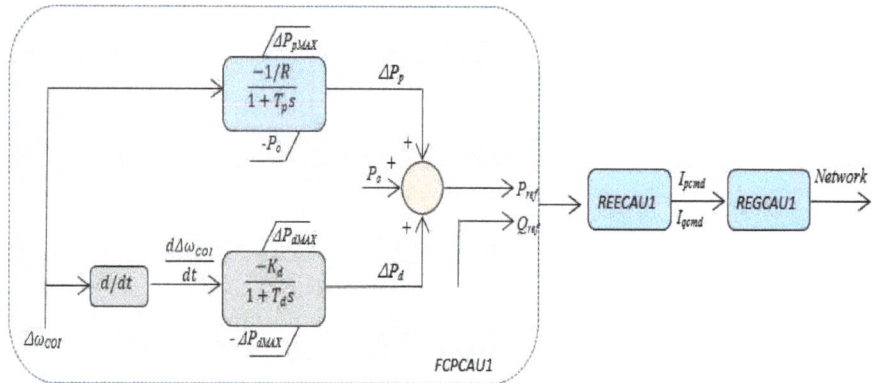

Figure 4. *FCPCAU1* wind inertia controller structure.

FCPCAU1 model storage, input and output parameters are presented with state variable values (*STATE*), real model parameters (*CON*) and real model variable values (*VAR*), as shown in Tables 2 and 3.

Table 2. *FCPCAU1* Model Storage Parameters.

Model Storage FCPCAU1	Parameters
STATE(K)	ΔP_p
STATE(K+1)	ΔP_d
VAR(L)	TOTALINERTIA
VAR(L+1)	TOTALFACTOR
VAR(L+2)	DELTAFCOI
VAR(L+3)	DERIVDELTAFCOI
VAR(L+4)	PSTART
VAR(L+5)	QSTART
CON(J)	R
CON(J+1)	K_d
CON(J+2)	T_p
CON(J+3)	T_d
CON(J+4)	ΔP_{pmax}
CON(J+5)	ΔP_{dmax}
CON(J+6)	DEFARENR

Table 3. *FCPCAU1* Model Input and Output Parameters.

Model Input FCPCAU1	Parameters
Hi	Model CONS
$\Delta \omega i$	SPEED
P_o	PELEC
Q_o	QELEC
Model Output FCPCAU1	**Parameters**
Pref	WPCMND
Qref	WQCMND

FCPCAU1 controller consists of two modules: the first one performs droop control, while the second one performs inertia emulation. These two modules combined results in the recovery of the frequency to a new steady-state value. Adjusting model storage parameters (e.g., *R*), the droop controller module can be disabled. The droop module produces an active power change proportional to the frequency deviation. It is reasonable that *R* should be tuned on a similar value as conventional synchronous generators, Figure 5. The wind turbine generators (WTGs) usually operate at the maximum power point tracking (MPPT). The active power increase (ΔP) during a sudden drop of system frequency must be obtained from the kinetic energy of the rotating parts (turbine and generator)

of WTG that causes a decrease of rotational speed. The droop control should be ended on time to avoid WT's stalling or coordinated with the de-loading control. For the de-loading mode of WT operation, the upper droop control limit ΔP_{pMAX} depends on the current wind power availability. FCPCAU1 operates assuming that the wind speed is constant. Tp (s) presents the time constant.

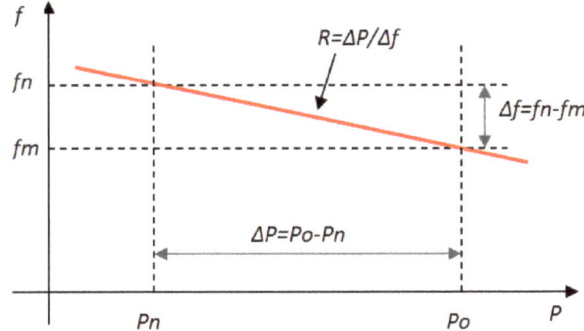

Figure 5. EPS frequency droop control.

The derivative module emulates conventional generating units' inherent property to support the grid frequency regulation, based on Equation (1). It is possible to consider the whole system or its part (one control area) and introduce ω_{COI}:

$$2H\frac{d\Delta\omega_{COI}}{dt} = P_m - P_e \quad (11)$$

where $\Delta\omega_{COI}$ is the rotational speed deviation of the COI from the nominal synchronous speed:
The wind power injected in the EPS is contained in P_e [30] according to:

$$2H\frac{d\Delta\omega_{COI}}{dt} = P_m - P_e - K_d\frac{d\Delta\omega_{COI}}{dt} \quad (12)$$

The controller provides an active power injection proportional to the derivative of the COI frequency of the control area in which WPP is connected, described by:

$$(2H + K_d)\frac{d\Delta\omega_{COI}}{dt} = P_m - P_e \quad (13)$$

where derivative gain K_d acts as synthetic inertial constant, increasing the total system inertia.
The derivative of the COI frequency deviation is calculated according to:

$$\frac{d\Delta\omega_{COI}(t)}{dt} = \frac{\Delta\omega_{COI}(t) - \Delta\omega_{COI}(t - \Delta t)}{\Delta t} \quad (14)$$

As a result, the frequency derivative effect has the same nature as the synchronous generators' inertia in the selected control area. When the frequency decreases, active power injection increases due to the negative derivative of frequency. It continues to change until the frequency variation becomes zero. Due to practical implementation, active power injection should be limited, and limitation band $\pm\Delta P_{dmax}$ band is introduced. Wind generator (REGCAU1) and wind electrical (REECAU1) models are controlled by the wind auxiliary control model's output reference signal (P_{ref}).

Having in the mind ongoing marketization of EPS inertial service, the proposed scheme can also be applied to selected generators belonging to balance responsible parties (BRPs). BRPs are financially responsible for continuously maintaining the balance between supply and demand within their generation portfolio.

5. EPS Modelling and Simulations

Traditionally, the generation's portfolio in EPS of this Southeast Europe region is based mainly on hydro and thermal power plants. The transmission network of these countries operates at 400 kV, 220 kV and 110 kV voltage levels. The high voltage network through the region is well meshed, providing robust interconnections illustrated in Figure 6. The nominal transmission capacity of interconnectors on average is above 40% of the regional average peak load. Some countries like BIH and Croatia have an unusually high capacity of interconnectors regarding their peak load.

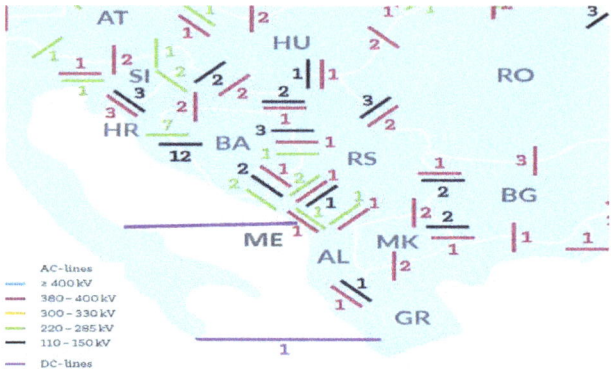

Figure 6. Southeast Europe (SEE) Region Electric Power System.

The SEE region's power system, presented in Figure 6, is modelled using PSS/E software. The PSS/E consists of a complete set of programs for the study of the EPS with both steady-state and dynamic simulations. The PSS/E software interfaced with Python programming language with the application program interface (API) is used to perform the dynamic simulations [31]. There are two automation processes in the PSS/E based on the API: the Python interpreter (Python) and the IPLAN simulator. Python has great flexibility in operation and has predefined modules for interacting with PSS/E [32]. In this paper, frequency response metrics, frequency of COI calculations and plots are performed with Python scripts.

The EPSs of Slovenia (SI), Croatia (HR), Bosnia and Herzegovina (BIH), Serbia (RS), North Macedonia, Kosovo, and Montenegro are modelled in detail, using complete models of generators, excitation systems and governors. The PSS/E library includes a family of generator models. GENROU (round rotor generator model) and IEEEG1 governor model is used to present thermal units [33]. IEEEG1 is the IEEE recommended general model for steam turbine speed governing systems. The hydro generators are presented with GENSAL (salient pole generator model) with HYGOV turbine speed governing systems. The simplified excitation system's dynamic model SEXS is used for all types of synchronous generators.

The classical generator model (GENCLS) is used to present all generators in EPSs of Hungary, Romania, Bulgaria, Albania, Greece, and Turkey. GENCLS is the classical constant voltage behind the transient reactance generator model (with $H = 4$ s, $D = 0.5$ p.u.).

The EPS is assumed to be working in a steady-state equilibrium before the active power disturbance's initialization. The simulated steady-state, network topology, generation units schedule and loads corresponding to the real scenario are recorded on 16 February 2020. Data are obtained from the ENTSO-E transparency platform. The total power generation in the analyzed system is 84.3 GW and consists of thermal and hydro as well as some wind power plants. The total system load is 82.6 GW. The FR analysis of each control area is out of the scope of this paper. This paper is focused on the FR performance of the ENTSO-E SCB control block (Slovenia, Croatia and BIH).

The power system data of ENTSO-E control block Slovenia, Croatia and BIH used in simulations are presented in Table 4.

Table 4. ENTSO-E Slovenia, Croatia and BIH (SCB) control block electric power system (EPS) data.

Control Area	Generation (MW)	E_{ksys} (MWs)	Load (MW)	Wind (MW)
BIH	1.990	9.273	1.385	87
Croatia	1.680	8.033	2.100	330
Slovenia	1.590	10.532	1.689	0
TOTAL (MW)	5.260	27.838	5.174	417

The aggregated wind turbine model is used to represent all wind turbines inside a wind farm. The wind speed remains constant during simulation. The user-written auxiliary model *FCPCAU1* is used to simulate WTGs in BIH EPS. There are two WPPs connected in the BIH control area, WPP Jelovaca (18 × 2 MW) and WPP Mesihovina (22 × 2.3 MW). The values of the parameters of the *FCPCAU1* are provided in Table 5.

Table 5. *FCPCAU1* parameter values.

Parameter	Value	Unit
R	0.1	-
K_d	3.3	-
T_p	0.3	s
T_d	0.5	s
ΔP_{pMAX}	0.5	-
ΔP_{dMAX}	0.5	-
DEFARENR	BIH [1]	

[1] Selected control area is Bosnia and Herzegovina (BIH).

5.1. Southeast Europe Region EPS Frequency Response Simulation

The active power disturbance $\Delta P_L = 2.75$ (p.u.), sudden loss of production of TPP Stanari (275 MW) in the BIH control area, is applied at $t = 2$ s after the simulation start. Based on imported data from PSS/E time-domain simulations, Python scripts have been developed to automate calculations and plots of EPS frequency response. For the simulated disturbance, the COI frequency is calculated for the SCB control block and the entire SEE region.

Power system frequency responses of Slovenia, Croatia, BIH and of FCOI of SEE region are presented in Figure 7.

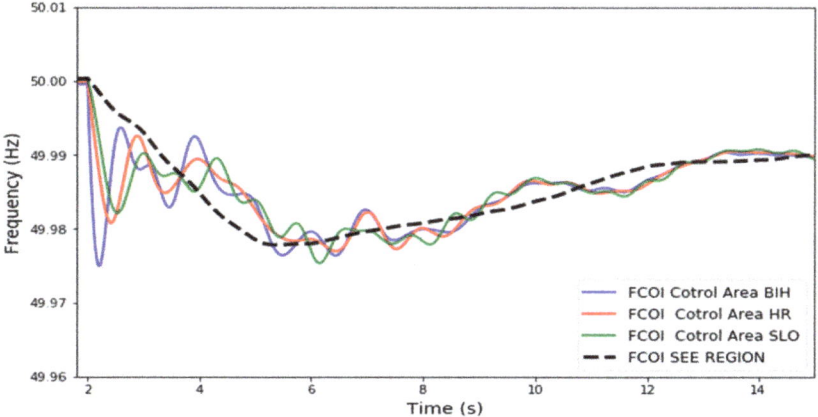

Figure 7. EPS frequency response, simulation time 15 s.

Following the disturbance, frequency excursions of COI Slovenia, Croatia and BIH are different and experienced inter-area electromechanical oscillations (EMOs).

These low frequency (0.1–0.8 Hz) EMOs are associated with coherent groups of synchronous generators from each control area (EPS) swinging against the other coherent groups of synchronous generators. Each coherent group oscillates differently than the others. In particular, at a given time, some groups locally accelerate, and others locally decelerate. Frequency oscillations caused by the active power imbalance are damped, and after 10–15 s, the system obtains a new steady-state.

At the end of the simulation (Figure 7), the system will not return to the nominal frequency (50 Hz) on its own. To compensate for the remaining frequency deviation, ENTSO-E Continental Europe interconnected system uses automatic secondary frequency control called automatic generation control (AGC). The AGC is not included in this simulation model.

5.2. Southeast Europe Region Frequency Response Evaluation

The frequency response (FR) performance indicators and inertia concentration are calculated for individual control areas and the entire SEE region. The results are presented in Table 6. The steady-state frequency deviation Δf_{ss} is the frequency deviation from the rated (nominal) system frequency (50 Hz) in a new steady-state, while Δf_{max} is the highest absolute frequency deviation (nadir). Generally, after the applied disturbance, the calculated frequency response metrics show that the system frequency remains in the prescribed range.

Table 6. Frequency response metrics of COI of control areas, SEE Region and WPP Jelovaca 110 kV PCC.

	Nadir	t_{min}	df/dt	Δf_{max}	Δf_{ss}	H_c
	Hz	s	Hz/s	mHz	mHz	s
BIH	49.9749	0.220	−0.238	25.1	11	6.7
Croatia	49.9808	0.245	−0.073	19.2	11	3.8
Slovenia	49.9821	0.255	−0.047	17.9	11	6.2
WPP Jel. 110 kV (BIH)	49.9849	0.228	−0.169	22.2	10	-
SEE Region	49.9778	3.38	−0.010	15.1	10	-

The lowest point of frequency drop (nadir) is recorded in EPS of BIH (49.9749 Hz) and the time it takes to reach the nadir t_{min} = 0.22 (s) (immediately after a disturbance occurred). The frequency decline of the SEE region is stopped at t_{min} = 5.38 (s) (3.38 s after a disturbance occurred) at a frequency value of 49.9778 Hz. The final steady-state frequency deviation in the SEE region EPS is Δf_{ss} = 10 (mHz).

Deviations from nominal frequency by more than 20 mHz are corrected by activating individual generators' primary frequency regulators and, if necessary, by activating the energy for frequency restore reserve FRR [34]. FRR is an operating reserve activated to restore frequency to the nominal value and power flows on the interconnected lines to the pre-fault scheduled values. It is also used for secondary and tertiary regulation. In Figure 8, the FCOI of SEE region, frequency measured at PCC of WPP Jelovaca and FCOI of EPS BIH are presented. The frequency at PCC of WPP Jelovaca (red line) is obtained from time-domain simulation, computing the numerical derivative of the bus voltage phase angle. It is evident that in the initial phase, the local bus frequency (red line) is much closer to the frequency of the total system frequency of the SEE region (black dotted line) than the frequency of COI BIH (blue line). It can be concluded that the frequency representation using the measured generator speeds (e.g., FCOI of the control area) more closely reflects the frequency dynamics of a local EPS.

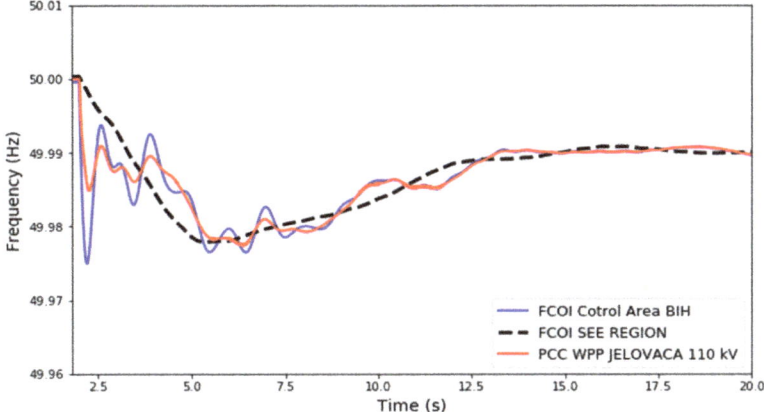

Figure 8. EPS frequency response, time simulation 20 s.

The initial slopes (ROCOF) of each control area and SEE region are presented in Figure 9. As expected, recorded ROCOFs of COI of Slovenia and Croatia are lower than the COI of BIH.

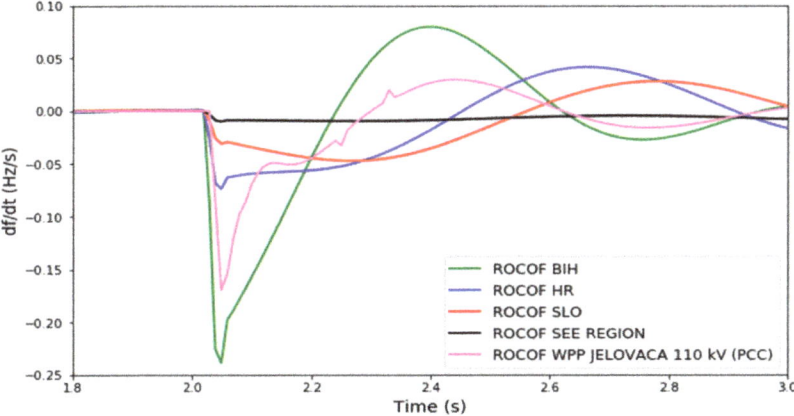

Figure 9. ROCOF of EPS.

Following the disturbance, the maximum ROCOF of COI BIH (−0.238 Hz/s) is significantly higher than the ROCOF of local frequency at PCC of WPP Jelovaca (−0.169 Hz/s). This different frequency behavior is used to design wind inertial controller *FCPCAU1* based on specifics of the FCOI of a control area. The performed simulation and calculations show that EPS of BIH, which is a significant exporter, has the largest inertia concentration 6.7 (s), due to low total system load and numerous synchronous generators connected to the grid. The inertia concentration of the EPS of Croatia is the smallest 3.8 (s) in the SCB LFC block. Croatia is a regional importer of electricity and its total load (2.100 MW) is high relative to the local generation (1.680 MW) and the number of connected SGs. The concentration of inertia indicates a potential need for WPPs located in EPS of Croatia to participate in the inertial response. It is an important finding because the EPS of Croatia has the largest installed wind capacity and the smallest kinetic energy in the SCB LFC block (Table 4).

5.3. WPP Inertial Controller FCPCAU1 Performance

Based on previous findings and calculated FR indicators, the WPPs' inertial contribution highly depends on used frequency input signals. In Figure 10, the power output of WPP Jelovaca (blue dotted line) during the analyzed disturbance is presented. When the frequency drop is detected (red line), the inertial controller *FCPCAU1* continuously follows df_{COI}/dt and forces the WPP to generate additional power. The additional power output from synthetic inertia depends on ROCOF, WPP inertia constant and WPP P_{max}. In the initial phase of active power response, WPP Jelovaca additionally generates 0.4 MW.

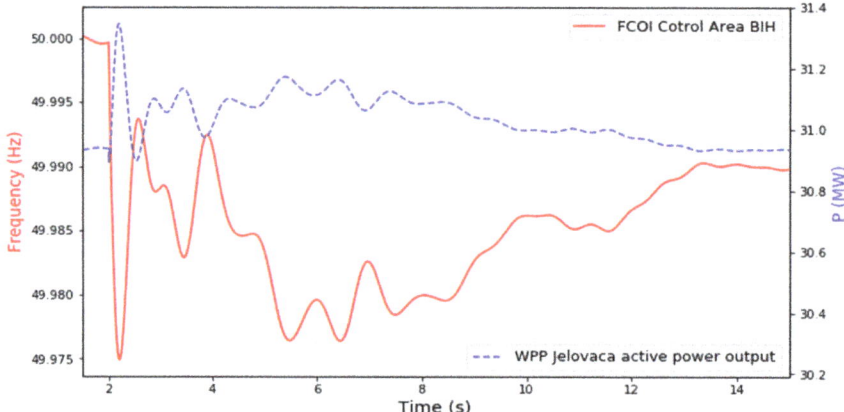

Figure 10. WPP Jelovaca (BIH) active power response.

In Figure 11, a comparison between WT active power responses considering different input signals for synthetic inertia control is presented. The initial peak of the inertial response triggered by the FCOI of the BIH control area (red line) results from fast releasing kinetic energy from WT. It is more significant than when local frequency measured at PCC of WPP (blue dotted line) is used.

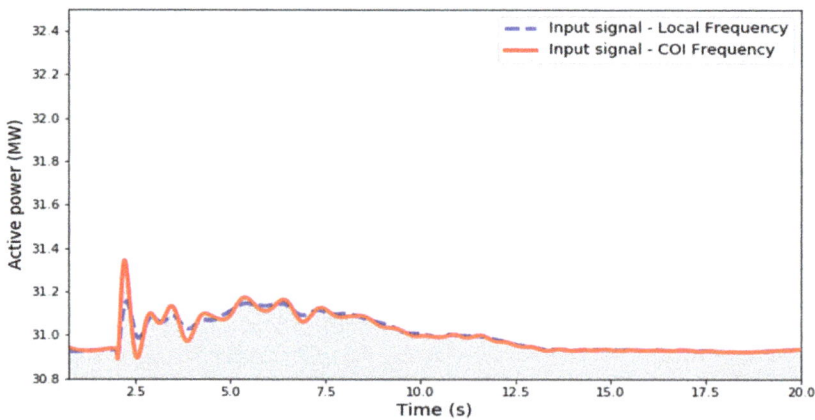

Figure 11. WPP Jelovaca (BIH) active power response comparison.

The value of derivative gain K_d highly influences *FCPCAU1* performance. Gain (K_d) excessive value could lead to WT rotor speed instability, because K_d scales change at every frequency. The inertial controller's parameters should be carefully determined to avoid eventual dangerous consequences on WT's mechanical parts.

In Figure 12, the FCOI of SEE region, FCOI of control area Croatia and frequency obtained from PSS/E simulation at PCC of WPP Ponikve (34 MW), WPP Velika Glava (43 MW) and WPP Vratarusa (42 MW) are presented. These wind farms are connected to the 110 kV transmission grid of the Croatia control area. WPP Ponikve and Velika Glava are electrically close to hydropower plants (HPP Dubrovnik and HPP Orlovac), so the frequency excursion at their PCC is significantly influenced by these HPP. WPP Vrataruša is electrically more distant from the fault location, so the frequency change at PCC is slightly different.

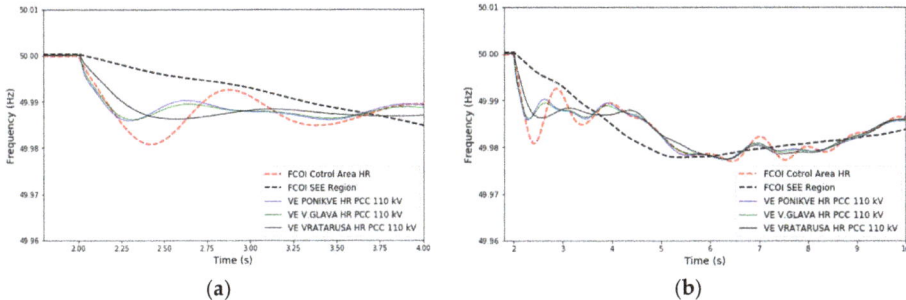

Figure 12. EPS frequency response. (a) simulation time 4 s, (b) simulation time 10 s.

In the first few seconds, the dynamics of the Croatian FCOI better characterizes the nature of the imbalance than the local PCC frequency at WPP Vratarusa. The frequency response indicates that the system frequency excursion in the inertial phase of disturbance is not the same in all system nodes. This finding will result in different synthetic inertial responses depending on the wind power plant (WPP) locations. Using FCOI of a control area instead of local frequency, all WPPs can provide a "firm frequency response" within the first few seconds after the disturbance, regardless of their locations.

6. Conclusions

The wind is expected to be a significant new electricity generation source in Southeast Europe, but WPP integration is slow and their share in total production is small. Generally, the frequency stability and control of EPS in this region are still based on the dynamic characteristics and controllability of conventional power plants.

Simulation results show that the modelling and the frequency calculation (COI or local frequency) significantly differ in a power system's transient behavior with high WPP integration. A new FR indicator, "concentration of inertia", was introduced. This indicator shows how inertia is distributed in real interconnected systems and can be relevant for the inertial response service's upcoming marketization. The inertial response of WT based on the COI frequency shows that WTTs can provide a "firm frequency response" in the first few seconds after the disturbance, regardless of their location. Some essential points like the impact of noise, time delay and measurement data quality on the estimation COI signal still require further research to enhance the WT inertial response based on the control area FCOI. In the future (low inertia) EPS, frequency control methods need to be redefined and new schemes need to be developed. The authors believe that this work presents a small step in that direction.

Author Contributions: Conceptualization, A.M. and M.K.; software, A.M. and E.N.; validation, A.M. and E.N.; formal analysis, A.M.; simulations, A.M.; writing—original draft preparation, A.M.; writing—review and editing, A.M., M.K. and E.N.; visualization, A.M.; supervision, M.K. All authors have read and agreed to the published version of the manuscript.

Funding: This research received no external funding.

Conflicts of Interest: The authors declare no conflict of interest.

References

1. Ulbig, A.; Borsche, T.S.; Andersson, G. Impact of Low Rotational Inertia on Power System Stability and Operation. *IFAC* **2014**, *47*, 7290–7297. [CrossRef]
2. Wind Europe. Wind energy in Europe in 2019—Trends and statistics. February 2020. Available online: https://windeurope.org/wp-content/uploads/files/about-wind/statistics/WindEurope-Annual-Statistics-2019.pdf (accessed on 1 September 2020).
3. Lalor, G.; Mullane, A.; O'Malley, M. Freq. control and wind turbine technologies. *IEEE Trans. Power Syst.* **2005**, *20*, 1905–1913. [CrossRef]
4. Doherty, R.; Mullane, A.; Nolan, G.; Burke, D.J.; Bryson, A.; O'Malley, M. An Assessment of the Impact of Wind Generation on System Frequency Control. *IEEE Trans. Power Syst.* **2010**, *25*, 452–460. [CrossRef]
5. Muljadi, E.; Gevorgian, V.; Singh, M.; Santoso, S. Understanding Inertial and Frequency Response of Wind Power Plants. In Proceedings of the 2012 IEEE Power Electronics and Machines in Wind Applications, Denver, CO, USA, 16–18 July 2012.
6. Nguyen, H.T.; Yang, G.; Nielsen, A.H.; Jensen, P.H. Challenges and Research Opportunities of Frequency Control in Low Inertia Systems. In Proceedings of the 2nd International Conference on Electrical Engineering and Green Energy, Rome, Italy, 28–30 June 2019.
7. Milano, F.; Dörfler, F.; Hug, G.; Hill, D.J.; Verbic, G. *Foundations and Challenges of Low-Inertia Systems*; IEEE: Piscataway, NJ, USA, 2018; pp. 1–25.
8. Gonzalez-Longatt, F.M. Impact of emulated inertia from wind power on under-frequency protection schemes of future power systems. *J. Mod. Power Syst. Clean Energy* **2016**, *4*, 211–218. [CrossRef]
9. AEMO. *Black system South Australia 28 September 2016*; Technical Report; Australian Energy Market Operator Limited: Sydney, Australia, 2017.
10. OFGEM. *Technical Report on the Event of 9 August 2019*; OFGEM: Canary Wharf, UK, 2019.
11. Tielens, T.; Hertem, D.V. The relevance of inertia in power systems. *Renew. Sustain. Energy Rev.* **2016**, *55*, 999–1009. [CrossRef]
12. Bevrani, H.; Ise, T.; Miura, Y. Virtual synchronous generators: A survey and new perspectives. *Int. J. Electr. Power Energy Syst.* **2014**, *54*, 244–254. [CrossRef]
13. Heydari, R.; Savaghebi, M.; Blaabjerg, F. Fast Frequency Control of Low-Inertia Hybrid Grid Utilizing Extended Virtual Synchronous Machine. In Proceedings of the 11th Power Electronics, Drive Systems, and Technologies Conference (PEDSTC), Tehran, Iran, 4–6 February 2020; pp. 1–5.
14. Wu, Z.; Gao, W.; Gao, T.; Ya, W.; Zhang, H.; Yan, S.; Wang, X. *State-of-the-Art Review on the Frequency Response of Wind Power Plants in Power Systems*; Springer: Berlin/Heidelberg, Germany, 2017.
15. Osmic, J.; Kusljugic, M.; Mujcinagic, A. Fuzzy Logic Controller for Inertial Support of Variable Speed Wind Generator. In Proceedings of the Medpower-10th Mediterranean Conference on Power Generation, Transmission, Distribution and Energy Conversion, Belgrade, Serbia, 6–9 November 2016.
16. Díaz-González, F.; Hau, M.; Sumper, A.; Gomis-Bellmunt, O. Participation of wind power plants in system frequency control: Review of grid code requirements and control methods. *Renew. Sustain. Energy Rev.* **2014**, *34*, 551–564. [CrossRef]
17. Eriksson, R.; Modig, N.; Elkington, K. Synthetic inertia versus fast frequency response: A definition. *IET Renew. Power Gener.* **2018**, *12*, 507–514. [CrossRef]
18. Gonzalez-Longatt, F.M. Activation Schemes of Synthetic Inertia Controller on Full Converter Wind Turbine (Type 4). In Proceedings of the IEEE Power & Energy Society General Meeting, Denver, CO, USA, 26–30 July 2015.
19. Mujcinagic, A.; Kusljugic, M.; Osmic, J. Frequency Response Metrics of an Interconnected Power System. In Proceedings of the 54th International Universities Power Engineering Conference (UPEC), Bucharest, Romania, 3–6 September 2019.
20. Agathokleous, C.; Ehnberg, J. A Quantitative Study on the Requirement for Additional Inertia in the European Power System until 2050 and the Potential Role of Wind Power. *Energies* **2020**, *13*, 2309. [CrossRef]
21. Thiesen, H.; Jauch, C.; Gloe, A. Design of a System Substituting Today's Inherent Inertia in the European Continental Synchronous Area. *Energies* **2016**, *9*, 582. [CrossRef]

22. Milano, F. Rotor Speed-Free Estimation of the Frequency of the Center of Inertia. *IEEE Trans. Power Syst.* **2018**, *33*, 1153–1155. [CrossRef]
23. ENTSO-E. Network Code on Load-Frequency Control and Reserves. 2013. Available online: https://eepublicdownloads.entsoe.eu/clean-documents/pre2015/resources/LCFR/130628-NC_LFCR-Issue1.pdf (accessed on 5 September 2020).
24. Grainger, J.J.; Stevenson, W.D. *Power System Analysis*; McGraw-Hill: New York, NY, USA, 1994.
25. ENTSO-E. Nordic Report Future System Inertia V2. 2018. Available online: https://docs.entsoe.eu/id/dataset/nordic-report-future-system-inertia/resource/6efce80b-2d87-48c0-b1fe-41b70f2e54e4 (accessed on 5 September 2020).
26. Anderson, P.; Fouad, M.A. *Power System Control and Stability*; John Willey & Sons: Hoboken, NJ, USA, 2003.
27. Kovaltchouk, T.; Debusschere, V.; Seddik, B.; Fiacchini, M.; Alamir, M. Assessment of the Impact of Frequency Containment Control and Synthetic Inertia on Intermittent Energies Generators Integration. In Proceedings of the Eleventh International Conference on Ecological Vehicles and Renewable Energies (EVER), Monte Carlo, Monaco, 6–8 April 2016.
28. González Rodriguez, A.G.; González Rodriguez, A.; Burgos Payán, M. Estimating wind turbines mechanical constants. *Renew. Energy Power Qual. J.* **2007**, *1*, 697–704. [CrossRef]
29. Krishnat, P.; Jay, S. *Creating Dynamics User Model Dynamic Linked Library (DLL) for Various PSS® Versions*; Siemens Energy Inc.: Erlangen, Germany, 2012.
30. Rodríguez-Bobada, F.; Ledesma, P.; Martínez, S.; Coronado, L.; Prieto, E. Simplified wind generator model for Transmission System Operator planning studies. In Proceedings of the 7th International Workshop on Large Scale Integration of Wind Power and on Transmission Networks for Offshore Wind Farms, Madrid, Spain, 26–27 May 2008.
31. SIEMENS. *Siemens PTI.PSS/E 34 Application Program Interface (API)*; SIEMENS: New York, NY, USA, 2015.
32. Khalid, M.U. *Python Based Power System Automation in PSS/E*; Lulu Enterprises, Inc.: Morrisville, NC, USA, 2014.
33. SIEMENS. *Siemens PTI.PSS/E 34 Model Library*; SIEMENS: New York, NY, USA, 2015.
34. Independent System Operator in BIH, Grid Code. January 2019. Available online: https://www.nosbih.ba/files/dokumenti/Legislativa/Mrezni%20kodeks/EN/GRID%20CODE%20%20January%202019.pdf (accessed on 18 September 2020).

Publisher's Note: MDPI stays neutral with regard to jurisdictional claims in published maps and institutional affiliations.

© 2020 by the authors. Licensee MDPI, Basel, Switzerland. This article is an open access article distributed under the terms and conditions of the Creative Commons Attribution (CC BY) license (http://creativecommons.org/licenses/by/4.0/).

Article

The LVRT Control Scheme for PMSG-Based Wind Turbine Generator Based on the Coordinated Control of Rotor Overspeed and Supercapacitor Energy Storage

Xiangwu Yan [1,2], Linlin Yang [1,*] and Tiecheng Li [3]

[1] State Key Laboratory of New Energy Power System, North China Electric Power University, Beijing 102206, China; xiangwuy@ncepu.edu.cn
[2] Key Laboratory of Distributed Energy Storage and Micro-grid of Hebei Province, North China Electric Power University, Baoding 071003, China
[3] State Grid Hebei Electric Power Research Institute, Shijiazhuang 050021, China; ltc8086@163.com
* Correspondence: yanglinlin@ncepu.edu.cn; Tel.: +86-1553-405-3644

Abstract: With the increasing penetration level of wind turbine generators (WTGs) integrated into the power system, the WTGs are enforced to aid network and fulfill the low voltage ride through (LVRT) requirements during faults. To enhance LVRT capability of permanent magnet synchronous generator (PMSG)-based WTG connected to the grid, this paper presents a novel coordinated control scheme named overspeed-while-storing control for PMSG-based WTG. The proposed control scheme purely regulates the rotor speed to reduce the input power of the machine-side converter (MSC) during slight voltage sags. Contrarily, when the severe voltage sag occurs, the coordinated control scheme sets the rotor speed at the upper-limit to decrease the input power of the MSC at the greatest extent, while the surplus power is absorbed by the supercapacitor energy storage (SCES) so as to reduce its maximum capacity. Moreover, the specific capacity configuration scheme of SCES is detailed in this paper. The effectiveness of the overspeed-while-storing control in enhancing the LVRT capability is validated under different levels of voltage sags and different fault types in MATLAB/Simulink.

Keywords: permanent magnet synchronous generator (PMSG); supercapacitor energy storage (SCES); rotor overspeed control; low voltage ride through (LVRT); capacity configuration of SCES

1. Introduction

Due to the abundant sources and advanced power generation technologies, the wind power is integrated into the grid on a large scale as a major green source. Currently, the low voltage ride through (LVRT) is one of the most important issues to the modern systems with high penetration of wind power. To overcome this difficulty, the grid-connected requirements for wind power have become stricter and stricter. The grid codes enforce the wind turbine generator (WTG) to keep grid-connected and provide reactive power support under faults.

At present, the doubly fed induction generator (DFIG) [1–6] and the permanent magnet synchronous generator (PMSG) [7–12] are two mainstream wind turbine generators (WTGs). On the one hand, the PMSG exempts the gearbox, which is easily broken, so as to reduce maintenance cost and improve the reliability of PMSG; on the other hand, the PMSG can complete isolation from the grid disturbances owing to the full-scale converter. In comparison with DFIG, the PMSG has a simpler structure, lower maintenance cost and higher LVRT capabilities [13].

Recently, the methods of the LVRT enhancement of PMSG have been divided into two primary types: inherent control modification and auxiliary device modification [14]. The inherent control modification was well covered in the following references [15–20]. The active power surplus was stored in the inertia of the turbine–generator system to keep the DC-link voltage stability in [15–17]. Moreover, in [16], a direct model predictive control is proposed for enhancing the dynamic response of the wind energy conversion systems. In addition, the control strategies of stator-side and grid-side were altered to provide reactive power for the grid under faults in [17]. Then, a new control structure is presented in [18,19], the machine-side converter (MSC) was utilized to regulate the DC-link voltage, while the grid-side converter (GSC) was used for fulfilling the maximum power point tracking (MPPT) of the wind turbine. Accordingly, when the voltage sags in the grid-side, the active power generated by the PMSG can be reduced, hence the surplus power of DC-side is decreased. Then, the voltage of DC-side is easily able to be stable. However, the performance of the control scheme in [18,19] is inaccurate in the normal condition. In [20], the pitch control was used to reduce the available wind power, and the excess energy is stored by the rotor and the DC-link capacitor. Moreover, the reactive power support was provided by the GSC. Whereas, the pitch control is a slow mechanical process and it cannot respond to the disturbances of system immediately. What is more, frequent change of the pitch angle results in the abrasion of equipment and the decrease of PMSG lifetime.

In addition, the studies on auxiliary device modification are also attention-attracting for its characteristics of a fast response and wide adjustable range under different levels of voltage sags. In [21,22], the excess energy of DC side was dissipated by the chopper circuit to avoid the overvoltage of the DC side and eliminate the mismatch between the input power of MSC and the output power of GSC. The method is utilized widely for its simple control strategy and low costs. However, it is noteworthy that the efficiency of PMSG is declining owing to the waste of energy, and the overheating problem occurs under severe faults. In [23], as a multiple-functional flexible alternating current transmission system (FACTS) device, the electronic power transformer (EPT) was combined with energy storage system to enhance the LVRT capability of PMSG. In [24], the supercapacitor energy storage (SCES) devices were installed on the DC-side, and they can absorb the surplus active power of DC-side to prevent the DC link capacitor from overvoltage. The effectiveness of the SCES is verified comparing with conventional current-limiting strategy. However, there was no specific capacity configuration scheme of the SCES in [24]. The superconducting magnetic energy storage (SMES) was presented in [25] to improve the LVRT capability and transient stability of PMSG, the superconducting fault current limiter (SFCL) was utilized to increase the output power of GSC, while the excess power was absorbed by the SMES so as to reduce its energy storage capacity.

In brief, the inherent control modification reduces the input power at the cost of the increasing mechanical tensions and faster aging, while the auxiliary device modification dissipates or absorbs the surplus power to prevent the DC capacitor from overvoltage at the expense of economic performance. In order to take full advantages of the two types of methods, based on the SCES control in [24], this paper presents a novel coordinated control scheme of rotor overspeed control and supercapacitor energy storage (SCES) control for PMSG-based WTG to improve the LVRT performance with comprehensive consideration of many factors. In this paper, the rotor speed is increased within allowable limits to reduce the input power of MSC under slight faults, and the rotor speed is set to the upper-limit to reduce the input power of MSC to the maximum extent under severe faults. The SCES is inoperative under slight faults to prevent the SCES from switching frequently, but the SCES can absorb the excess energy to prevent the DC link capacitor from overvoltage under severe faults. The specific capacity of the SCES is calculated, compared and verified. In addition, the GSC is utilized to maintain the DC voltage stability under the normal condition and provide reactive power support under faults.

The rest of this paper is structured as follows: The grid codes and technical principle of LVRT are introduced in detail in Section 2. The rotor overspeed control scheme of PMSG-based WTG is described in Section 3. The SCES control scheme is given in Section 4. The two kinds of coordinated control scheme for PMSG-based WTG are presented in

Section 5. The simulation results and analyses of the proposed coordinated control scheme are shown in Section 6. Finally, the conclusions are drawn in Section 7.

2. The Grid Codes and Technical Principle of LVRT

2.1. The Grid Codes of China

According to the grid codes of China [26], the PMSG-based WTG should have sufficient LVRT capabilities. Here, the LVRT requirements of China are shown in Figure 1. It is stated that the PMSG-based WTG should keep connected to the power system for the grid voltage above the curve and the trip of PMSG-based WTG occurs otherwise. In the worst case, the voltage sagged to 0.2 p.u. and lasted for 0.625 s at most, and the voltage should recover to 0.9 p.u. within 2 s.

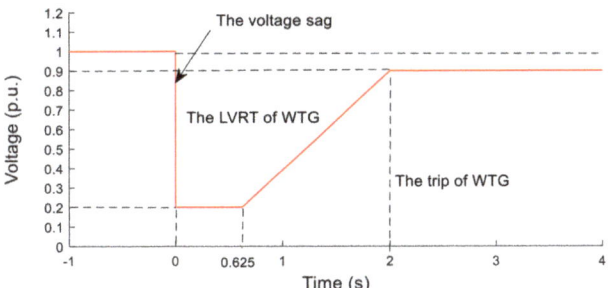

Figure 1. The grid codes of China.

2.2. The Technical Principle of LVRT

In the steady state, Equation (1) holds if the losses of the PMSG-based WTG and full-scale converters are ignored.

$$P_m = P_s = P_g \tag{1}$$

where P_m is the mechanical power captured by the wind turbine, P_s is the input power of MSC and P_g is the output power of GSC.

The output power P_g is decreasing on account of voltage sags and current limiting measures of GSC during faults. However, the PMSG is incapable of responding the grid faults due to the complete decoupling from grid for adopting the full-scale converter. Consequently, a mismatch between the input power of MSC P_s and the output power of GSC P_g is produced, and then it leads to the unbalance energy on the DC-side.

According to the above principle, this paper mainly takes measures to fulfill the LVRT requirements of PMSG-based WTG from the following three aspects in Figure 2:

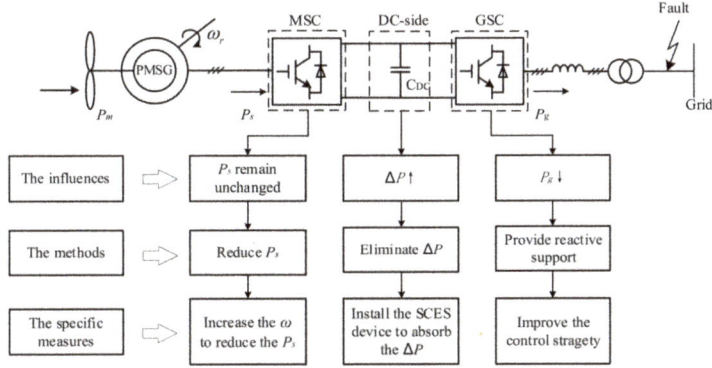

Figure 2. The technical principle of low voltage ride through (LVRT).

(1) The MSC side: It can reduce the input power P_s by increasing the rotor angular speed ω. However, this method threatens the stability of system when the rotor angular speed out-of-limit. Therefore, it generally needs to be combined with other methods.
(2) The GSC side: The control strategy of GSC should switch to reactive power compensation mode during faults, so as to provide reactive power support to hold the grid voltage.
(3) The DC side: In order to dissipate the unbalance power ΔP, the SCES is installed on the DC link and absorbs the excess energy to prevent the DC link capacitor from overvoltage.

3. The Overspeed Control Scheme

The mechanical power captured by the wind turbine can be described as Equation (2):

$$P_m = \frac{1}{2}\pi\rho R^2 V^3 C_p \qquad (2)$$

where ρ is the air density, R is the radius of the wind turbine blade, V is the wind speed, and C_p is the wind power coefficient [27], which is the function of the tip speed radio λ and the pitch angle β. Generally, in order to maximize the use of wind energy, the C_p is set to maximum value C_{pmax}, and tip speed radio λ is set to the optimal value λ_{opt}:

$$C_p(\lambda, \beta) = 0.5176(\frac{116}{\lambda_i} - 0.4\beta - 5)\exp(\frac{-21}{\lambda_i}) + 0.0068\lambda \qquad (3)$$

$$\frac{1}{\lambda_i} = \frac{1}{\lambda + 0.08\beta} - \frac{0.035}{1 + \beta^3} \qquad (4)$$

where λ is given as Equation (5).

$$\lambda = \frac{\omega R}{V} \qquad (5)$$

It is assumed that the wind speed V is constant during the short time interval for the electromagnetic transient analysis. In addition, we did not take the pitch control into account due to its slow mechanical response. Therefore, the wind speed V was set to V_N (i.e., $V = V_N$), and the pitch angle β was set to 0 (i.e., $\beta = 0$). According to Equations (2)–(5), the input power curve is shown as Figure 3:

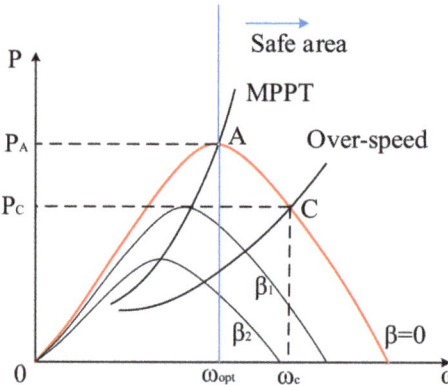

Figure 3. The input power curve of the permanent magnet synchronous generator (PMSG).

The unbalance power of the DC-side produced by the mismatch between the input power and the output power during faults can be expressed as:

$$\Delta P = P_s - P_g \qquad (6)$$

Thus, to eliminate the mismatch between the input power of MSC P_s and the output power of GSC P_g, the input power Ps needs to be reduced by ΔP. Then the deloading rate is defined as $d = \Delta P/P_A$, where P_A is the maximum power captured by the wind turbine in the MPPT mode. Due to the static instability problem caused by under-speed control [28], then overspeed control is generally adopted to keep the rotor working in the safe area (i.e., the right area of the optimal speed ω_{opt}) [29]. The power of the overspeed point P_C can be represented as:

$$P_C = P_A - \Delta P = (1-d)P_A \qquad (7)$$

The wind power coefficient of the overspeed point can be expressed as:

$$C_{pc} = (1-d)C_{pmax} \qquad (8)$$

Under the condition of $\beta = 0$, C_p is given according to Equation (8), and then λ can be obtained by Equations (3) and (4). Furthermore, the rotor speed reference ω_{ref} can be calculated easily according to Equation (5). However, consider the strong nonlinearity of the C_p-λ curve, it is hard to find the concrete expression of inverse function of $C_p = f(\lambda)$. To address this issue, in this paper, the least square method is was to fit the inverse function of $C_p = f(\lambda)$.

$$\lambda = f(C_p) = a_0 + a_1 C_p + a_2 C_p^2 + \cdots + a_n C_p^n \qquad (9)$$

When the order of polynomial $n = 3$, the equation is shown as:

$$\lambda = 13.5594 - 11.5724 C_p + 29.3383 C_p^2 - 56.6625 C_p^3 \qquad (10)$$

Comparing with the actual curve, the fitting curve with the allowable errors is shown in Figure 4.

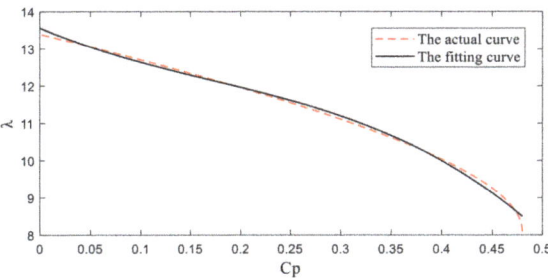

Figure 4. Actual curve and fitting curve of λ-Cp.

The overspeed control strategy of MSC is shown in Figure 5. The outer loop is the speed control loop while the inner loop is the current control loop. The d axis current reference i_{dref} is set to 0 both in the steady state and faulted state, while the q axis outer speed reference ω_{ref} switches between the normal mode and faulted mode according to the state of the system. When the system is steady, the switch is work at Mode 1, the speed reference ω_{ref} is set to ω_{opt} (i.e., $\omega_{ref} = \omega_{opt}$), the PMSG is working at point A, the input power of MSC is equivalent to the maximum mechanical power captured by the wind turbine P_A. When the voltage sags, the switch is working at Mode 2, the speed reference ω_{ref} is set to ω_c. The reference control variables of the MSC are altered so as to reduce the input power, and eliminate the mismatch between the input power and the output power.

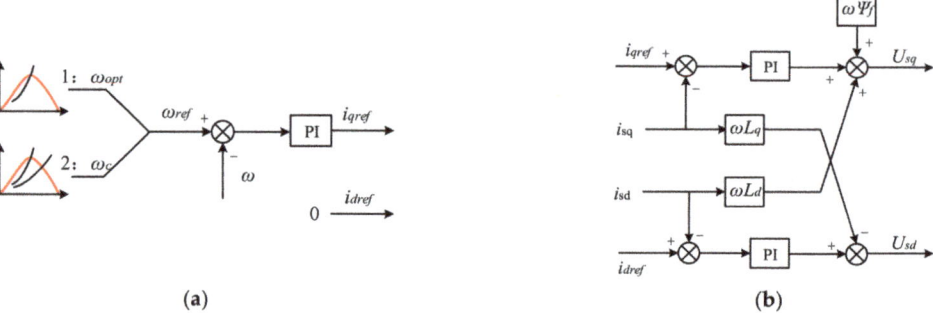

Figure 5. The control strategy of the machine-side converter (MSC): (**a**) the outer loop control and (**b**) the inner loop control.

According to Figure 1, under the symmetrical fault (i.e., the worst case of voltage sags), the unbalanced energy generated during faults can be expressed as:

$$E = \int_0^2 \Delta P dt \tag{11}$$

where the integral time t means the time that PMSG keeps connected to the grid under faults, and ΔP is the deviation between the output power during faults and the rated power.

The unbalanced power is generated due to the voltage sags and current limiting measures of GSC. Thus, the output current is keeping at the upper limit under voltage sags, that is to say, the current is constant during faults. Consequently, the output power P_g is proportional to the grid voltage u_g.

By substituting the data of Figure 1 into Equation (11) [30], the unbalanced energy generated during faults can be calculated as:

$$E = P_N[(1-0.2) \times 0.625 + 0.5 \times (0.8+0.1) \times (2-0.625)] \tag{12}$$

where P_N is the rated power of PMSG. The total unbalance energy of the DC-side is 28 kJ in the worst case of voltage sags. Hence, to eliminating the unbalanced energy 28 kJ within 2 s, the average changed power by regulating the rotor is 14 W, and the corresponding speed is $\omega = 1.4\omega_N$ according to Figure 3. However, the rotor speed of the wind turbine, generally, should not exceed 1.2 ω_N [31]. Thus, the deloading rate d should keep between 0 and 12% without the pitch control.

The rotor overspeed control scheme can enhance the inherent LVRT performance of PMSG by regulating the rotor speed without auxiliary devices. Apparently, the advantage of the overspeed control is low cost. However, the rotor overspeed control scheme is only applicable to the condition of slight faults due to the limit of maximum rotor speed.

4. The Supercapacitor Energy Storage (SCES) Control Scheme

To cope with the issue existing in the rotor overspeed control scheme, the SCES control scheme is proposed to improve the LVRT capability of PMSG under all conditions of voltage sags [24].

The unbalanced power on the DC-side under faults leads to overvoltage as the following:

$$\Delta P = P_s - P_g = CU_{dc}\frac{dU_{dc}}{dt} \tag{13}$$

To maintain the stability of DC voltage during faults, the energy storage systems, which consist of the SCES and the bidirectional DC–DC converter, are installed on the DC-side. In comparison with other energy storage, the SCES is more attractive and suitable for LVRT occasions owing to its higher power density, more cycle times, and shorter charge–discharge time. Consequently, The SCES is selected to store or release the energy in this paper.

In addition, the bidirectional DC–DC converter is mainly used to fulfill the charge–discharge control of the SCES and improve the stability of DC voltage. If the DC voltage rises, the bidirectional DC–DC converter works at the buck mode. Otherwise, the bidirectional DC–DC converter works at the boost mode.

Thus, the control strategy of the bidirectional DC–DC converter is displayed in Figure 6. The voltage control outer loop achieves the voltage stability by tracking DC voltage, while the current control inner loop improves the response speed. Furthermore, the working mode of the bidirectional DC–DC converter is shown in the Figure 7, when the S1 is triggered, the converter works in the buck mode and the SCES absorbs the energy from the DC-side; when the S2 is triggered, the converter works in the boost mode and the SCES transfers the stored energy to the DC link. It should be noted that S1 and S2 cannot be triggered concurrently.

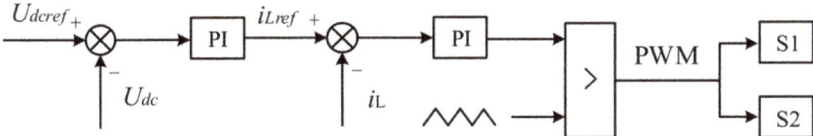

Figure 6. The control strategy of the bidirectional DC–DC converter.

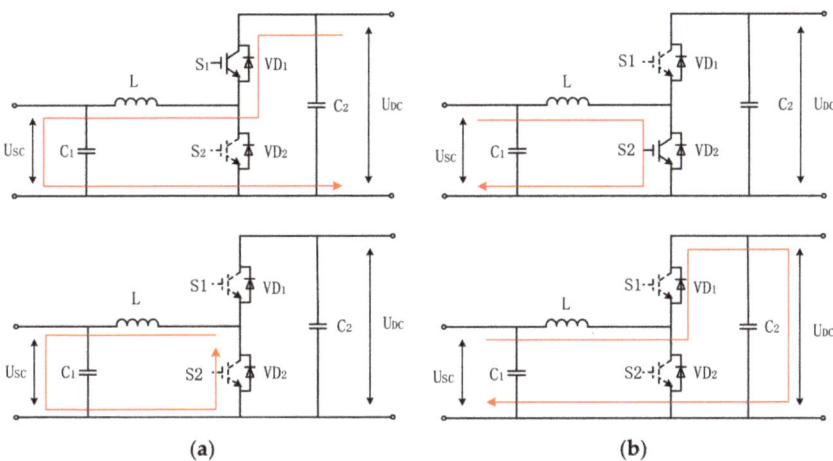

Figure 7. The bidirectional DC–DC working mode: (**a**) buck mode and (**b**) boost mode

In practice, consider the impacts of the series equivalent resistance of the SCES, the losses caused by the large number of supercapacitors integrated in series or in parallel, and the possibility of grid faults occurring multiple times during a period of time. Thus, the actual capacity of the SCES should be multiplied by a larger reliability coefficient based on the theoretical value. In addition, the SCES usually works at a middle voltage U_0 so as to charge or discharge. In this paper, we set the reliability coefficient of unidirectional energy transmission to 1.5, and then the reliability coefficient of bidirectional energy transmission was set to 3 [30]. According to Equation (12), the theoretical value of unbalanced energy is 28 kJ. Therefore, the triple energy of SCES (i.e., 84 kJ) should be configured in this paper.

This paper adopted the single supercapacitor and the specifications were: Cs = 300 F, Us = 2.5 V and Rs = 200 mΩ. Then 120 supercapacitors were connected in series to form the SCES. Here, define discharge depth $h = U_{min}/U_{max}$, and the energy absorbed or released by the SCES can be described as:

$$W_{sc} = \frac{1}{2}C_{sc}(U_{max}^2 - U_{min}^2) = W_{max}(1 - h^2) \quad (14)$$

The state of charge (SOC) of the SCES can be expressed as:

$$SOC = \frac{W_{max} - W_{sc}}{W_{max}} = \frac{W_{max} - W_{max}(1 - h^2)}{W_{max}} = h^2 \quad (15)$$

Generally, the discharge depth was set to 50% (i.e., h = 50%), then the SOC of SCES was calculated between 25 and 100%, and the chargeable and dischargeable energy of the SCES was 75%. Apparently, the utilization rate of the SCES was relatively high.

From Equation (14), the maximum energy absorbed by the SCES can be calculated as follows:

$$W_{sc} = \frac{1}{2}C_{sc}U_{max}^2(1 - h^2) = 84kJ \quad (16)$$

Thus, the value meets the requirement of reliability coefficient. In addition, the initial middle voltage U_0 of the SCES can be obtained from Equation (17) [32]:

$$\frac{1}{2}C_{sc}U_{max}^2 - \frac{1}{2}C_{sc}U_0^2 = \frac{1}{2}C_{sc}U_0^2 - \frac{1}{2}C_{sc}U_{min}^2 \quad (17)$$

Therefore, the SCES control scheme is applicative under the all conditions of voltage sags as long as the capacity configuration is appropriate. However, the separate utilization of the SCES control scheme is not conducive to the economic performance of the system owing to its high cost.

5. The Coordinated Control of Rotor Overspeed and SCES

According to the aforementioned control scheme in Section 3, the rotor overspeed control scheme under severe faults is prone to result in the rotor speed of the PMSG out-of-limit and affects the safety and stability of the power system. Additionally, in Section 4, the cost for separate utilization of the SCES control is high.

Thus, in order to take the advantages of the rotor overspeed control and the SCES control, and overcome drawbacks of either of the two control methods, now we are in a position to propose the coordinated control strategies, in which two innovative schemes are discussed as follows.

5.1. Coordinated Control Scheme I: Overspeed-Before-Storing

A reliable and effective coordinated control scheme named overspeed-before-storing is proposed (i.e., Scheme I). The rotor overspeed control is adopted under the slight faults; while the SCES control is adopted under the severe faults.

Introduce the voltage sag depth $k = \Delta U/U_N$. Noteworthily, when k is equal to 30%, the rotor speed is up to the upper bound limit (i.e., 1.2 p.u.). Therefore, we set the critical value as 30% to avoid the rotor speed out-of-limit.

The control strategy of Scheme I is as follows: When the voltage sag depth $k \leq 30\%$, the rotor overspeed control is adopted. Contrarily, the SCES control is adopted. The specific control flow chart is shown in Figure 8.

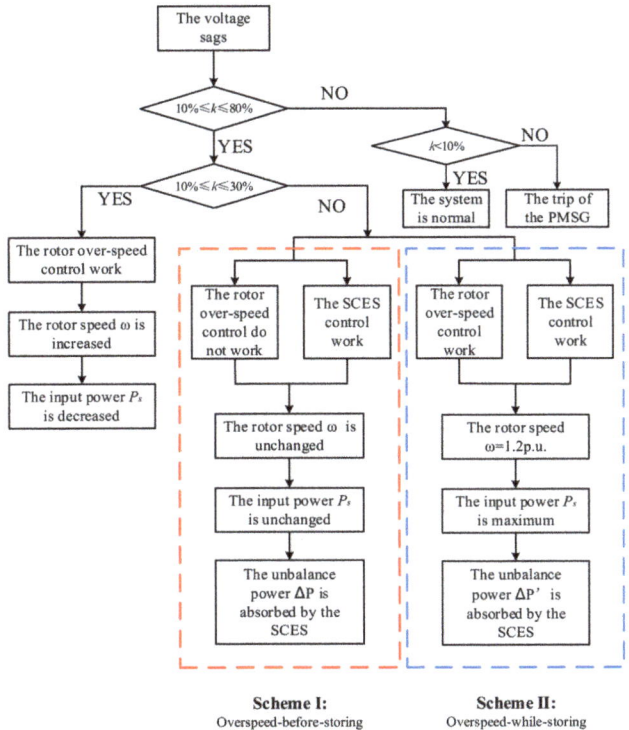

Figure 8. The flow chart of the two coordination control schemes.

Theoretically speaking, Scheme I can fulfill the LVRT requirements of PMSG under the all voltage sags conditions. Furthermore, Scheme I can avoid frequent switching of SCES under slight faults. Whereas, in Scheme I, the capacity configuration of SCES is the same as the SCES control scheme in Section 4, and the economic performance of the system is not improved.

5.2. Coordinated Control Scheme II: Overspeed-While-Storing

Based on further optimization and improvement of Scheme I, the coordinated control scheme named overspeed-while-storing (i.e., Scheme II) is presented. When the voltage sag depth $k \leq 30\%$, the rotor overspeed control scheme was adopted, and it was the same as in the Scheme I. However, if $k > 30\%$, unlike Scheme I, the rotor overspeed control and the SCES control worked together. The rotor angular speed increased and maintained at maximum (i.e., $\omega = 1.2$ p.u.), and the surplus energy was absorbed by the SCES to stabilize the DC-side voltage. The detailed control flow chart is represented in Figure 8.

In Scheme II, the unbalance power of DC-side under the severe faults is given as:

$$\Delta P' = \Delta P - \Delta P_s \tag{18}$$

where ΔP is the unbalance power between MSC and GSC without rotor overspeed control, and ΔP_s is the variable quantity of input power by setting the rotor speed to maximum.

Substitute (18) into Equations (11) and (12), it can be calculated that the total unbalance power of DC-side under voltage sags is 23 kJ. To absorb the excess energy, the 100 supercapacitors are installed in series to form the SCES.

The maximum energy absorbed by the SCES is calculated as:

$$W'_{sc} = \frac{1}{2}C'_{sc}U'^2_{max}(1-d^2) = 70 \text{ kJ} \qquad (19)$$

In practice, consider the limit of voltage transformation range of the DC–DC converter, the bound limits of voltage magnitudes on the SCES can be designed between 195 and 584 V [33], that is, the corresponding voltage transformation ratio is between 0.25 and 0.75. Consequently, the maximum working voltage of the SCES meets the requirements under the three control schemes (i.e., the SCES control scheme, the coordinated control Scheme I and Scheme II).

Comparing with the other three control schemes mentioned above, the coordinated control Scheme II could enhance the LVRT capability under the all conditions of voltage sags. Meanwhile, the rotor speed could stay in the safe limit. In addition, due to the impact of the regulation of rotor speed, the capacity configuration of SCES in coordinated control Scheme II is smaller than that in the SCES control scheme. In conclusion, the coordinated control Scheme II improves the stability and economy of the system simultaneously.

For all of the four control schemes mentioned above, the control strategy of the GSC is identical and it is presented as follows:

In the normal condition, the GSC works at the unity power factor mode (i.e., i_{qref} = 0), in other words, there is no reactive power injected to the grid.

When the voltage sags, the reactive current reference can be written as:

$$i_{gq} \geq 1.5 \times (0.9 - u_g)I_N \qquad (20)$$

$$i_{gdref2} = \sqrt{i^2_{max} - i^2_{gqref}} \qquad (21)$$

where u_g is the grid voltage and I_N is the rated current. According to Equation (20), instead of the unity power factor control, the GSC is utilized to provide reactive power support, in other words, the GSC is set to the Q-priority mode [34]. Accordingly, the active current should be limited by Equation (21) during faults because of the current limit of GSC, and the active current reference adopts the smaller one between i_{dref1} and i_{dref2}. The reference control variables of the GSC are altered so as to provide reactive support to the grid under faults. The control strategy of GSC is shown in Figure 9.

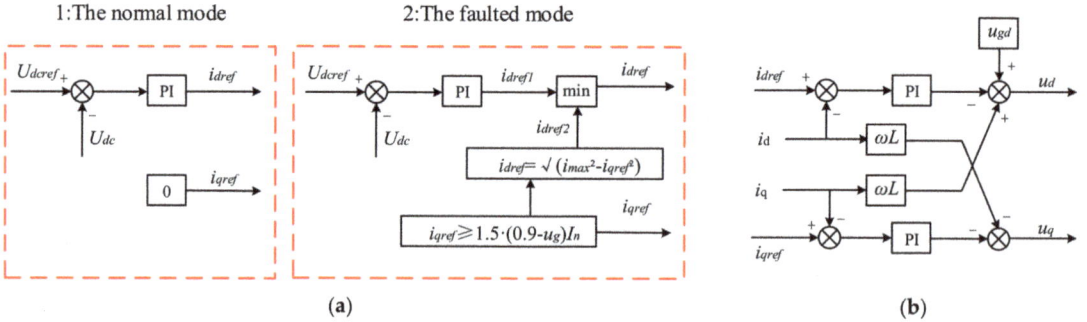

Figure 9. The control strategy of the grid-side converter (GSC): (**a**) the outer loop control and (**b**) the inner loop control.

6. Simulation Validation

The simplified modeling of the grid-connected wind system in MATLAB/Simulink is depicted in Figure 10. The wind system consists of a PMSG-base WTG, the MSC, the GSC, the SCES and the bidirectional DC–DC converter. The parameters of the modeling are listed in Table 1.

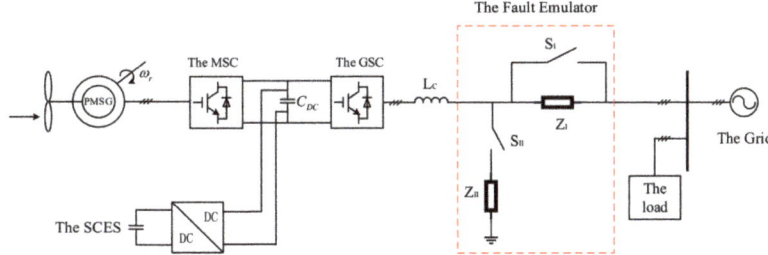

Figure 10. The simplified modeling of the grid-connected wind system.

Table 1. Simulation parameters of the wind system.

	Name	Value
The parameters of PMSG	Rated power	25 kW
	Stator resistor	0.05 Ω
	Stator inductance	2×10^{-3} H
	Permanent magnet flux	2 Wb
	Pole pairs of PMSG	8
	Rotational inertia	12 kg·m^2
The parameters of system	DC-link rated voltage	778 V
	System frequency	50 Hz
The parameters of SCES under the SCES control and overspeed-before-storing control	Equivalent capacitance of SCES	2.5 F
	SCES rated voltage	300 V
	Equivalent resistor of the SCES	2.4 Ω
The parameters of SCES underoverspeed-while-storing control	Equivalent capacitance of SCES	3 F
	SCES rated voltage	250 V
	Equivalent resistor of the SCES	2 Ω

The parameters of the below simulation analysis diagram are as follows: u_a is the grid voltage, ω is the rotor angular speed, P_s is the input power of MSC, U_{dc} is the DC-side voltage, P_g is the output power of GSC, Q is the reactive power generated by the GSC, U_{sc} is the voltage of the SCES and SOC is the state of charge of the SCES.

6.1. The Simulation Results of the Four Control Scheme under Symmetrical Faults

6.1.1. The Simulation Results of the Rotor Overspeed Control Scheme

The response characteristics of PMSG-connected system with the rotor overspeed control scheme under faults are shown in Figure 11. According to the voltage sag depth, the rotor speed ω is regulated to reduce the input power of MSC, so that it can maintain the power balance of the system and keep the DC voltage fluctuated in the range within permission. The GSC can provide the reactive power to the grid under the voltage sags. When the voltage sag depth $k > 30\%$, the rotor speed has exceeded the upper limit of wind turbine, which is not allowable in practice.

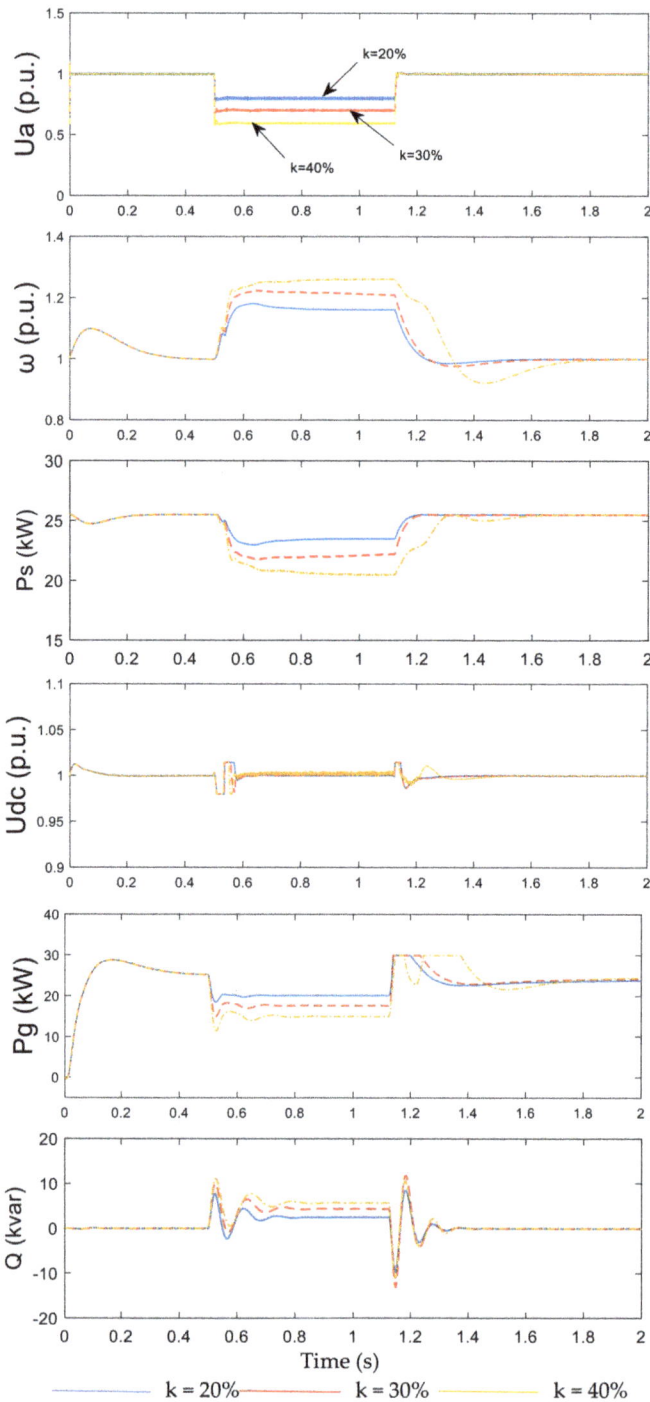

Figure 11. The simulation results of the rotor overspeed control scheme.

6.1.2. The Simulation Results of the SCES Control Scheme

The response characteristics of PMSG-connected system with the SCES control scheme under faults are presented in Figure 12. The input power of the MSC remained unaltered under any voltage sags because there no measures were adopted in the machine side. The unbalance power of the DC-side leads to the fluctuation of DC voltage. The SCES was utilized to absorb the unbalanced power on the DC side and stabilized the DC voltage. The GSC was used to provide the reactive power to enhance the LVRT capacity. When the worst fault occurred (i.e., the voltage sag depth k was 80%), the voltage of the SCES U_{sc} increased from 230 to 279 V, SOC increased from 76% to 93%, the two values of the SCES were both in the allowable limit. The SCES control scheme could enhance the LVRT capacity of the PMSG under the all levels of faults.

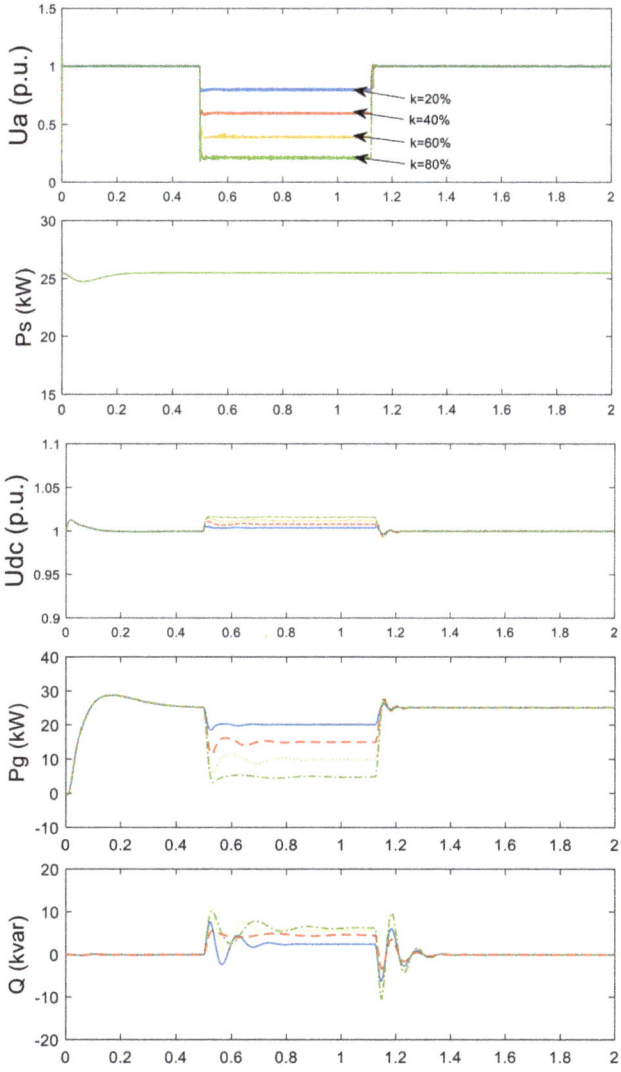

Figure 12. *Cont.*

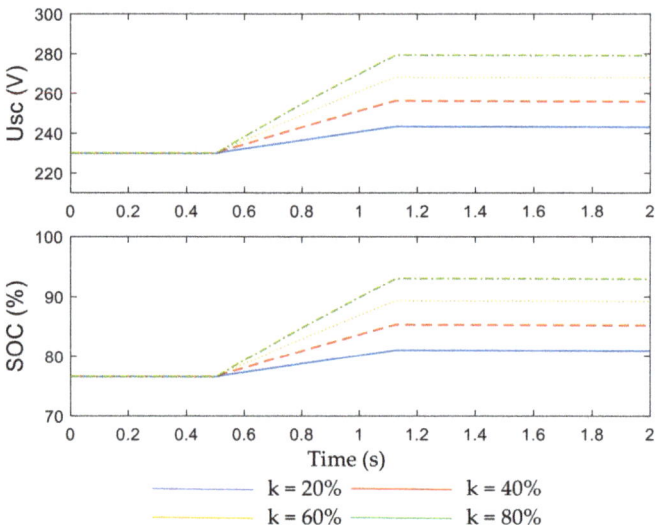

Figure 12. The simulation results of the supercapacitor energy storage (SCES) control scheme.

6.1.3. The Simulation Results of Coordinated Control Scheme I: Overspeed-Before-Storing

The rotor overspeed control and the SCES control are combined to enhance the LVRT capacity under all voltage sags and maintain that the rotor is not out-of-limit. When the voltage sag depth $k \leq 30\%$, the rotor overspeed control is adopted; otherwise, the SCES control is adopted. Since the two control schemes are still used separately, the simulation results were the same as those in the Sections 6.1.1 and 6.1.2.

6.1.4. The Simulation Results of Coordinated Control Scheme II: Overspeed-While-Storing

The response characteristics of PMSG-connected system with the coordinated control Scheme II under faults are described in Figure 13. When the voltage sag depth $k \leq 30\%$, the input power was declining by regulating the rotor speed, while, the SCES was inoperative. When the voltage sag depth $k > 30\%$, the rotor speed was set to the maximum 1.2 p.u. to reduce the input power. Meanwhile, the bidirectional DC–DC converter worked in the buck mode and the SCES absorbed the excess energy of the DC side. The GSC provided the reactive power to support the LVRT of PMSG. Comparing with the SCES control scheme, the capacity configuration of the SCES was decreased. The coordinated control Scheme II improved the stability and economy of the system simultaneously. As seen, even though the worst fault occurred (i.e., the voltage sag depth k was 80%), the two values of the SCES were in the allowable limit.

6.2. The Simulations Results of Overspeed-while-Storing Control under Asymmetrical Faults

Actually, only 12% of grid dips are the symmetrical fault. Therefore, it is necessary to verify the performance of the overspeed-while-storing control proposed in this paper under asymmetrical faults. In this section, the overspeed-while-storing control under the two asymmetrical faults, which are single line-to-ground fault and double line-to-ground fault, were simulated and analyzed in detail.

6.2.1. The Simulations Results under the Single Line-to-Ground Fault

In Figure 14, it can be observed that, when the voltage of A-phase sagged to 20% (i.e., the maximum degree of voltage sags), the rotor speed was increased but did not reach to the upper limit, that is the unbalanced power can be eliminated only by regulating the rotor speed, and then the SCES was inoperative. In addition, the reactive power generated by the GSC contained 2 ω oscillations due to the negative component of the grid, and the voltage of DC-side also consisted of 2 ω oscillations. However, the oscillation of DC voltage was within the allowable range. Apparently, the overspeed-while-storing control fulfilled the LVRT of PMSG and prevented the DC capacitor from overvoltage under single line-to-ground fault.

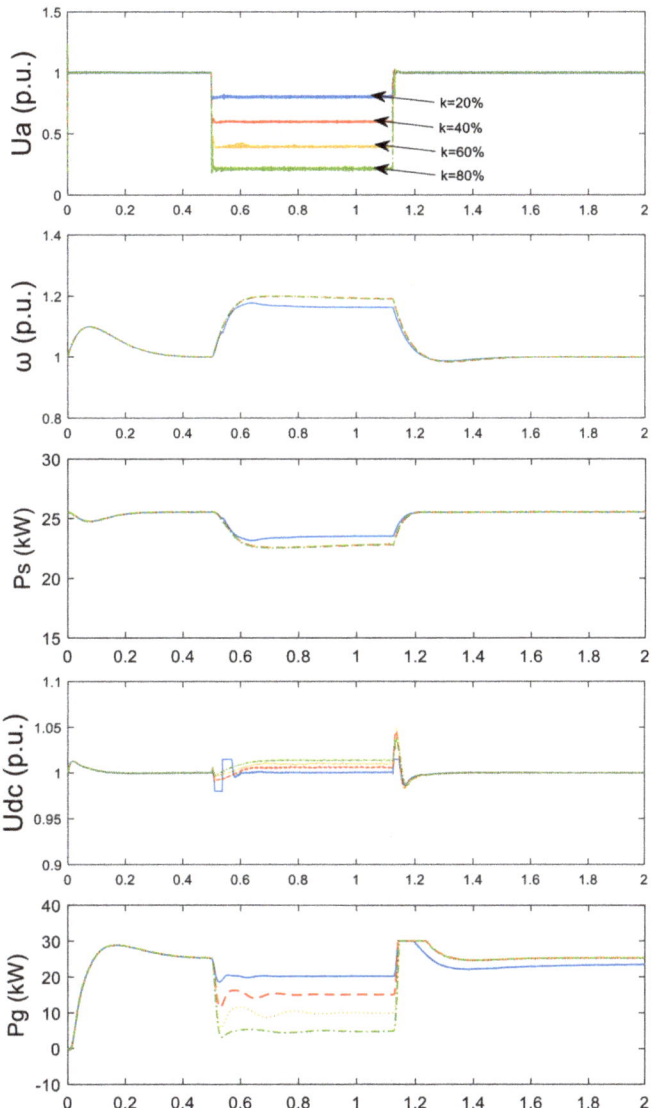

Figure 13. *Cont.*

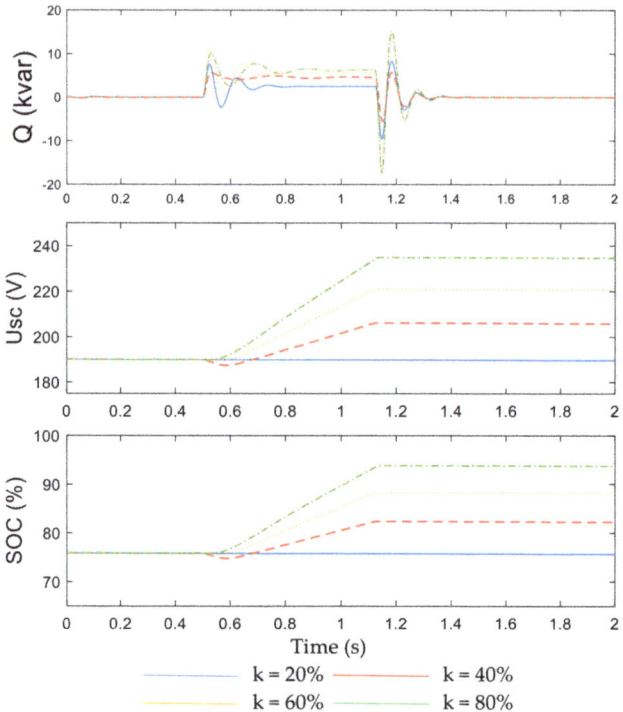

Figure 13. The simulation results of the overspeed-while-storing control.

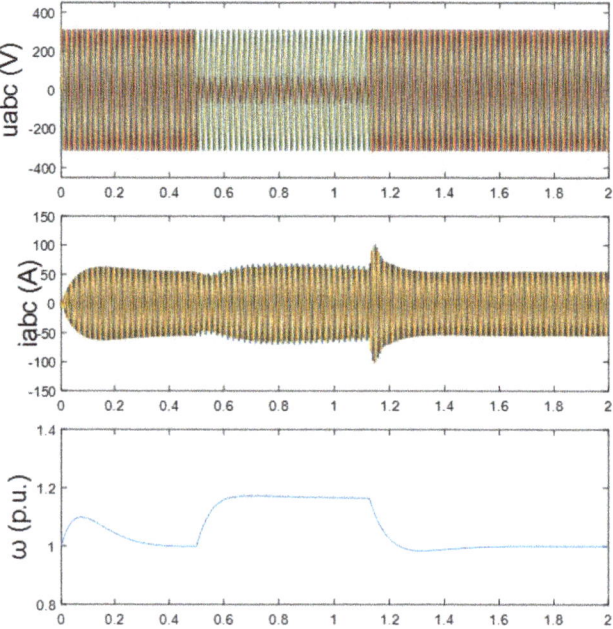

Figure 14. Cont.

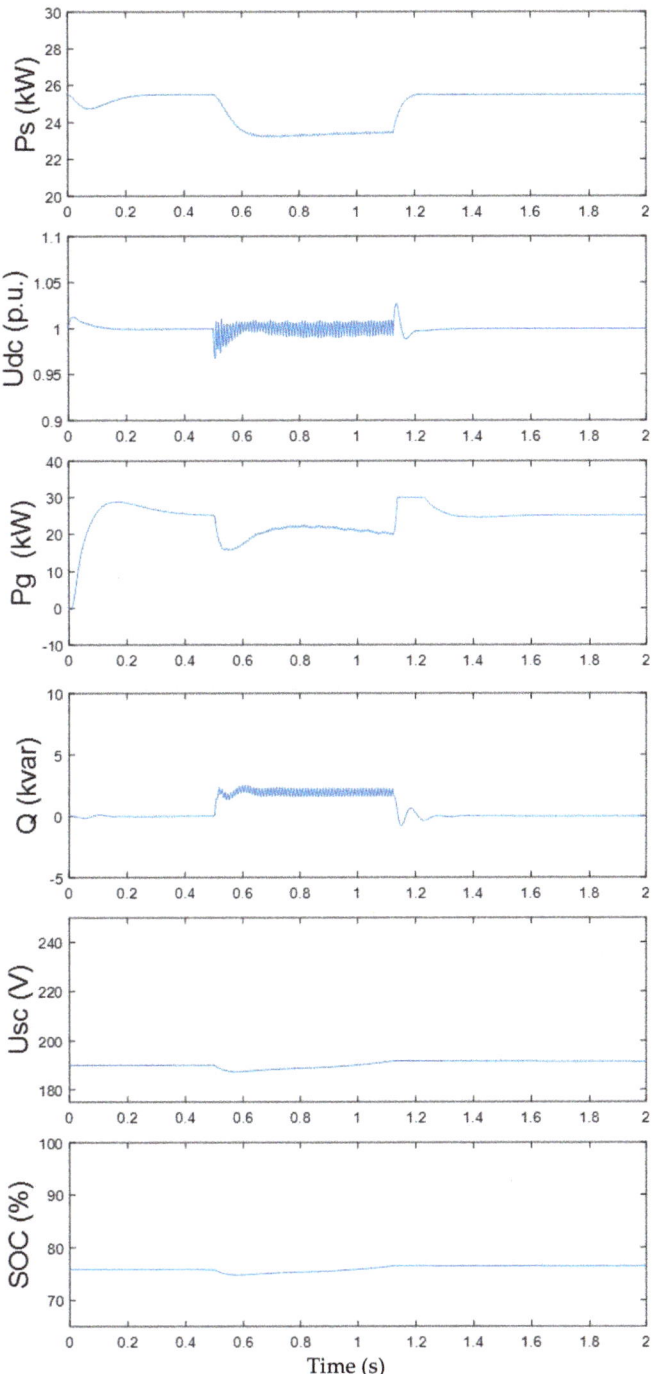

Figure 14. The simulation results of the overspeed-while-storing control under a single line-to-ground fault.

6.2.2. The Simulations Results of the Double Line-to-Ground Fault

Another simulation of the overspeed-while-storing under the double line-to-ground is shown in Figure 15, when the voltage of B-phase and C-phase sagged to 20%, the rotor was increased to the maximum value (i.e., 1.2 p.u.) and the surplus power of DC-side was absorbed by the SCES. The values of SCES were in the allowable range due to the capacity being configured under the most severe fault (i.e., symmetrical fault). It can be seen that the DC voltage and reactive power suffered 2ω oscillation during faults, as a result of the negative component of the unbalanced voltage. However, this 2ω oscillation was small enough not to affect the stability of the system. Thus, the overspeed-while-storing control fulfilled the LVRT of PMSG and prevented the DC capacitor from overvoltage under the double line-to-ground fault.

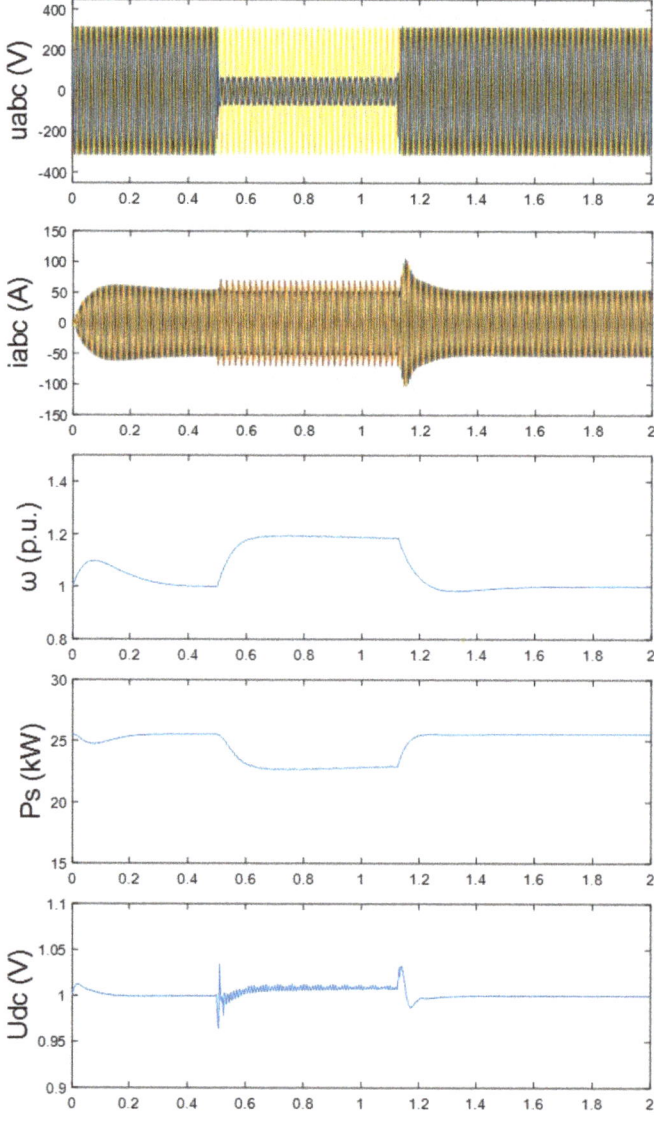

Figure 15. *Cont.*

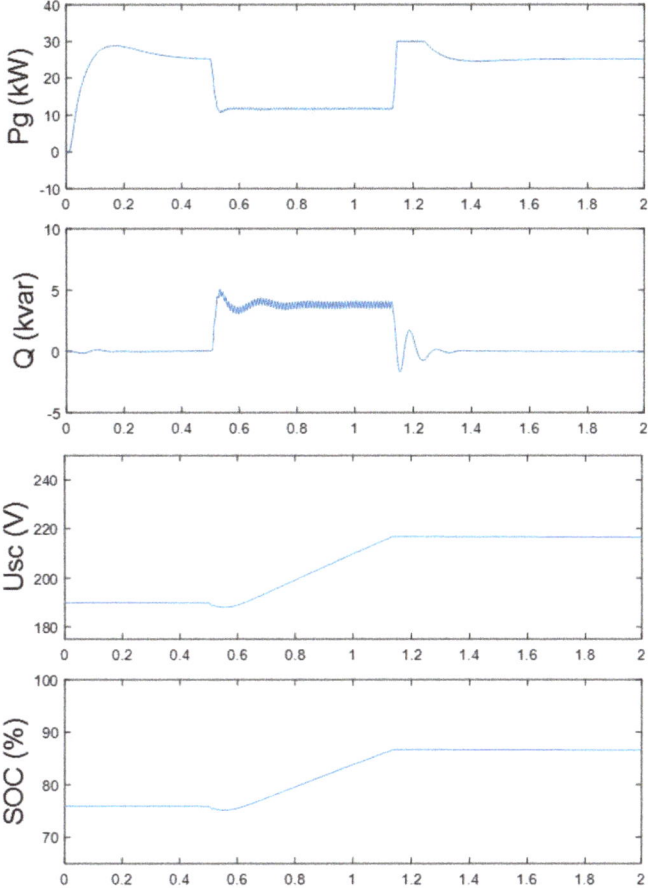

Figure 15. The simulation results of the overspeed-while-storing control under the double line-to-ground fault.

6.3. The Comparison between the Conventional Control and the Overspeed-while-Storing Control under Faults

In conventional control [24], the 1.5 times current-limiting strategy and unity power factor control were adopted in GSC. As shown in Figure 16, when the voltage sagged to 20%, the voltage of DC-side could be reached to triple of rated value and there was no reactive power generated by the GSC in conventional control. Whereas, the DC voltage could be stabilized in the allowable limit and the GSC could provide reactive power under faults by using overspeed-while-storing control. According to the grid codes of the power system, the overspeed-while-storing control could fulfill the LVRT requirements effectively.

7. Conclusions

A novel LVRT control strategy for the PMSG-based wind turbine generator (WTG) based on the coordinated control named overspeed-while-storing is presented in this paper. Particularly, when the voltage sags were slight (i.e., the voltage sag depth $k \leq 30\%$), the mismatch issue between the input power and the output power could be solved by only regulating the rotor speed. Otherwise, the rotor speed was set to the maximum to reduce the input power to the maximum extent. Meanwhile, the supercapacitor energy storage (SCES) was utilized to absorb the excess energy of the DC side to maintain the DC voltage stability. The coordinated control method proposed in this paper, on the one hand, could fulfill the LVRT requirements of PMSG under all conditions of voltage sags; on the other hand, it requires the smaller capacity configuration than that in the SCES control scheme. Consequently, the proposed coordinated control method could improve the stability and economy of the system simultaneously. All these analytical results were validated in the single PMSG-connected system.

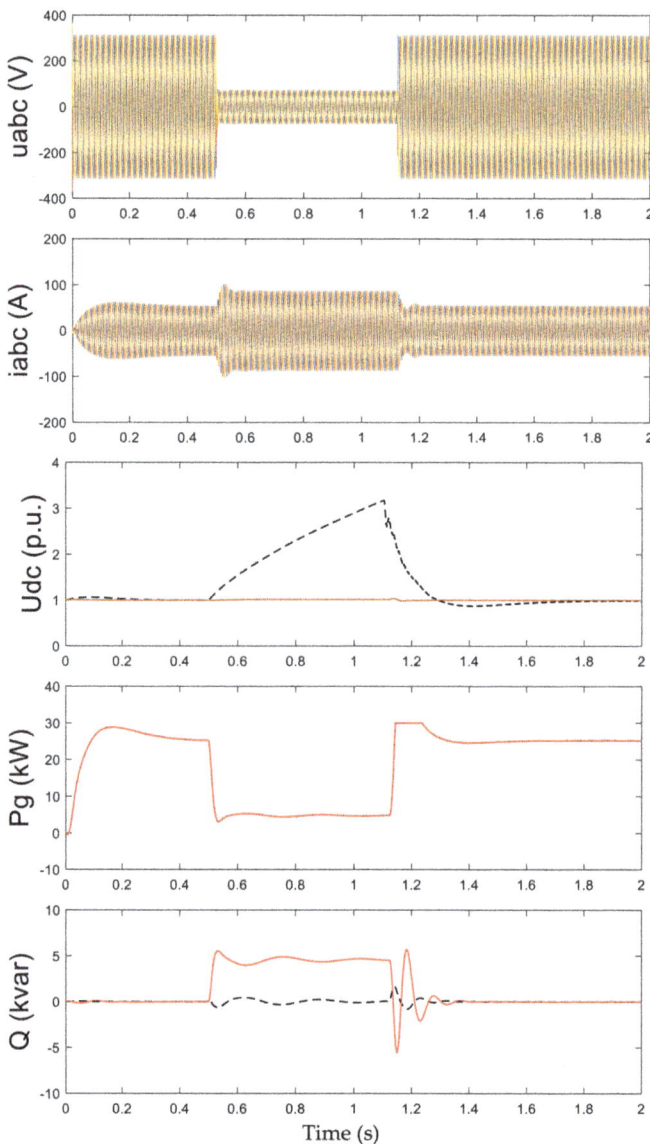

Figure 16. The simulation results comparison between the conventional control and the overspeed-while-storing control.

In this paper, only the LVRT under symmetrical faults and asymmetrical faults were discussed. Future works will focus on the condition of HVRT and the frequency regulation of the PMSG.

Author Contributions: Conceptualization, X.Y., L.Y. and T.L.; Methodology, X.Y.; Software, L.Y.; Validation, L.Y.; Formal analysis, X.Y.; Investigation, L.Y.; Writing—original draft preparation, L.Y.; Writing—review and editing, X.Y.; Visualization, L.Y. and T.L.; Supervision, X.Y., T.L.; Project administration, X.Y. and T.L.; Funding acquisition, X.Y. and T.L. All authors have read and agreed to the published version of the manuscript.

Funding: This research was funded by the Science and Technology Project of State Grid Corporation of China, grant number: SGHEDK00DYJS1900061.

Institutional Review Board Statement: Not applicable.

Informed Consent Statement: Not applicable.

Data Availability Statement: Data available in a publicly accessible repository.

Conflicts of Interest: The authors declare no conflict of interest.

References

1. Zhen, Y.Z.; Su, N.S.; Li, M.L. Research on Doubly-Fed Induction Generators Synergetic Control Strategy and Stability Technology for High/Low Voltage Ride Through. *Power Syst. Technol.* **2020**, 1–11. [CrossRef]
2. Zhang, R.W.; Qin, B.Y.; Li, H.Y. Low Voltage Ride-Through Control Strategy for DFIG-based Wind Turbine based on Disturbance Attenuation. *Autom. Electr. Power Syst.* **2020**, *44*, 112–120.
3. Liu, R.K.; Yao, J.; Wang, X.W.; Sun, P.; Pei, J.X.; Hu, J.B. Dynamic Stability Analysis and Improved LVRT Schemes of DFIG-based Wind Turbines During a Symmetrical Fault in a Weak Grid. *IEEE Trans. Power Electron.* **2020**, *35*, 303–318. [CrossRef]
4. Nadour, M.; Essadki, A.; Nasser, T. Improving Low-Voltage Ride-Through Capability of a Multimegawatt DFIG based Wind Turbine under Grid Faults. *Protec. Contr. Modern. Power Syst.* **2020**, *5*, 370–382. [CrossRef]
5. Xiao, X.Y.; Yang, R.H.; Zheng, Z.X.; Wang, Y. Cooperative Rotor-Side SMES and Transient Control for Improving the LVRT Capability of Grid-Connected DFIG-Based Wind Farm. *IEEE Trans. Appl. Supercond.* **2019**, *29*, 1–5. [CrossRef]
6. Nian, H.; Cheng, P.; He, Y.K. Review on Operation Techniques for DFIG-Based Wind Energy Conversion Systems under Network Faults. *Proc. CSEE* **2015**, *35*, 4184–4197.
7. Wang, D.; Liu, C.R.; Li, G.Y. Research on the Fault Ride-Through Optimal Control Strategy of PMSG-based Wind Turbine. *Power Syst. Protec. Contr.* **2015**, *43*, 83–89.
8. Ren, Y.F.; Hu, H.B.; Xue, Y. Low Voltage Ride-Through Capability Improvement of PMSG based on Chopper Circuit and Reactive Priority Control. *High Vol. Eng.* **2016**, *42*, 11–18.
9. Arani, M.F.M.; Mohamed, Y.A.I. Assessment and Enhancement of a Full-Scale PMSG-Based Wind Power Generator Performance under Faults. *IEEE Trans. Energy Convers.* **2016**, *31*, 728–739. [CrossRef]
10. Jahanpour-Dehkordi, M.; Vaez-Zadeh, S.; Mohammadi, J. Development of a Combined Control System to Improve the Performance of a PMSG-Based Wind Energy Conversion System under Normal and Grid Fault Conditions. *IEEE Trans. Energy Convers.* **2019**, *34*, 1287–1295. [CrossRef]
11. Nasiri, M.; Mohammadi, R. Peak Current Limitation for Grid Side Inverter by Limited Active Power in PMSG-Based Wind Turbines during Different Grid Faults. *IEEE Trans. Sustain. Energy* **2017**, *8*, 3–12. [CrossRef]
12. Yao, J.; Guo, L.; Zhou, T.; Xu, D.; Liu, R. Capacity Configuration and Coordinated Operation of a Hybrid Wind Farm With FSIG-Based and PMSG-Based Wind Farms during Grid Faults. *IEEE Trans. Energy Convers.* **2017**, *32*, 1188–1199. [CrossRef]
13. Ren, Y.F.; Peng, W.; Liu, H.T. Enhancement of Low Voltage Ride Through Ability of Permanent Magnet Synchronous Wind Power Generation Sets based on Hybrid Energy Storage System Composed of Vanadium Redox Battery and Super Capacitor. *Power Syst. Technol.* **2014**, *38*, 3016–3023.
14. Nasiri, M.; Milimonfared, J.; Fathi, S.H. A Review of Low-Voltage Ride-Through Enhancement Methods for Permanent Magnet Synchronous Generator based Wind Turbines. *Renew. Sustain. Energy Rev.* **2015**, *47*, 399–415. [CrossRef]
15. Yousef, A.; Nasiri, A.; Abdelbaqi, O. Wind Turbine Level Energy Storage for Low Voltage Ride Through (LVRT) Support. In Proceedings of the 2014 IEEE Symposium on Power Electronics and Machines for Wind and Water Applications, Milwaukee, WI, USA, 24–26 July 2014; pp. 1–6. [CrossRef]
16. Abdelrahem, M.; Kennel, R. Fault-Ride through Strategy for Permanent-Magnet Synchronous Generators in Variable-Speed Wind Turbines. *Energies* **2016**, *9*, 1066. [CrossRef]
17. Tripathi, S.M.; Tiwari, A.N.; Singh, D. Low-Voltage Ride-Through Enhancement with the ω and T Controls of PMSG in a Grid-integrated Wind Generation System. *IET Gener. Transm. Distrib.* **2019**, *13*, 1979–1988. [CrossRef]
18. Kim, K.; Jeung, Y.; Lee, D.; Kim, H. LVRT Scheme of PMSG Wind Power Systems Based on Feedback Linearization. *IEEE Trans. Power Electron.* **2012**, *27*, 2376–2384. [CrossRef]

19. Alepuz, S.; Calle, A.; Busquets-Monge, S.; Kouro, S.; Wu, B. Use of Stored Energy in PMSG Rotor Inertia for Low-Voltage Ride-Through in Back-to-Back NPC Converter-Based Wind Power Systems. *IEEE Trans. Ind. Electron.* **2013**, *60*, 1787–1796. [CrossRef]
20. Liu, Z.Y.; Liu, C.R.; Li, G.Y. Coordinated Power Control Method for Improving Low Voltage Ride Through Capability of Wind Turbines with Permanent Magnet Synchronous Generators. *Autom. Electr. Power Syst.* **2015**, *39*, 23–29.
21. Zhang, Q.; Li, F.T.; Jiang, Y.M. Comprehensive Control Strategy for Improving Low Voltage Ride Through Capability of Permanent Magnet Synchronous Generator. *Power Syst. Protec. Contr.* **2017**, *45*, 62–67.
22. Wang, Y.Q.; Yu, J.Z.; Feng, J.G. Research on Low/High Voltage Ride through of Permanent Magnet Synchronous Wind Turbine. *Power Syst. Protec. Contr.* **2018**, *46*, 34–42.
23. Huang, H.; Mao, C.; Lu, J.; Wang, D. Electronic Power Transformer Control Strategy in Wind Energy Conversion Systems for Low Voltage Ride-through Capability Enhancement of Directly Driven Wind Turbines with Permanent Magnet Synchronous Generators (D-PMSGs). *Energies* **2014**, *7*, 7330–7347. [CrossRef]
24. Liu, S.H.; Zhou, Y.S.; Xu, Z.H. Research on Low-Voltage Ride Through Capability of Permanent Magnetic Synchronous Offshore Wind Power based on Super-capacitor Energy Storage. *Power Syst. Protec. Contr.* **2018**, *46*, 9–15.
25. Huang, C.; Xiao, X.Y.; Zheng, Z.; Wang, Y. Cooperative Control of SFCL and SMES for Protecting PMSG-Based WTGs under Grid Faults. *IEEE Trans. Appl. Supercond.* **2019**, *29*, 1–6. [CrossRef]
26. Wang, P.; Wang, G.H.; Zhang, J.W. Design and Application of Supercapacitor Energy Storage System used in Low Voltage Ride through of Wind Power Generation System. *Proc. CSEE* **2014**, *34*, 1528–1537.
27. Huang, S.; Wang, H.; Liao, W. Control Strategy based on VSC-HVDC Series Topology Offshore Wind Farm for Low Voltage Ride Through. *Trans. China Electrotech. Soc.* **2015**, *30*, 362–369.
28. Bian, X.Y.; Jiang, Y.; Zhao, Y. Coordinated Frequency Regulation Strategy of Wind, Diesel and Load for Microgrid with High-penetration Renewable Energy. *Autom. Electr. Power Syst.* **2018**, *42*, 102–109.
29. Ding, L.; Yin, S.Y.; Wang, T.X. Integrated Frequency Control Strategy of DFIGs based on Virtual Inertia and Over-speed Control. *Power Syst. Technol.* **2015**, *39*, 2385–2391.
30. Zhang, M.R.; Li, N.; Wang, Z.X. LVRT Ability of PMSG Wind Power System. *Electr. Power Autom. Equip.* **2014**, *34*, 128–134.
31. Meng, Y.Q.; Wang, J.; Li, L. New Low Voltage Ride-through Coordinated Control Schemes of Permanent Magnet Synchronous Generator Considering Rotor Speed Limit and DC Discharging Circuit Optimization. *Proc. CSEE* **2015**, *35*, 6283–6292.
32. Zhang, M.R.; Li, Y.H.; Ouyang, L. DC Bus Voltage Stability Control of DC-DC Converter in the Permanent Mmagnet Wind Power Grid-connected System based on the Hybrid System. *Trans. China Electrotech. Soc.* **2015**, *30*, 62–69. [CrossRef]
33. Zhang, C. Research on Fault Ride through and Power Smooth Control of Direct-drive Wind Power Generator with Supercapacitor. *Inner Mongolia Univ. Tech.* **2015**, *20*, 22–25.
34. Yuan, L.; Meng, K.; Huang, J.; Dong, Z.Y.; Zhang, W.; Xie, X. Development of HVRT and LVRT Control Strategy for PMSG-Based Wind Turbine Generators. *Energies* **2020**, *13*, 5442. [CrossRef]

Article

Analysis and Mitigation of Sub-Synchronous Resonance for Doubly Fed Induction Generator under VSG Control

Yingzong Jiao [1], Feng Li [2], Hui Dai [2] and Heng Nian [1,*]

1. College of Electrical Engineering, Zhejiang University, Hangzhou 310027, China; jyzzju@zju.edu.cn
2. State Grid Huaian Power Supply Company, Huaian 223002, China; lifeng.ha@js.sgcc.com.cn (F.L.); daihui.ha@js.sgcc.com.cn (H.D.)
* Correspondence: nianheng@zju.edu.cn

Received: 2 March 2020; Accepted: 24 March 2020; Published: 1 April 2020

Abstract: This paper presents the analysis and mitigation of sub-synchronous resonance (SSR) for doubly fed induction generators (DFIG) under virtual synchronous generator (VSG) control, based on impedance methods. VSGs are considered to have grid-supporting ability and good stability in inductance-based weak grids, and are implemented in renewable power generations, including DFIG systems. However, stability analyses of VSGs for DFIG connecting with series capacitor compensation are absent. Therefore, this paper focuses on the analysis and mitigation of SSR for DFIG under VSG control. Impedance modeling of DFIG systems is used to analyze SSR stability. Based on impedance analysis, the influence of VSG control parameters and the configuration of damping factor of reactive power are discussed. Next, a parameter configuration method to mitigate SSR is proposed. Finally, time-domain simulation and fast fourier transform (FFT) results are given to validate the correctness and effectiveness of the impedance model and parameter configuration methods.

Keywords: virtual synchronous generator; doubly fed induction generator; sub-synchronous resonance; impedance modeling

1. Introduction

Recently, renewable power generation has developed rapidly. Wind power generation systems (WPGS) based on doubly fed induction generator (DFIG), which has the advantages of relatively low cost and variable speed constant frequency operation, have been widely installed worldwide [1,2]. With the penetration of WPGS in power grids increasing continuously, conventional power sources like synchronous generators (SG) are decreasing; frequency stability and lack of inertia are becoming main concerns in power grids [3]. It is necessary to enhance the ability of WPGS to support power systems like conventional SGs. Based on this consideration, the virtual synchronous generator (VSG) control is proposed [4,5].

VSG is a grid-friendly control strategy which emulates the operating principle of SGs. VSG is first introduced to stabilize the grid frequency by adding virtual rotational inertia to distributed generation systems in [4]. The concepts of "Self-Synchronized Synchronverters" is proposed in [5], in which phase-locked loop can be neglected. The method to stabilize the power system based on VSG control with alternating moment of inertia is introduced in [6]. The comparison of VSG control and droop control are given in [7,8]. According to [8], VSG control inherits the advantages of droop control and provides inertia support for the system. In [9], "virtual synchronized control" for DFIG is presented to increase the inertia support capability and frequency stability of the weak grid. Furthermore, VSG control shows the superior small signal stability better than conventional vector control when a DFIG is connected to a weak grid [10,11].

Wind farms are usually located far away from load center. And wind farms are connected to weak grids through long transmission lines with high impedance. Therefore, series capacitor compensation is usually used to increase power transmission capability among AC lines [12]. However, series capacitors in transmission lines, which connects weak grid with DFIG based wind farms, may cause sub-synchronous resonance (SSR). In recent years, several accidents in DFIG based wind farm caused by SSR have been reported worldwide [13,14].

Currently, the studies of modeling, analyzing and mitigating control strategy for DFIG-based wind farms interconnected with series compensation have been reported in [14–20]. Based on modal analysis, the impacts of series compensation level, wind speed and current loop gains on SSR are studied in [15,16]. According to [15], the higher compensation level, the lower wind speed and the larger control loop gains, the more possible SSR occurs. As for mitigation strategies of SSR, there are two main methods: (1) auxiliary damping hardware. FACTS devices, such as static var compensator (SVC), thyristor-controlled series capacitor (TCSC) and gated-controlled series capacitors (GCSC), can be used to mitigate SSR [17,18]; (2) damping control strategies. Sub-synchronous resonance damping control is implemented in grid-side converters (GSC) [19] and control performance of different control signals, including rotor speed, capacitor voltage and current magnitude, is analyzed. Intelligent algorithms can be also adopted in suppressing SSR, such as the improved particle swarm optimization algorithm [20].

The works mentioned above focus on the SSR of DFIG under conventional vector control. However, the stability analysis of VSG control in DFIG interconnected with series compensation is absent. Furthermore, VSG control is quite different with vector control, especially in power-synchronization and voltage control loop. Therefore, the stability of DFIG under VSG control in series compensation network should be analyzed in detail.

Impedance modeling is an effective stability analysis method [21,22], which has been used in studying SSR [23,24]. The impedance modeling of VSG control has been implemented in [25], in which VSG control shows the better stability than the vector control in ultraweak inductance-based grid. However, the reactive power and voltage control loops are neglected during VSG modeling, which are very important in VSG control and SSR analysis. Therefore, the impedance model of VSG control for DFIG including voltage control loop should be developed.

The major contributions of this paper can be concluded as: (1) building the detailed impedance model of DFIG under VSG control including DFIG model, swing equation, voltage-reactive power control, frame transformation and rotor voltage calculation; (2) analyzing the influence of VSG control for DFIG on SSR stability and configuring the VSG control parameter based on impedance analysis.

The rest of this paper is organized as follows. In Section 2, VSG control for DFIG is introduced briefly. In Section 3, the impedance modeling of DFIG under VSG control is illustrated. In Section 4, the impedance model is verified and the SSR analysis is also discussed, followed by the influence of VSG control parameters and configuration method are investigated. In Section 5, simulation studies are implemented to verify the correctness and effectiveness of the impedance model and configuration method. Finally, the conclusion is drawn in Section 6.

2. VSG Control for DFIG

In this section, the VSG control for DFIG is introduced. Different with voltage source converters (VSC), the stator of DFIG is connected to the grid directly and outputs the most part of power. Thus, in order to introduce the VSG control for DFIG, the equivalent circuit of DFIG is given in Figure 1. It should be noted that the control target of grid-side converters (GSC) is to keep the constant DC voltage on the DC bus. Therefore, GSC still works in the conventional current vector control [2]. Moreover, GSC has little influence on SSR stability, therefore the analysis of GSC will be neglected in the following [19,23,24].

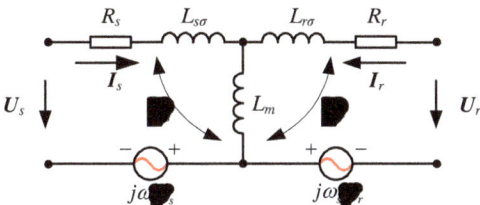

Figure 1. Equivalent circuit of the doubly fed induction generator (DFIG) model.

According to Figure 1, the voltage and flux equations of DFIG in synchronous rotating frame (SRF) can be expressed as:

$$U_{sdq} = R_s I_{sdq} + d\psi_{sdq}/dt + j\omega\psi_{sdq} \tag{1}$$

$$U_{rdq} = R_r I_{rdq} + d\psi_{rdq}/dt + j\omega_s\psi_{rdq} \tag{2}$$

$$\psi_{sdq} = L_s I_{sdq} + L_m I_{rdq} \tag{3}$$

$$\psi_{rdq} = L_m I_{sdq} + L_r I_{rdq} \tag{4}$$

$$\psi_{mdq} = L_m I_{sdq} + L_m I_{rdq} \tag{5}$$

where I_{sdq} and I_{rdq} are the stator and rotor current respectively, U_{sdq} and U_{rdq} are the stator and rotor voltage respectively, ψ_{sdq}, ψ_{rdq} and ψ_{mdq} are the stator, rotor and air-gap flux respectively, L_m, L_s and L_r are the mutual, stator and rotor inductance, in which $L_s = L_m + L_{s\sigma}$, $L_r = L_m + L_{r\sigma}$, $L_{s\sigma}$ and $L_{r\sigma}$ are the stator and rotor self-inductance respectively, ω is the SRF angular frequency, ω_s is the slip angular frequency, $\omega_s = \omega - \omega_r$, ω_r is the rotor angular frequency. All variables are referenced to the stator.

Under steady-state conditions, neglecting differential terms of stator flux ψ_{sdq} in (1), the stator voltage model can be represented as:

$$E_{\text{DFIG}} = j\omega\psi_{mdq} = U_{sdq} + I_{sdq}(R_s + j\omega_1 L_{s\sigma}) \tag{6}$$

Similar with the steady stator circuit equation of SG in [26], the term $j\omega\psi_{mdq}$ in (6) can be defined as the internal voltage E_{DFIG} of DFIG. In this way, the VSG control for DFIG can emulate a conventional SG by controlling the phase and magnitude of ψ_m.

The phase and magnitude of can be controlled by active power and reactive power, respectively. The phase and frequency reference of ψ_m can be calculated by the rotor swing equation [5–11,27]:

$$\omega = \frac{1}{J_P} \int (P_s^* - P_s + D_P(\omega_0 - \omega)) \, dt + \omega_0 \tag{7}$$

$$\theta = \int \omega \, dt \tag{8}$$

where, P_s is the stator active power, P_s^* is stator active power reference, which can be calculated by the methods of the maximum power point tracking (MPPT) or de-loading control, ω_0 is rated angular frequency, J_P and D_P are the inertia and damping coefficients of active power respectively, θ is the phase reference of E_{DFIG}, which is used in the frame transformation. In this way, the phase-locked-loop (PLL) can be canceled.

The magnitude reference of internal voltage can be calculated by the feedback control of stator voltage and stator reactive power. The reference of reactive power can be set by the grid operators or given by the voltage-drooping, which can be written as:

$$\left|E_{\text{DFIG}}\right|^* = \frac{1}{J_Q} \int (Q_s^* - Q_s + D_Q(U_0 - |U|)) + U_0 \tag{9}$$

where, Q_s and Q_s^* are the stator reactive power and its reference, J_Q are D_Q are the inertia and damping coefficient of reactive power control loop respectively, U_0 is the rated value of grid voltage, $|U|$ is equals to the magnitude of U_{sdq}.

VSG control for DFIG works in the virtual synchronous rotating frame (VSRF), in which d-axis is aligned to E_{DFIG}. And flux feedback control is implemented in control scheme to track the flux reference, which can be expressed as:

$$E_d^* = |E_{DFIG}|^* \quad E_q^* = 0 \tag{10}$$

$$U_{rdq}^* = (k_P + k_I \int)(E_{DFIG}^* - \omega \psi_{mdq}) \tag{11}$$

where, U_{rdq}^* is the output of VSG control loop, k_P and k_I are the proportional and integral coefficients of $\omega \psi_{mdq}$ feedback controller and $\omega \psi_{mdq}$ can be calculated based on (5), respectively.

Based on (7)–(11), the control block diagram of VSG for DFIG is presented in Figure 2. The active and reactive power can be controlled by the phase and magnitude of internal voltage E_{DFIG}, respectively. The inertia and damping characteristics can also be implemented by the rotor swing equation in the VSG control of DFIG.

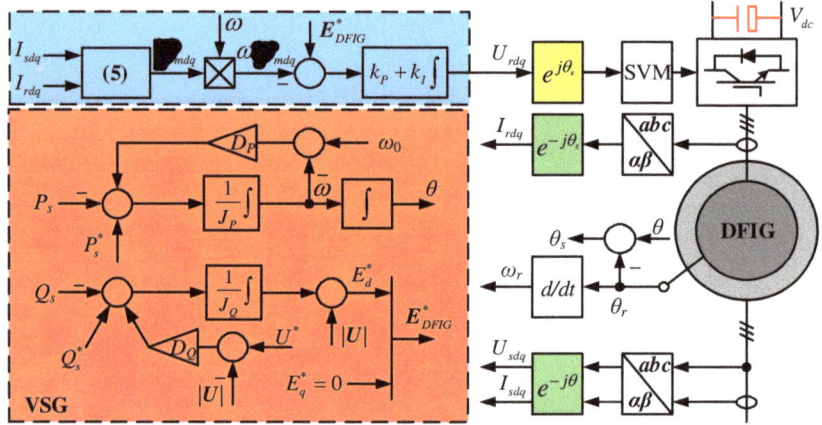

Figure 2. Control block diagram of virtual synchronous generator (VSG) for DFIG.

3. Impedance Modeling of VSG Control for DFIG

In this section, the impedance modeling of VSG control for DFIG is presented. For clarity, bold letters are used in this paper to denote complex space vectors, e.g., $X = Xd + jXq$; bold letters also denote complex transfer functions or transfer matrix, e.g., $X(s) = Xd(s) + jXq(s)$. Impedance modeling is based on the small signal analysis [21,22], therefore, the state variables with a small-signal perturbation can be written as:

$$x = X_0 + \Delta x \tag{12}$$

where x is the state variable, X_0 is the steady-state value, Δx is the small-signal perturbation, x can denote voltages, currents and other state variables in VSG control of DFIG.

3.1. Modeling of DFIG

Submitting (3) and (4) into (1) and (2) and taking the Laplace transformation of (1) and (2), the small signal model of DFIG can be expressed as:

$$\Delta U_{sdq} = G_1(s)\Delta I_{sdq} + G_2(s)\Delta I_{rdq} \tag{13}$$

$$\Delta U_{rdq} = G_3(s)\Delta I_{sdq} + G_4(s)\Delta I_{rdq} \tag{14}$$

In the impedance model, the stator voltage and rotor voltage are the inputs of the model; the stator current is the output. Thus, the small-signal model of DFIG can rewritten as:

$$\Delta I_{sdq} = G_{us\_is}(s)\Delta U_{sdq} + G_{ir\_is}(s)\Delta I_{rdq} \tag{15}$$

$$\Delta I_{rdq} = G_{us\_ir}(s)\Delta U_{sdq} + G_{ur\_ir}(s)\Delta U_{rdq} \tag{16}$$

Figure 3 illustrates the block diagram of transfer matrices of DFIG in the synchronous dq-frame. As can be seen from Figure 3, the stator current depends on the stator voltage and rotor voltage. The transfer functions in (13)–(16) are expressed as:

$$G_1(s) = \begin{bmatrix} R_s + sL_s & -\omega L_s \\ \omega L_s & R_s + sL_s \end{bmatrix} \tag{17}$$

$$G_2(s) = \begin{bmatrix} sL_m & -\omega L_m \\ \omega L_m & sL_m \end{bmatrix} \tag{18}$$

$$G_3(s) = \begin{bmatrix} sL_m & -\omega_s L_m \\ \omega_s L_m & sL_m \end{bmatrix} \tag{19}$$

$$G_4(s) = \begin{bmatrix} R_r + sL_r & -\omega_s L_r \\ \omega_s L_r & R_r + sL_r \end{bmatrix} \tag{20}$$

$$\begin{cases} G_{us\_is} = G_1^{-1} \\ G_{ir\_is} = -G_1^{-1}G_2 \\ G_{us\_ir} = -\left(-G_3 G_1^{-1} G_2 + G_4\right)^{-1} G_3 G_1^{-1} \\ G_{ur\_ir} = \left(-G_3 G_1^{-1} G_2 + G_4\right)^{-1} \end{cases} \tag{21}$$

where the superscript −1 means the inverse matrix and the subscript us_is means the transfer matrix from stator voltage to stator current, the meanings of other subscripts are similar to this.

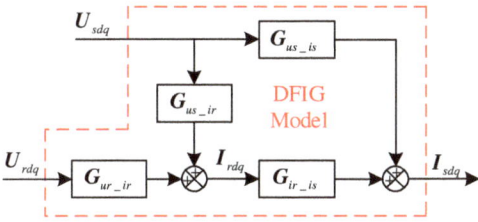

Figure 3. The model of DFIG in synchronous dq-frame.

3.2. Modeling of Power and Voltage Amplitude

In synchronous dq-frame, the power and voltage amplitude can be calculated as:

$$\begin{cases} P_s = -1.5(u_{sd}i_{sd} + u_{sq}i_{sq}) \\ Q_s = -1.5(u_{sd}i_{sq} - u_{sq}i_{sd}) \end{cases} \tag{22}$$

$$|u| = \sqrt{u_{sd}^2 + u_{sq}^2} \tag{23}$$

By linearizing the (22) and (23), the small-signal-model can be expressed as:

$$\begin{cases} \begin{bmatrix} \Delta P_s \\ 0 \end{bmatrix} = \overbrace{\frac{3}{2}\begin{bmatrix} -u_{sd0} & -u_{sq0} \\ 0 & 0 \end{bmatrix}}^{G_{i\_P}} \begin{bmatrix} \Delta i_{sd} \\ \Delta i_{sq} \end{bmatrix} + \overbrace{\frac{3}{2}\begin{bmatrix} -i_{sd0} & -i_{sq0} \\ 0 & 0 \end{bmatrix}}^{G_{u\_P}} \begin{bmatrix} \Delta u_{sd} \\ \Delta u_{sq} \end{bmatrix} \\ \begin{bmatrix} \Delta Q_s \\ 0 \end{bmatrix} = \underbrace{\frac{3}{2}\begin{bmatrix} -u_{sq0} & u_{sd0} \\ 0 & 0 \end{bmatrix}}_{G_{i\_Q}} \begin{bmatrix} \Delta i_{sd} \\ \Delta i_{sq} \end{bmatrix} + \underbrace{\frac{3}{2}\begin{bmatrix} i_{sq0} & -i_{sd0} \\ 0 & 0 \end{bmatrix}}_{G_{u\_Q}} \begin{bmatrix} \Delta u_{sd} \\ \Delta u_{sq} \end{bmatrix} \end{cases} \quad (24)$$

$$\begin{bmatrix} \Delta |U| \\ 0 \end{bmatrix} = \overbrace{\frac{1}{U_0}\begin{bmatrix} u_{sd0} & u_{sq0} \\ 0 & 0 \end{bmatrix}}^{G_U} \begin{bmatrix} \Delta u_{sd} \\ \Delta u_{sq} \end{bmatrix} \quad (25)$$

3.3. Modeling of VSG Control

The VSG control of DFIG consists two power control loops. The phase and amplitude of internal voltage of DFIG are controlled by active and reactive power loops, respectively. According to (7)–(9), the small signal model of power loops can be written as:

$$\Delta \theta = \frac{1}{s}\Delta \omega = \frac{1}{s}\frac{-1}{J_P s + D_P}\Delta P_s \quad (26)$$

$$\Delta |E_{DFIG}|^* = \frac{-1}{J_Q s}\Delta Q_s - D_Q \Delta |u| \quad (27)$$

By submitting (24) and (25) into (26) and (27), the model can be expressed as:

$$\begin{bmatrix} \Delta \theta \\ 0 \end{bmatrix} = G_{P\_\theta}\begin{bmatrix} \Delta P \\ 0 \end{bmatrix} = G_{P\_\theta}G_{i\_P}\begin{bmatrix} \Delta i_{sd} \\ \Delta i_{sq} \end{bmatrix} + G_{P\_\theta}G_{u\_P}\begin{bmatrix} \Delta u_{sd} \\ \Delta u_{sq} \end{bmatrix} \quad (28)$$

$$\begin{bmatrix} \Delta E_d^* \\ 0 \end{bmatrix} = G_{Q\_E}\begin{bmatrix} \Delta Q_s \\ 0 \end{bmatrix} + G_{U\_E}\begin{bmatrix} \Delta |U| \\ 0 \end{bmatrix} \\ = G_{Q\_E}G_{i\_Q}\begin{bmatrix} \Delta i_{sd} \\ \Delta i_{sq} \end{bmatrix} + G_{Q\_E}G_{u\_Q}\begin{bmatrix} \Delta u_{sd} \\ \Delta u_{sq} \end{bmatrix} + G_{U\_E}G_U\begin{bmatrix} \Delta u_{sd} \\ \Delta u_{sq} \end{bmatrix} \quad (29)$$

The transfer functions in (28) and (29) are expressed as:

$$G_{P\_\theta} = \begin{bmatrix} \frac{1}{s}\frac{-1}{J_P s + D_P} & 0 \\ 0 & 0 \end{bmatrix} \quad (30)$$

$$G_{Q\_E} = \begin{bmatrix} \frac{-1}{J_Q s} & 0 \\ 0 & 0 \end{bmatrix} \quad (31)$$

$$G_{U\_E} = \begin{bmatrix} -D_Q & 0 \\ 0 & 0 \end{bmatrix} \quad (32)$$

Based on the (28)–(32), the small signal model of VSG control can be obtained.

3.4. Modeling of Frame Transformation

Since the calculation of rotor voltage are applied in virtual synchronous reference frame (VSRF), the park transformation and inverse park transformation are used in VSG control (green and yellow blocks in Figure 2, respectively). The small-signal perturbation in the angular reference (28) affects the VSG control for DFIG the through the frame transformation. Therefore, the frame transformation should be involved in impedance modeling. It should be noted that the frame transformation does not affect the modeling of DFIG, power calculation and power control loops.

The park transformation (green blocks in Figure 2) of stator and rotor currents can be expressed as:

$$\begin{aligned} I^v_{dq} &= I_{\alpha\beta} e^{-j\theta} = (I_{dq0} + \Delta I^s_{dq}) e^{-j\Delta\theta} \\ &\approx (I_{dq0} + \Delta I^s_{dq})(1 - j\Delta\theta) \\ \Rightarrow \Delta I^v_{dq} &= \Delta I^s_{dq} - j I_{dq0} \Delta\theta \end{aligned} \quad (33)$$

where the superscript v means the VSRF and the superscript s means the synchronous dq-frame.

The inverse park transformation (yellow block in Figure 2) of rotor voltage can be expressed as:

$$\begin{aligned} U^s_{r\alpha\beta} &= U^v_{rdq} e^{j(\theta_s + \Delta\theta)} = (U_{rdq0} + \Delta U^v_{rdq}) e^{j\Delta\theta_s} \\ &\approx (U_{rdq0} + \Delta U^v_{rdq})(1 + j\Delta\theta) \\ \Rightarrow \Delta U^s_{rdq} &= \Delta U^v_{rdq} + j U_{rdq0} \Delta\theta \end{aligned} \quad (34)$$

3.5. Modeling of Rotor Voltage

Based on the (5) and (11), the small signal model of rotor voltage can be obtained:

$$\Delta U^v_{rdq} = G_{PI} \Delta |E_{DFIG}|^* + G_{PI\_i}(\Delta I^v_{sdq} + \Delta I^v_{rdq}) \quad (35)$$

By submitting (33) and (34) into (35), the rotor voltage in synchronous dq-frame can be expressed as:

$$\begin{aligned} \Delta U^s_{rdq} &= G_{urdq0} \begin{bmatrix} \Delta\theta \\ 0 \end{bmatrix} + G_{PI} \Delta |E_{DFIG}|^* \\ &+ G_{PI\_i}(\Delta I^s_{sdq} + G_{isdq0} \begin{bmatrix} \Delta\theta \\ 0 \end{bmatrix}) + G_{PI\_i}(\Delta I^s_{rdq} + G_{irdq0} \begin{bmatrix} \Delta\theta \\ 0 \end{bmatrix}) \end{aligned} \quad (36)$$

The transfer matrixes in (36) are expressed as:

$$G_{urdq0} = \begin{bmatrix} -u_{rq0} & 0 \\ u_{rd0} & 0 \end{bmatrix} G_{isdq0} = \begin{bmatrix} i_{sq0} & 0 \\ -i_{sd0} & 0 \end{bmatrix} G_{irdq0} = \begin{bmatrix} i_{rq0} & 0 \\ -i_{rd0} & 0 \end{bmatrix} \quad (37)$$

$$G_{PI} = \begin{bmatrix} \frac{1}{2}\frac{k_P s + k_I}{s} & 0 \\ 0 & \frac{1}{2}\frac{k_P s + k_I}{s} \end{bmatrix} G_{PI\_i} = -\omega_0 L_m G_{PI} \quad (38)$$

Based on the (22)–(36), the block diagram of transfer matrixes of VSG control in synchronous dq-frame is presented in Figure 4.

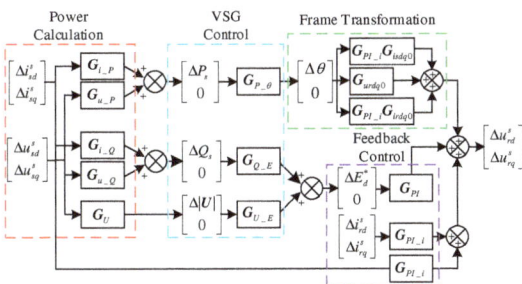

Figure 4. Block diagram of small signal transfer matrices for VSG control in synchronous dq-frame.

In Figure 4, the currents and stator voltage are the inputs of the VSG and the rotor voltage is the output. Thus, the model can be expressed as:

$$\Delta U^s_{rdq} = G_{us\_ur} \Delta U^s_{sdq} + G_{is\_ur} \Delta I^s_{sdq} + G_{ir\_ur} \Delta I^s_{rdq} \tag{39}$$

$$\begin{cases} G_{us\_ur} = (G_{urdq0} + G_{PI\_i}(G_{isdq0} + G_{irdq0}))G_{P\_\theta}G_{u\_P} \\ \qquad + G_{PI}G_{Q\_E}G_{u\_Q} + G_{PI}G_{U\_E}G_U \\ G_{is\_ur} = (G_{urdq0} + G_{PI\_i}(G_{isdq0} + G_{irdq0}))G_{P\_\theta}G_{i\_P} \\ \qquad + G_{PI}G_{Q\_E}G_{i\_Q} + G_{PI\_i} \\ G_{ir\_ur} = G_{PI\_i} \end{cases} \tag{40}$$

3.6. Sequence Impedance of VSG Control for DFIG

The small signal models of DFIG and VSG control have been obtained in Figures 3 and 4, respectively. The detailed model of VSG control for DFIG based on transfer matrices can be presented as Figure 5, in which "∆" is omitted for simplicity.

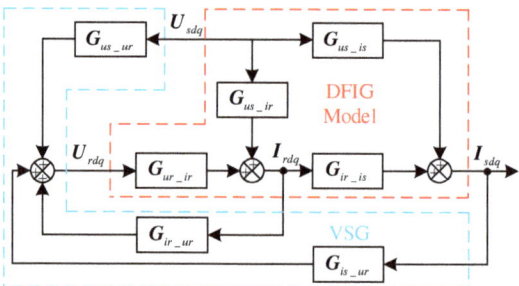

Figure 5. The whole block diagram of transfer matrices of DFIG with VSG control in synchronous dq-frame.

Therefore, the stator current can be expressed as:

$$U_{sdq} = -Z_{dq} I_{sdq} \tag{41}$$

$$Z_{dq} = \frac{G_{ur\_ir} G_{ir\_ur} + G_{ur\_ir} G_{ir\_is} G_{is\_ur} - E}{G_{us\_is} + G_{us\_ir} G_{ir\_is} + G_{us\_ur} G_{ur\_ir} G_{ir\_is}} \tag{42}$$

where the **E** is the identity matrix. $Z_{dq}(s)$ is the dq-frame impedance model, which is used to reveal the mathematical relations between the models in the dq-domain.

A general dq-frame impedance matrix $Z_{dq}(s)$ is expressed as:

$$\mathbf{Z}_{dq}(s) = \begin{bmatrix} Z_{dd}(s) & Z_{dq}(s) \\ Z_{qd}(s) & Z_{qq}(s) \end{bmatrix} \tag{43}$$

Since the dq-domain impedance model cannot be directly used in the practical situation, the $\alpha\beta$-domain or the sequence-domain model should be obtained. Based on the unifying approach in [22], the sequence-domain model in stationary frame can be expressed as:

$$\mathbf{Z}_{\pm}(s) = \begin{bmatrix} Z_{+dq}(s - j\omega_1) & Z_{-dq}(s - j\omega_1) \\ Z^*_{-dq}(s - j\omega_1) & Z^*_{+dq}(s - j\omega_1) \end{bmatrix} \tag{44}$$

$$\begin{aligned} Z_{+dq}(s) &= \frac{Z_{dd}(s) + Z_{qq}(s)}{2} + j\frac{Z_{qd}(s) - Z_{dq}(s)}{2} \\ Z_{-dq}(s) &= \frac{Z_{dd}(s) - Z_{qq}(s)}{2} + j\frac{Z_{qd}(s) + Z_{dq}(s)}{2} \end{aligned} \tag{45}$$

where the subscript ± means the positive-sequence and negative sequence components of impedance and the superscript * means the complex conjugate of the transfer functions.

4. Impedance Validation and SSR Analysis

In this section, the proposed sequence impedance model of VSG control for DFIG will be validated. Then, based on the impedance model, the SSR analysis of VSG control for DFIG interconnected with series compensation is presented. The simulation study is implemented to verify the SSR analysis; several conclusions can be obtained from the impedance analysis. Finally, the influence of VSG parameters on impedance is illustrated in detail. The parameters configuration of VSG is also discussed.

4.1. Impedance Validation

To verify the correctness of the impedance modeling in Section 3, a corresponding simulation model based on Simulink is built. The parameters of DFIG and VSG control for simulation are listed in the Appendix A Table A1; the DFIG is connected to an ideal grid, for the sake of excluding the influence from the grid impedance. The results of impedance model and simulation are given in the case of 0.7 pu rotor speed.

Figure 6 shows the model validation results of impedance in (44) and simulation frequency scanning. As can be seen, the asterisks are all located in the solid lines. It verifies that the impedance model in (44) is accurate in describing the impedance characteristics of VSG control for DFIG.

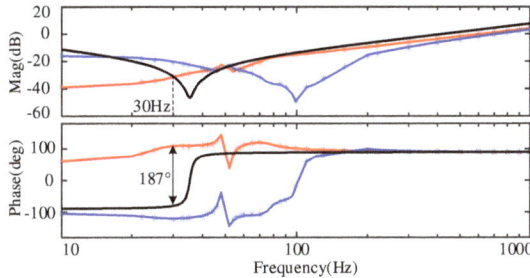

Figure 6. Model validation by frequency scanning, the lines are calculated by the impedance matrix (44), the asterisks are obtained by simulation results. Red and blue lines are positive and negative impedance, respectively. The black line is the impedance of network at 50% compensation level.

Since VSG control emulates the operating principle of SG, the impedance of DFIG under VSG is similar with the output impedance of SG [25]. The sequence impedance model of VSG control for DFIG is basically inductive. As can be seen from phase curve, the influence of VSG control can be found mainly around 50 Hz.

4.2. SSR Analysis

To analyze the SSR stability of VSG control for DFIG, a simulation case study of weak grid with series compensation net is introduced first. The simulation study system is derived from IEEE first benchmark model [15,16,23]; its schematic diagram is shown in Figure 7. The DFIG-based wind farm is a 100 MW aggregated equivalent system, which is aggregated from 50 2-MW DFIGs in the Appendix A Table A1. The transformer is 690 V/161 kV. The main parameters of the network system are given in the Appendix A Table A2.

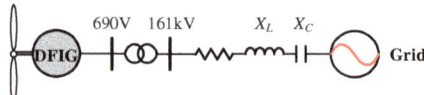

Figure 7. Diagram of a DFIG-based wind farm connecting the grid with a series-compensated transmission line.

The compensation level can be defined as [12]:

$$K = \frac{X_C}{X_L} \times 100\% \tag{46}$$

where X_L is the inductive reactance of the network including transmission line and transformer, X_C is the capacitive reactance in the transmission line.

Based on the impedance theory, instability happens when the impedance ratio curve encircles (−1,0). Therefore, bode plots can be used to analyze resonance stability [23,25]. When the compensation level is 50%, the impedance curve of series compensation network is shown as the black line in Figure 6. As can be seen, there is a magnitude curve intersection at 30 Hz; the phase difference is 187° (more than 180° is unstable), which indicates that there is an SSR instability in the situation; the resonant frequency is 30 Hz.

Figure 8 shows the bode plots of the impedance model developed in [23]. Compared with the proposed model in this paper, only the current control is considered in the impedance model in [23]. VSG control loops and frame transformation are neglected.

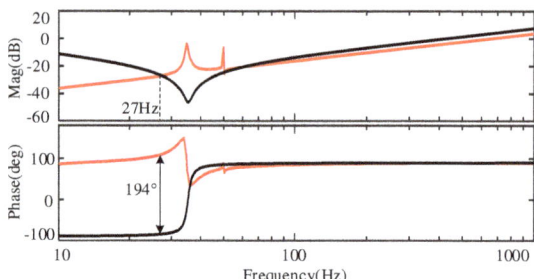

Figure 8. Bode diagram. Red line is the impedance in [23]. The black line is network impedance.

As can be seen from Figure 8, when the compensation level is 50%, there is a curve intersection at 27 Hz and the phase difference is 194°. The impedance model in [23] indicates that there is an SSR instability in the situation; the resonant frequency is 27 Hz.

To verify the SSR analysis, a simulation model is built in Matlab/Simulink. The simulation initially operates at 25% compensation level. Then, at $t = 1$ sec, additional capacitors are switched in, which imitates the transmission line fault and makes the compensation level reach 50%. Figure 9 shows the stator current of DFIG under VSG control; the fast fourier transform (FFT) analysis result of currents is given in Figure 10. As can be seen, SSR occurs at the resonant frequency 30 Hz. Therefore, the following conclusions can be obtained:

1. The DFIG with VSG control also has the SSR problem, when the weak grid reaches a high compensation level.
2. The correctness of proposed impedance model is validated based on the frequency scanning and SSR prediction.

3. Compared with the impedance model only considering current control, the proposed impedance model is more accurate, which indicates that VSG control has an important influence on the SSR.

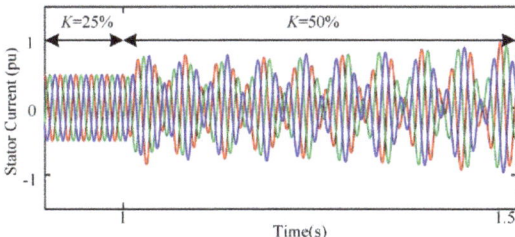

Figure 9. Waveform of stator currents, at the 50% series-compensation level.

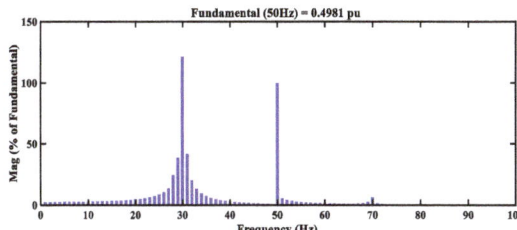

Figure 10. Fast Fourier transform (FFT) analysis of stator currents, at 50% series-compensation level.

4.3. Influence and Configuration of VSG Control Parameters

The inertia and damping are the key parameters in VSG control. Based on the SSR analysis above, VSG control plays an important role in SSR stability. Therefore, the influence of VSG control parameters is introduced in this section.

Figure 11 shows the bode plots of impedance with different inertia of active power control. The three impedance curves almost overlap each other. As can be seen, the inertia of active power control has little influence on impedance. Thus, inertia of active power cannot be used to mitigate the SSR.

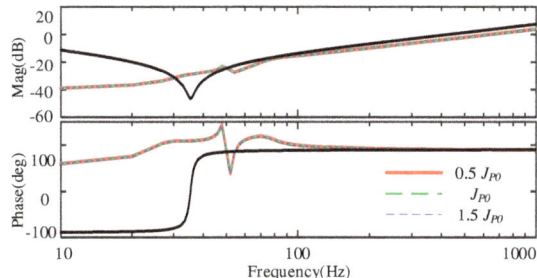

Figure 11. Bode diagram of different inertia of active power. Red line: 0.5 J_{P0}; green dash line: J_{P0}; blue dash line: 1.5 J_{P0}.

Figure 12 shows the bode plots of impedance with different damping of active power control. As can be seen, the phase decreases when the damping of active power control increases. It means that the phase difference of DFIG and network decreases, which will decrease the possibility of SSR. However, the damping of active power control is associated with primary frequency control [26,27], which configuration is limited according to grid code. Moreover, with the value increasing continuously, the phase variation is small.

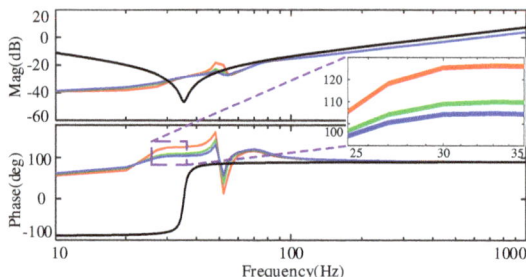

Figure 12. Bode diagram of different damping of active power. Red line: 0.5 D_{P0}; green line: D_{P0}; blue line: 1.5 D_{P0}.

Figure 13 shows the bode plots of impedance with different inertia of reactive power control. As can be seen, with the inertia changing, the phase varies without obvious regularity. Therefore, the inertia of reactive power is not suitable for mitigating SSR.

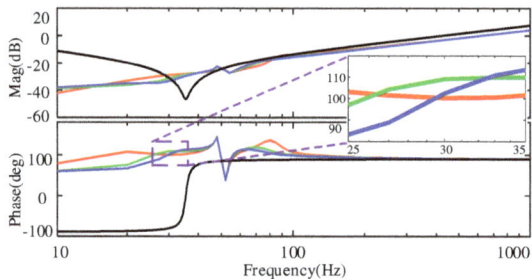

Figure 13. Bode diagram of different inertia of reactive power. Red line: 0.5 J_{Q0}; green line: J_{Q0}; blue line: 1.5 J_{Q0}.

Figure 14 shows the bode plots of impedance with different damping of reactive power D_Q. As can be seen, as D_Q increases, the phase decreases significantly. It indicates that the damping of reactive power has the ability to mitigate the SSR. When the damping of reactive power increase more than 1.5 D_{Q0}, the phase difference is less than 180 degree, which means that the small SSR disturbance is stable. And it should be noted that phase variation is small when D_Q increases continuously. Moreover, with the D_Q increasing, the magnitude of impedance intersection decreases, which may slow down the recovery of SSR. With the consideration of phase margin and impedance magnitude, 2 D_{Q0} is a proper value to mitigates the SSR. The zoom figure is shown in Figure 15.

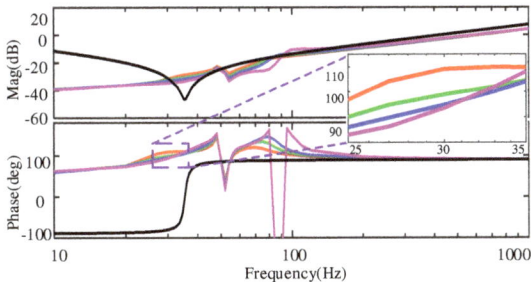

Figure 14. Bode diagram of different damping of reactive power. Red line: D_{Q0}; green line: 1.5 D_{Q0}; blue line: 2 D_{Q0}; pink line: 4 D_{Q0}.

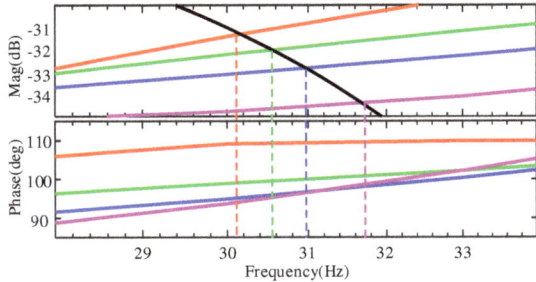

Figure 15. Zoom Bode diagram of different damping of reactive power.

Table 1 shows the summary of the influence of VSG control parameters. As can be seen, the damping factor of reactive power D_Q is suitable for mitigating the sub-synchronous resonance (SSR); the value should be set to 2 D_{Q0}.

Table 1. Influence of VSG Parameters Increasing.

Inertia of Active Power	Damping of Active Power	Inertia of Reactive Power	Damping of Reactive Power
Has little influence on impedance.	Phase decreases. But limited by the grid code.	Phase varies without obvious regularity.	Phase decreases.
×	×	×	√

5. Simulation Results

To validate the correctness and effectiveness of the proposed impedance model and parameters configuration method, the simulation results are given in this section. The simulation situation is the same as Figure 9 in Section 4.2. At the 1 sec, the series compensation level increases from 25% to 50%.

Figure 16 shows the results of stator currents with different damping factor D_Q. As can be seen, at the 1 s, there is the SSR disturbance caused by series compensation changing. Compared with Figure 9, the SSR disturbance in Figure 16 is mitigated well. When the D_Q increases more than 1.5 D_{Q0}, the phase difference is less than 180°, which means that the SSR disturbance is stable and validates the correctness of the SSR analysis in Section 4.3. According to the FFT analysis, when D_Q increases to 1.5 D_{Q0}, 2 D_{Q0} and 4 D_{Q0}, the resonant frequency is 30.5 Hz, 31 Hz and 32 Hz, respectively. The FFT results also coincide with the SSR analysis in Figure 15.

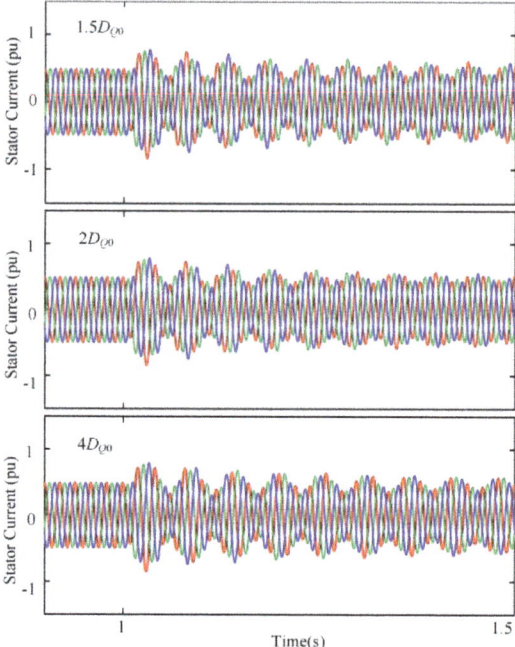

Figure 16. Waveform of stator currents.

Figure 17 shows the simulation results of electromagnetic torque; the enlarged figure of Figure 17 is given in Figure 18. Compared with the conditions of 1.5 D_{Q0} and 4 D_{Q0}, the attenuation of torque with 2 D_{Q0} is faster. To illustrate the results more intuitively, the damping time is defined as the length of time that the disturbance in electromagnetic torque damps into 0.05 pu (dashed lines in Figures 17 and 18). The damping time of different parameter's value is given in the Table 2. When the parameter is 2 D_{Q0}, the damping time is less than 0.5 s. By contrast, the damping time is more than 0.8 s, when the parameter is 1.5 D_{Q0} and 4 D_{Q0}. The simulation results verify the effectiveness of parameter configuration method.

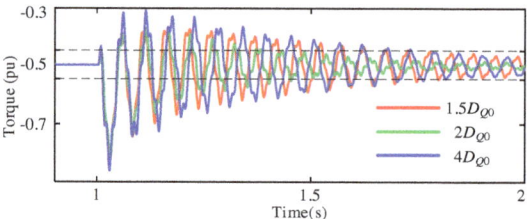

Figure 17. Waveform of electromagnetic torque.

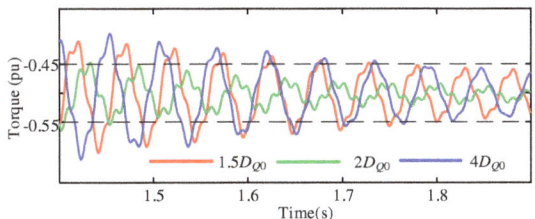

Figure 18. Enlarged waveform of electromagnetic torque.

Table 2. SSR damping time.

Damping of reactive power	1.5 D_{Q0}	2 D_{Q0}	4 D_{Q0}
Damping time	>0.8 s	<0.5 s	>0.8 s

6. Conclusions

This paper analyzes the SSR stability of DFIG under VSG control based on impedance method. Accurate impedance model of DFIG under VSG control, which considers DFIG model, swing equation, voltage-reactive power control loop, frame transformation and rotor voltage calculation, is developed. It can be found that DFIG with VSG control also has the SSR problem, when the weak grid reaches a high series-compensation level. The simulation results of stator current and FFT result coincide the impedance model analysis well, which indicates that the impedance modeling is an effective way to analyze the SSR in DFIG under VSG control. The influence of VSG control parameters on SSR stability is also analyzed. Based on the impedance analysis, the damping factor of reactive power is suitable for mitigating SSR. With the damping of reactive power increasing, the phase difference between DFIG and network decreases, which means that the damping of reactive power can be designed to mitigate SSR disturbance. Simulation results validate the correctness and effectiveness of the proposed impedance model and parameters configuration method. The robust control to damp SSR disturbance of DFIG under VSG control will be investigated based on the impedance model in the future.

Author Contributions: Conceptualization, Y.J., H.N.; methodology, Y.J., H.N.; software, Y.J., H.N.; validation, Y.J., H.N.; formal analysis, Y.J., H.N.; investigation, Y.J., H.N.; writing—original draft preparation, Y.J.; writing—review and editing, Y.J., H.N.; visualization, Y.J., H.N.; supervision, F.L., H.D.; project administration, F.L., H.D.; funding acquisition, F.L., H.D. All authors have read and agreed to the published version of the manuscript.

Funding: This paper was supported by the National Natural Science Foundation of China under Grant 51977194.

Conflicts of Interest: The authors declare no conflict of interest.

Appendix A

Table A1. Parameters of DFIG and VSG Control.

Parameters	Value
Rated power	2 MVA
Rated voltage	690 V
Rated frequency	50 Hz
Stator/Rotor ratio	0.34
Mutual inductance (p.u.)	3.90
Stator leakage inductance (p.u.)	0.171
Rotor leakage inductance (p.u.)	0.167
Stator resistance (p.u.)	0.0127
Rotor resistance (p.u.)	0.0127
DC voltage	1200 V
Inertia of active power J_{P0}	100
Damping of active power D_{P0}	$318310(50 \times 2 \times 10\hat{\ }6/(100\pi))$
Inertia of reactive power J_{Q0}	100
Damping of reactive power D_{Q0}	$17750(5 \times 2 \times 10\hat{\ }6/(\sqrt{2/3} \times 690))$
Flux control coefficient k_P, k_I	1, 10

Table A2. Parameters of Network System.

Parameters	Value
Transformer ratio	690 V/161 KV
Rated power	100 MVA
Line resistance	0.02 pu
Line inductance	0.5 pu
Line capacitive reactance at 50% compensation level	64.8 Ω

References

1. Chen, Z.; Guerrero, J.M.; Blaabjerg, F. A review of the state of the art of power electronics for wind turbines. *IEEE Trans. Power Electron.* **2009**, *24*, 1859–1875. [CrossRef]
2. Tapia, A.; Tapia, G.; Ostolaza, J.X.; Saenz, J.R. Modeling and control of a wind turbine driven doubly fed induction generator. *IEEE Trans. Energy Convers.* **2003**, *18*, 194–204. [CrossRef]
3. Blaabjerg, F.; Ma, K. Future on power electronics for wind turbine systems. *IEEE J. Emerg. Sel. Top. Power Electron.* **2013**, *1*, 139–152. [CrossRef]
4. Driesen, J.; Visscher, K. Virtual synchronous generators. In Proceedings of the 2008 IEEE Power and Energy Society General Meeting—Conversion and Delivery of Electrical Energy in the 21st Century, Pittsburgh, PA, USA, 20–24 July 2008; pp. 1–3.
5. Zhong, Q.; Nguyen, P.; Ma, Z.; Sheng, W. Self-synchronized synchronverters: Inverters without a dedicated synchronization unit. *IEEE Trans. Power Electron.* **2014**, *29*, 617–630. [CrossRef]
6. Alipoor, J.; Miura, Y.; Ise, T. Power system stabilization using virtual synchronous generator with alternating moment of inertia. *IEEE J. Emerg. Sel. Top. Power Electron.* **2015**, *3*, 451–458. [CrossRef]
7. D'Arco, S.; Suul, J.A. Equivalence of virtual synchronous machines and frequency-droops for converter-based MicroGrids. *IEEE Trans. Smart Grid* **2014**, *5*, 394–395. [CrossRef]
8. Liu, J.; Miura, Y.; Ise, T. Comparison of dynamic characteristics between virtual synchronous generator and droop control in inverter-based distributed generators. *IEEE Trans. Power Electron.* **2016**, *31*, 3600–3611. [CrossRef]
9. Wang, S.; Hu, J.; Yuan, X.; Sun, L. On inertial dynamics of virtual-synchronous-controlled dfig-based wind turbines. *IEEE Trans. Energy Convers.* **2015**, *30*, 1691–1702. [CrossRef]
10. Wang, S.; Hu, J.; Yuan, X. Virtual synchronous control for grid-connected dfig-based wind turbines. *IEEE J. Emerg. Sel. Top. Power Electron.* **2015**, *3*, 932–944. [CrossRef]

11. Huang, L.; Xin, H.; Zhang, L.; Wang, Z.; Wu, K.; Wang, H. Synchronization and frequency regulation of dfig-based wind turbine generators with synchronized control. *IEEE Trans. Energy Convers.* **2017**, *32*, 1251–1262. [CrossRef]
12. Reader's guide to subsynchronous resonance. *IEEE Trans. Power Syst.* **1992**, *7*, 150–157. [CrossRef]
13. Irwin, G.D.; Jindal, A.K.; Isaacs, A.L. Sub-synchronous control interactions between type 3 wind turbines and series compensated AC transmission systems. In Proceedings of the 2011 IEEE Power and Energy Society General Meeting, San Diego, CA, USA, 24–28 July 2011; pp. 1–6.
14. Wang, L.; Xie, X.; Jiang, Q.; Liu, H.; Li, Y.; Liu, H. Investigation of SSR in Practical DFIG-based wind farms connected to a series-compensated power system. *IEEE Trans. Power Syst.* **2015**, *30*, 2772–2779. [CrossRef]
15. Fan, L.; Kavasseri, R.; Miao, Z.L.; Zhu, C. Modeling of DFIG-based wind farms for SSR Analysis. *IEEE Trans. Power Deliv.* **2010**, *25*, 2073–2082. [CrossRef]
16. Fan, L.; Zhu, C.; Miao, Z.; Hu, M. Modal analysis of a DFIG-based wind farm interfaced with a series compensated network. *IEEE Trans. Energy Convers.* **2011**, *26*, 1010–1020. [CrossRef]
17. Varma, R.K.; Auddy, S.; Semsedini, Y. Mitigation of subsynchronous resonance in a series-compensated wind farm using FACTS controllers. *IEEE Trans. Energy Convers.* **2008**, *23*, 1645–1654. [CrossRef]
18. Mohammadpour, H.A.; Santi, E. Modeling and control of gate-controlled series capacitor interfaced with a dfig-based wind farm. *IEEE Trans. Ind. Electron.* **2015**, *62*, 1022–1033. [CrossRef]
19. Fan, L.; Miao, Z. Mitigating SSR using DFIG-based wind generation. *IEEE Trans. Sustain. Energy* **2012**, *3*, 349–358. [CrossRef]
20. Yao, J.; Wang, X.; Li, J.; Liu, R.; Zhang, H. Sub-synchronous resonance damping control for series-compensated DFIG-based wind farm with improved particle swarm optimization algorithm. *IEEE Trans. Energy Convers.* **2019**, *34*, 849–859. [CrossRef]
21. Cespedes, M.; Sun, J. Impedance modeling and analysis of grid-connected voltage-source converters. *IEEE Trans. Power Electron.* **2014**, *29*, 1254–1261. [CrossRef]
22. Wang, X.; Harnefors, L.; Blaabjerg, F. Unified impedance model of grid-connected voltage-source converters. *IEEE Trans. Power Electron.* **2018**, *33*, 1775–1787. [CrossRef]
23. Miao, Z. Impedance-model-based SSR analysis for type 3 wind generator and series-compensated network. *IEEE Trans. Energy Convers.* **2012**, *27*, 984–991. [CrossRef]
24. Liu, H.; Xie, X.; Zhang, C.; Li, Y.; Liu, H.; Hu, Y. Quantitative SSR analysis of series-compensated DFIG-based wind farms using aggregated RLC circuit model. *IEEE Trans. Power Syst.* **2017**, *32*, 474–483. [CrossRef]
25. Wu, W.; Chen, Y.; Zhou, L.; Luo, A.; Zhou, X.; He, Z.; Yang, L.; Xie, Z.; Liu, Z.; Zhang, M. Sequence impedance modeling and stability comparative analysis of voltage-controlled vsgs and current-controlled VSGs. *IEEE Trans. Ind. Electron.* **2019**, *66*, 6460–6472. [CrossRef]
26. Kundur, P.; Balu, N.J.; Lauby, M.G. *Power System Stability and Control*; McGraw-Hill: New York, NY, USA, 1994.
27. Nian, H.; Jiao, Y. Improved virtual synchronous generator control of DFIG to ride-through symmetrical voltage fault. *IEEE Trans. Energy Convers.* **2019**. [CrossRef]

© 2020 by the authors. Licensee MDPI, Basel, Switzerland. This article is an open access article distributed under the terms and conditions of the Creative Commons Attribution (CC BY) license (http://creativecommons.org/licenses/by/4.0/).

Article

Influence of Active Power Output and Control Parameters of Full-Converter Wind Farms on Sub-Synchronous Oscillation Characteristics in Weak Grids

Yafeng Hao [1], Jun Liang [1,2,*], Kewen Wang [1], Guanglu Wu [3], Tibin Joseph [2] and Ruijuan Sun [1]

1. School of Electrical Engineering, Zhengzhou University, Zhengzhou 450001, China; haoyafeng_1@163.com (Y.H.); kwwang@zzu.edu.cn (K.W.); sunruijuan_1@163.com (R.S.)
2. School of Engineering, Cardiff University, Cardiff CF24 3AA, UK; JosephT@cardiff.ac.uk
3. State Key Laboratory of Power Grid Safety and Energy Conservation (China Electric Power Research Institute), Beijing 100192, China; wuguanglu1989@126.com
* Correspondence: LiangJ1@cardiff.ac.uk

Received: 22 August 2020; Accepted: 1 October 2020; Published: 7 October 2020

Abstract: Active power outputs of a wind farm connected to a weak power grid greatly affect the stability of grid-connected voltage source converter (VSC) systems. This paper studies the impact of active power outputs and control parameters on the subsynchronous oscillation characteristics of full-converter wind farms connected weak power grids. Eigenvalue and participation factor analysis was performed to identify the dominant oscillation modes of the system under consideration. The impact of active power output and control parameters on the damping characteristics of subsynchronous oscillation is analysed with the eigenvalue method. The analysis shows that when the phase-locked loop (PLL) proportional gain is high, the subsynchronous oscillation damping characteristics are worsened as the active power output increases. On the contrary, when the PLL proportional gain is small, the subsynchronous oscillation damping characteristics are improved as the active power output increases. By adjusting the control parameters in the PLL and DC link voltage controllers, system stability can be improved. Time-domain results verify the analysis and the findings.

Keywords: weak grids; full-converter wind; active power output; control parameters; subsynchronous oscillation; eigenvalue analysis

---

1. Introduction

In recent years, as a clean, renewable and relatively proven technology, wind power generation has grown significantly in order to tackle the climate change and replace fossil fuels generators. By the end of 2019, the cumulative installed capacity of wind power worldwide reached 650 GW, of which 60.4 GW was newly added [1]. With the development of wind power and high voltage direct current transmission system (HVDC), subsynchronous interaction (SSI) has attracted the attention of academia and industry. The SSI is generally classified into the following three types: subsynchronous resonance (SSR), subsynchronous control interaction (SSCI) and subsynchronous torsional interaction (SSTI) [2]. In 2009, an SSI incident occurred in southern Texas, USA. A doubly fed induction generator (DFIG)-based wind farm was integrated into the grids via a high-series compensation transmission line. This caused a subsynchronous control interaction, resulting in a large number of wind turbine trips [3,4]. In 2012, the Guyuan wind farm in China also experienced the interaction between the control of DFIG and series compensation devices, causing the SSI event.

With the increase of grid-connected wind power capacity and the use of long-distance transmission lines, the support from the grids for wind farm connection is weakened [5]. There is a different type of interaction observed recently between full-converter wind farms and weak AC networks. In 2015, the permanent magnet synchronous generators based wind farm in Xinjiang, China, suffered from a severe oscillation event without series compensation. The oscillation frequency coincided with the torsional vibration frequency of the nearby thermal power unit, which led to the torsional vibration protection action of the thermal power unit resulting in generator shut down [6]. This type of interaction between full-converter wind farms and weak AC networks is also called subsynchronous oscillation (SSO), which is the topic investigated in this paper.

The dominant elements that affect the subsynchronous interaction characteristics in different scenarios of wind farms connected to the grids [6–25]. References [7–9] established DFIG-based wind farms interconnected with the grids and analysed the influence of the number of wind turbine generators (WTGs), wind speed, series compensation, line resistance, and outer and inner loop control parameters on subsynchronous interaction. For instance, reference [8] points out that when the DFIG-based wind farm is connected to series capacitive compensated transmission systems, the system damping decreases with the rise of series compensation or the decrease of total line resistance. Meanwhile, the variations of series compensation also affect the oscillation frequency of subsynchronous interaction. As for SSO in the full-converter wind farm or the VSC connected to AC grid system, the eigenvalue analysis, impedance-based analysis and the complex torque coefficient approach are conducted in [6,10–20] to research the dominant elements that affect the SSO characteristics. AC system strengths, WTGs number, wind speed, converter control parameters, PLL parameters and aggregation characteristics of wind farms are considered in these works. The work in [6] indicates the SSO will occur with the decrease of the AC system strengths. And control parameters also have great impacts on the SSO characteristics. In addition, SSO caused by the interconnection of direct-drive wind farms via voltage source converter based high voltage direct current (VSC-HVDC) transmission system has been studied in the references [21–28]. These elements, including wind farm control parameters, HVDC control parameters, PLL parameters and filter parameters are analysed and the coordinated controller is designed.

Until recently, there were very few papers specifically analysing the impact of the active power output of wind farms on the SSO characteristics. However, the change of active power output during the operation of wind farms will have a more significant impact on system stability. References [17–20] established the model of full-converter wind farm integrated into the grids or the VSC connected to AC grid. The SSO characteristics of the system are studied, and the impact of active output is analysed. The works in [17–19] pointed out that as the active power output of wind farms increases, the damping of the SSO mode decreases. However, it is revealed in [20] that increasing the active power output of wind farms will increase the damping of the SSO mode and increase system stability. When the active power output is too small, the system will result in diverging oscillation and loss of system stability.

In the view of the impact of active power output on SSO characteristics, some studies identified that the greater the active power output, the worse the SSO damping characteristics will be [17–19]. However, some studies that found that the higher the active output, the better the SSO damping characteristics will be [20]. Meanwhile, the existing researches are based on a certain set of control parameter without considering the influence of different control parameters. Therefore, it is necessary to study further the relationship between active power output and damping characteristics of SSO mode.

This paper investigates the impact of active power output on SSO characteristics by a small-signal analysis based on analytical models. The correlation between the active power output and the damping of the SSO mode with different control parameters is analysed through dynamic modelling and linear system analysis. First, the critical factor that determines the correlation is identified. Then, based on the eigenvalue analysis results, the strategy to increase the damping of SSO mode and improve system stability is proposed. Case studies and time-domain simulation verify the analysis result.

The rest of the paper is organized as follows: Section 2 builds the dynamic model of the system with full converter wind farm connected to the AC grids. In Section 3, both eigenvalue analysis and calculation of participation factors are carried to study the impact of active power output on SSO characteristics. The correlation between the active power output and the damping of the SSO mode is analysed with different control parameters and the critical factors that affect the SSO characteristics are presented. Meanwhile, the strategy to improve the stability of the system is proposed. Section 4 presents case studies and time-simulation results. Finally, the brief conclusions are given in Section 5.

2. System Modeling

A full-converter wind model including wind turbine, synchronous generator (SG), machine-side converter (MSC), DC link, grid-side converter (GSC), phase-locked loop (PLL), and converter control system is considered. It is assumed that wind farms usually consist of the same type of wind turbines with similar control parameters and operating conditions. Therefore, a wind farm is represented by an equivalent wind turbine. The schematic diagram of grid-connected wind power system structure is shown in Figure 1. L_{f1} and R_{f1} represent the filter inductance and filter resistance, respectively. C_1 represents the reactive power compensation parallel capacitor. R_2+jX_2 represents the equivalent impedance of both 25 kV line and 220 kV line. R_3+jX_3 represents the equivalent impedance of the transmission line near the grids. v_{pcc} denotes the voltage of point of common coupling (PCC). v_{grid} denotes the infinite grid voltage. i_1 and i_2 are the grid-side output current and transmission line current, respectively. Since the grid-connected dynamics of full-converter mainly depends on the control features of GSC, this paper ignores the machine-side dynamics. The wind turbine, SG and MSC are simplified as constant power sources [6].

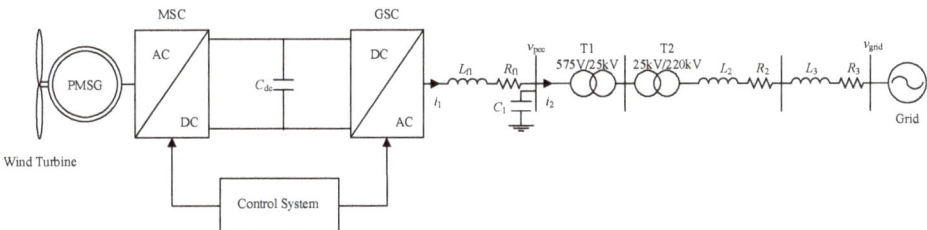

Figure 1. The diagram of the grid-connected wind farm.

The following section will establish a dynamic mathematical model of the grid-connected system. There are two dq reference frames in the dynamic mathematical model, namely the PLL-based dq frame and the grid-based dq frame. The PLL-based reference frame aligns its d-axis with the PCC voltage space vector v_{pcc} through the PLL output phase. Meanwhile, the grid-based reference frame has its d-axis aligned with the grid voltage space vector v_{grid} [10,17]. Superscripts 'c' and 'g' represent variables in the PLL-based reference frame and the grid-based reference frame, respectively. Phasor diagram of the component in different reference frames is shown in Figure 2.

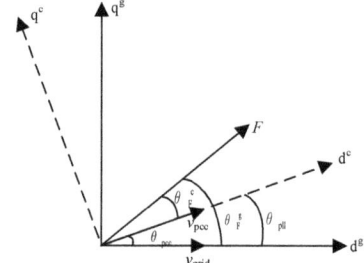

Figure 2. Phasor diagram of component in different reference frame.

2.1. Modeling of DC-Link

Since the machine-side dynamics are ignored, it is assumed that the active power output of the generator remains constant and is represented by P_{wind}. The dynamic mathematical model can be obtained from the DC link active power balance equation as Equation (1).

$$C_{dc} V_{dc} \frac{dV_{dc}}{dt} = P_{\text{wind}} - P_g \tag{1}$$

$$\frac{C_{dc} V_{dc,base}^2}{2 P_{base}^2} \frac{d(V_{dc}^{pu})^2}{dt} = P_{\text{wind}}^{pu} - P_g^{pu} \tag{2}$$

$$P_g = v_{pcc,d}^c i_{1d}^c + v_{pcc,q}^c i_{1q}^c \tag{3}$$

P_g and P_{base} are the GSC power delivered to the grids and base power, respectively. V_{dc} and $V_{dc,base}$ are expresses as DC voltage and rated DC voltage, respectively. Superscript 'pu' represents per unit variables. Subscripts 'd' and 'q' respectively notate the d-axis and q-axis components of variables. Hereafter the dc-link dynamic mathematical model is expressed by Equation (2). For convenience, the superscript 'pu' is omitted.

2.2. Outer and Inner Control Loop of GSC

The GSC control block diagram is shown in Figure 3. DC-link voltage control (DVC) and reactive power control are adopted for GSC, which contributes to balancing the power flow through DC link, maintaining DC-link voltage and operating at unit power factor for wind farm. The dynamic mathematical model of the outer and inner loop can be expressed as

$$\begin{cases} \frac{dx_1}{dt} = K_{idc}(V_{dc}^2 - V_{dc,ref}^2) \\ \frac{dx_2}{dt} = K_{ii}(i_{1d,ref}^c - i_{1d}^c) \\ \frac{dx_3}{dt} = K_{ii}(i_{1q,ref}^c - i_{1q}^c) \end{cases} \tag{4}$$

$$\begin{cases} v_d^c = K_{pdc} K_{pi}(V_{dc}^2 - V_{dc,ref}^2) + K_{pi} x_1 - K_{pi} i_{1d}^c + \\ \qquad x_2 - \omega L_{f1} i_{1q}^c + v_{pcc,d}^c \\ v_q^c = K_{pi}(i_{1q,ref}^c - i_{1q}^c) + x_3 + \omega L_{f1} i_{1d}^c \end{cases} \tag{5}$$

where x_1, x_2 and x_3 represent intermediate state variables. K_{pdc} and K_{idc} notate DVC proportional gain and integral gain, respectively. K_{pi} and K_{ii} are the proportional gain and integral gain of the inner current control loop, respectively. Subscript 'ref' denotes the system reference value.

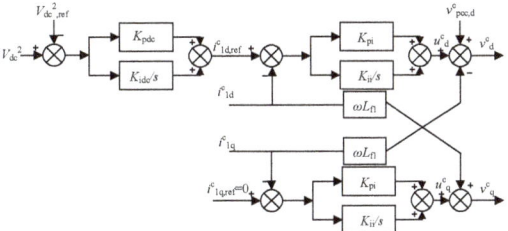

Figure 3. The control block diagram of GSC

2.3. Phase-Locked Loop Model

PLL uses the three-phase voltage at PCC bus as inputs to obtain the phase of the PCC voltage to achieve synchronization between the wind farm and the grids. The control block diagram of the PLL is illustrated in Figure 4. The PLL principle has been well documented [26] and will not be discussed here. ω_0 represents the rated angular frequency of the grids. $\Delta\omega$ notates the frequency deviation. θ_{pll} is the voltage phase of the PLL output. K_{ppll} and K_{ipll} denote the PLL proportional gain and integral gain, respectively. PLL dynamic mathematical model can be expressed as

$$\begin{cases} \frac{dx_{pll}}{dt} = K_{iPLL} v^c_{pcc,q} \\ \frac{d\Delta\theta_{pll}}{dt} = \omega_0 + \Delta\omega \end{cases} \quad (6)$$

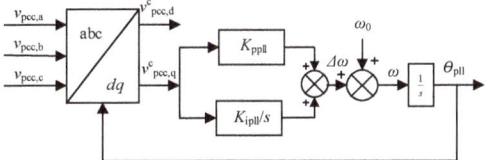

Figure 4. The control block diagram of the PLL.

2.4. Grid Dynamics

The grid dynamics mainly include shunt capacitor dynamics, filter inductance dynamics, and transmission line equivalent inductance dynamics. The dynamic mathematical model of the grid is established in the grid-based reference frame. The grid dynamic mathematical model can be written as Equation group (7):

$$\begin{cases} \frac{di^g_{1d}}{dt} = \frac{1}{L_{f1}}(v^g_d - v^g_{pcc,d} - R_{f1}i^g_{1d} + \omega_0 L_{f1} i^g_{1q}) \\ \frac{di^g_{1q}}{dt} = \frac{1}{L_{f1}}(v^g_q - v^g_{pcc,q} - R_{f1}i^g_{1q} - \omega_0 L_{f1} i^g_{1d}) \\ \frac{di^g_{2d}}{dt} = \frac{1}{L_g}(v^g_{pcc,d} - v^g_{grid,d} - R_g i^g_{2d} + \omega_0 L_g i^g_{2q}) \\ \frac{di^g_{2q}}{dt} = \frac{1}{L_g}(v^g_{pcc,q} - v^g_{grid,q} - R_g i^g_{2q} - \omega_0 L_g i^g_{2d}) \\ \frac{dv^g_{pcc,d}}{dt} = \frac{1}{C_1}(i^g_{1d} - i^g_{2d} + \omega_0 C_1 v^g_{pcc,q}) \\ \frac{dv^g_{pcc,q}}{dt} = \frac{1}{C_1}(i^g_{1q} - i^g_{2q} - \omega_0 C_1 v^g_{pcc,d}) \end{cases} \quad (7)$$

R_g and L_g denote the total equivalent resistance and inductance of the grid, including the transformers and the transmission lines. The impedance from the PCC to the grid can be represented as a single impedance $R_g + j\omega_0 L_g$, $R_g = R_{T1} + R_{T2} + R_2 + R_3$, $L_g = L_{T1} + L_{T2} + L_2 + L_3$.

3. Eigenvalue Analysis

3.1. Analysis of the Dominant Oscillation Mode

In this paper, a wind farm consisting of fifty 2 MW wind turbines connected to the AC grid through long-distance transmission lines is used as the target test system. The parameters of the system are listed in Table 1. The short circuit ratio (SCR) of this system is 1.53, which indicates that the wind farm is connected to a very weak AC grid [27]. The parameters of the wind generator are shown in Table 2.

Table 1. Parameters of the grid-connected system.

Parameter	Value (pu, S_B = 100 MVA)
Transformer T1(575 V/25 kV)	$X_{T1} = 0.06$, $R_{T1} = 0.006$
Transformer T2(25 kV/220 kV)	$X_{T2} = 0.065$, $R_{T2} = 0.0065$
Long-distance transmission line	$X_2 = 0.525$, $R_2 = 0.0525$
Short-distance transmission line	$X_3 = 0.01$, $R_3 = 0.001$

Table 2. The parameters of a single wind generator.

Parameter	Value (pu, S_B = 2 MVA)
Rated power	2 MW
Rated frequency	50 Hz
GSC filter	$X_{f1} = 0.15$, $R_{f1} = 0.003$, $y_{c1} = 0.25$
DC capacitor	0.09 F
Rated DC voltage	1100 V
DVC	$K_{pdc} = 1.1$, $K_{idc} = 137.5$
Current control	$K_{pi} = 0.4758$, $K_{ii} = 3.28$
PLL	$K_{ppll} = 314$, $K_{ipll} = 24{,}700$

In the system dynamic mathematical model established in this paper, the state variables are $x = [\ i^g_{1d},\ i^g_{1q},\ i^g_{2d},\ i^g_{2q},\ v^g_{pcc,d},\ v^g_{pcc,q},\ x_{pll},\ \Delta\theta_{pll},\ V_{dc},\ x_1,\ x_2,\ x_3]$. By linearizing the dynamic mathematical model at an operating condition x_0, the small signal model of the system can be established as Equation (8) shows.

$$\frac{d\Delta x}{dt} = A\Delta x \qquad (8)$$

In Equation (8), A represents the eigenmatrix of the small signal model as shown in Appendix A and Δx denotes incremental state vector.

When the active power output of the wind farm is maintained at 0.8 pu, the eigenmatrix is used to calculate the eigenvalues of the system as shown in Table 3. It can be observed that there are four oscillation modes in the target system, of which $\lambda_{6,7}$ and $\lambda_{9,10}$ belong to the SSO mode. However, the real parts of the eigenvalues $\lambda_{6,7}$ are positive, which indicates that the mode exhibits negative damping and the system is unstable. For this mode, the participation factors of state variables are shown in Figure 5. In Figure 5, the first six state variables represent the dynamics of the grids and the last six state variables represent the dynamics of the wind farm. Therefore, this mode is related to both the grid dynamics and the wind farm dynamics and reflects the subsynchronous interaction between the AC grids and the wind farm. As far as the control loops are concerned, the participation factors of these state variables ($\Delta\theta_{pll}$, x_{pll}, V^2_{dc}, x_1) are higher. That is, PLL and DVC have a greater impact on this mode.

Table 3. Eigenvalues of the weak grids-connected wind power system.

Mode	Eigenvalue
$\lambda_{1,2}$	$-569.33 \pm j1764.69$
$\lambda_{3,4}$	$-87.53 \pm j836.15$
λ_5	-976.21
$\lambda_{6,7}$	$2.62 \pm j199.47$
λ_8	-91.51
$\lambda_{9,10}$	$-15.89 \pm j55.99$
λ_{11}	-6.90
λ_{12}	-6.89

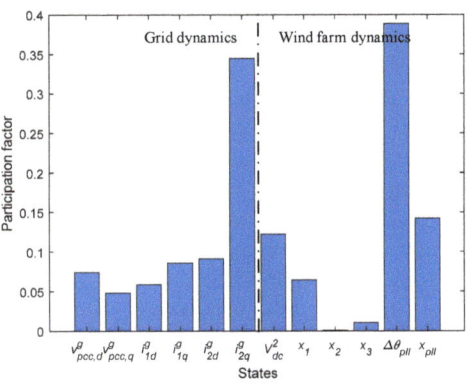

Figure 5. Participation factors of the state variables in the dominant oscillation mode

3.2. Impacts of the Active Power Outputs of the Wind Farm on Subsynchronous Oscillation Characteristics with Different Control Parameters

There are two main factors that affect the eigenvalues in the weak grids: one is active power output (operating condition), and the other is the control structure and control parameters. By calculating the participation factors, it can be seen that the PLL and the DVC loop have a greater impact on the dominant oscillation mode. In this section, the eigenvalue method will be used to analyse the impact of active power output on SSO characteristics with different control parameters. For convenience of expression, the following sections will use comparative gain to express the control parameters. The comparative gain represents a multiple of the pre-set value of the parameters given in Table 2.

3.2.1. Impacts of Active Power Outputs with Different PLL Proportional Gains

To evaluate this case, K_{ppll} is selected between 0.1 and 1.2 times of its pre-set value. When the active power output increases from 0.6 pu to 1.0 pu, the variations of the dominant eigenvalues with different K_{ppll} are shown in Figure 6 (only those parts are shown where the imaginary part is positive). When the value of K_{ppll} is large (e.g., when the factors are larger than 0.3 times), the eigenvalues move toward the right half plane (RHP) with the increase of the active power output, the mode damping decreases, and the system stability decreases. The active power output is negatively related to the mode damping. When the value of K_{ppll} is small (e.g., when the factors are smaller than 0.3 times), the eigenvalues move towards the left half plane (LHP) as the active power output increases. The active power output is positively correlated with the mode damping. There are only slight changes of the frequency of the SSO modes with different active power outputs.

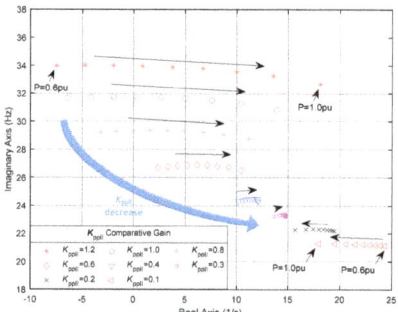

Figure 6. The impact of the active power output on the dominant SSO modes with different K_{ppll}.

When K_{ppll} takes these intermediate values, the correlation between the active power output and the mode damping will change from negative correlation to positive correlation with the decrease of K_{ppll}. When K_{ppll} takes the critical value, the real part of the dominant eigenvalues changes with the active power output as shown in Figure 7. Moreover, as depicted in Figure 7, the real part of the dominant eigenvalues gradually increases when the active power output increases from 0.6 pu to 0.75 pu, while the real part of the dominant eigenvalues decreases when the active power output increases from 0.75 pu to 1.0 pu. It can be found that when K_{ppll} takes the critical value, the mode damping decreases first and then increases as the active power output increases.

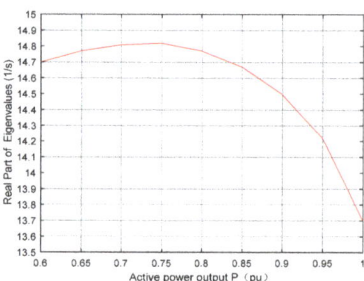

Figure 7. The impact of the active power output on the dominant SSO modes with the critical value of K_{ppll}.

In addition, it can also be seen from Figure 6 that when the active power output is negatively correlated with the mode damping, the larger the value of K_{ppll}, the greater the variation of the mode damping with the active power output will be. That is, the stability of the system is more affected by the active power output. Conversely, when the active power output is positively correlated with the mode damping, the smaller the value of K_{ppll}, the stability of the system is more affected by the active power output.

From the results above, a conclusion can be drawn that when selecting a larger K_{ppll}, the active power output is negatively correlated with the damping of this SSO mode, while when selecting a smaller K_{ppll}, the active power output is positively correlated with the damping of this SSO mode. Moreover, there is a critical value K_{ppll} for correlation. Meanwhile, the closer K_{ppll} is to the critical value, the less the system stability is affected by the active power output.

3.2.2. Impacts of Active Power Outputs with Different PLL Integral Gain

In the two cases where K_{ppll} is selected to be larger (negative correlation) and smaller (positive correlation), the impact of the active power outputs on the mode damping with different K_{ipll} is observed. When the active power output increases from 0.6 pu to 1.0 pu, the dominant eigenvalue is plotted as shown in Figure 8. As shown in Figure 8a, with different K_{ipll}, the dominant eigenvalues move towards the RHP as the active power output increases and in effect decreasing the mode damping. At the same time, Figure 8b shows response with smaller K_{ppll} value. With different K_{ipll}, the dominant eigenvalues move towards the LHP as the active power output increases and the mode damping increases. It can be observed that adjusting K_{ipll} does not affect the correlation between the active power output and the damping of this SSO mode. However, under the same active power output condition, the damping of the SSO mode increases when K_{ipll} decreases. This is because the typical control parameters of a PLL are designed to ensure good phase tracking responses. However, in a weak grid, a fast PLL response will enlarge the interaction between the weak grid and the wind turbine converter, which will reduce the system stability. Therefore, a smaller integral gain is selected to improve the stability by compromising the PLL response characteristics.

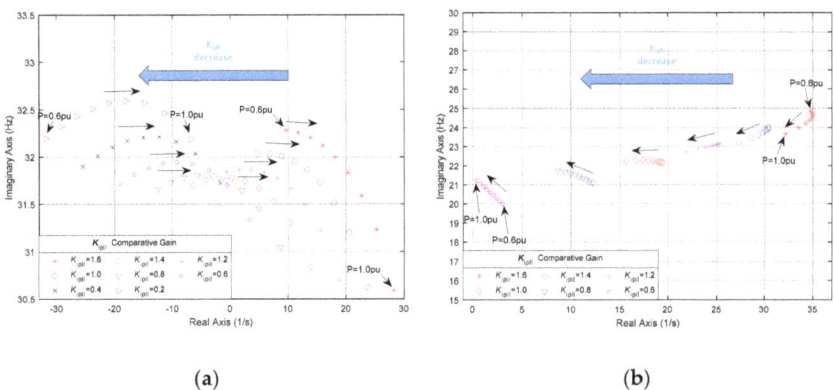

Figure 8. The impact of the active power output on the dominant SSO modes with different Kipll. (a) At large K_{ppll}. (b) At small K_{ppll}.

3.2.3. Impacts of the Active Power Outputs with Different DVC Proportional Gain.

According to the above analysis on PLL parameters, four representative PLL parameters are selected as shown in Table 4. The impact of K_{pdc} on the correlation between the active power output and the damping of the dominant SSO mode is analysed with the four different PLL parameters. When the active power output increases from 0.6 pu to 1.0 pu, the variations of the dominant eigenvalues with different K_{pdc} are presented in Figure 9.

Table 4. Four different PLL parameters.

	K_{ppll}	K_{ipll}
Case 1	314 (the pre-set value)	24,700 (the pre-set value)
Case 2	314	24,700 × 0.8
Case 3	314 × 0.2	24,700
Case 4	314 × 0.2	24,700 × 0.8

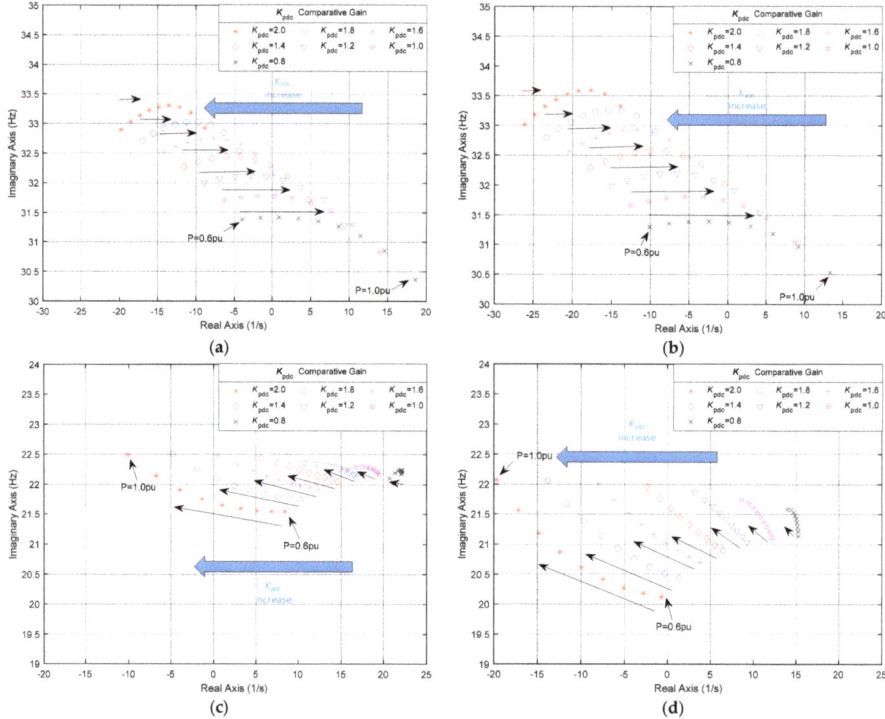

Figure 9. The impact of the active power output on the dominant SSO modes with different K_{pdc}. (**a**) Case 1. (**b**) Case 2. (**c**) Case 3. (**d**) Case 4.

The K_{ppll} of PLL is selected to be larger in Figure 9a,b. Figure 9a,b show that with different K_{pdc}, the dominant eigenvalues move towards the RHP as the active power output increases, and the mode damping decreases. The K_{ppll} of PLL is selected to be smaller in Figure 9c,d and shows that with different K_{pdc}, the dominant eigenvalues move towards the LHP as the active power output increases, and the mode damping increases. Therefore, adjusting K_{pdc} does not change the correlation between the active power output and the mode damping. However, when the active power output is negatively correlated with the mode damping, the smaller the value of K_{pdc}, the greater the variation of the mode damping with the active power output. That is, system stability is more affected by the active power output (as shown in Figure 9a,b). Conversely, when the active power output is positively correlated with the mode damping, the greater the value of K_{pdc}, the greater the system stability affected by the active power output (as shown in Figure 9c,d).

Meanwhile, it can be found that the increase in K_{pdc} leads to increase the mode damping under the same active power output condition. When the damping of the SSO mode is small, the stability can be improved by increasing K_{pdc}. Comparing Figure 9a,c with Figure 9b,d, it can be seen that better and improved system stability can be achieved by simultaneously decreasing K_{ipll} and increasing K_{pdc}.

A conclusion can be drawn from the analysis that the correlation between the active power output and the damping of the dominant SSO mode mainly depends on K_{ppll}. When K_{ppll} is large, the active power output is negatively correlated with the damping of this SSO mode. When K_{ppll} is small, the active power output is positively correlated with the damping of the dominant SSO mode. Moreover, there is a critical range for K_{ppll}, in which SSO damping is near consistent irrespective to the change of active power variation. Meanwhile, the system stability can be improved by appropriately decreasing K_{ipll} or increasing K_{pdc}.

4. Case Study and Simulation Verifications

To validate the effectiveness of the conclusions in Section 3, the impact of the active power output on the eigenvalues of the system is analysed with different control parameters shown in Figure 1. At the same time, the detailed simulation model of the studied system is developed in Matlab/Simulink (2018a, MathWorks, Natick, MA, USA) for validation.

4.1. Verification of the Negative Correlation when the PLL Proportional Gain is Large

When the active power output is 0.6pu, the system has good stability through trial-and-error and adjustment of control parameters. The control parameters in this case are called the based-case as shown in Table 5. When the control parameters of the based-case in Table 5 are used (with the larger K_{ppll} selected), the eigenvalue locus of the two SSO modes with the increase in active power output are plotted in Figure 10a. It is found that under the control parameters of the based-case, the eigenvalues $\lambda_{6,7}$ move to the RHP with the increase of active power output. The mode damping decreases continuously, and the system stability is weakened. When the active power output reaches 0.75 pu, $\lambda_{6,7}$ first crosses the imaginary axis and enters the RHP. The system becomes unstable. That is, there is a negative correlation between the active power output and the damping of the $\lambda_{6,7}$ mode. The results proved that when K_{ppll} is large, the active power output is negatively correlated with the damping of this SSO mode.

Table 5. Four different control parameters

	Parameters	Based-Case	Group 1	Group 2	Group 3
PLL	K_{ppll}	314	314	314 × 0.2	314 × 0.2
	K_{ipll}	24,700	24,700 × 0.8	24,700 × 0.8	24,700 × 0.7
DVC	K_{pdc}	1.1	1.1 × 1.6	1.1 × 1.6	1.1 × 2
	K_{idc}	137.5	137.5	137.5	137.5
Inner current control loop		$K_{pi} = 0.4758$		$K_{ii} = 3.28$	

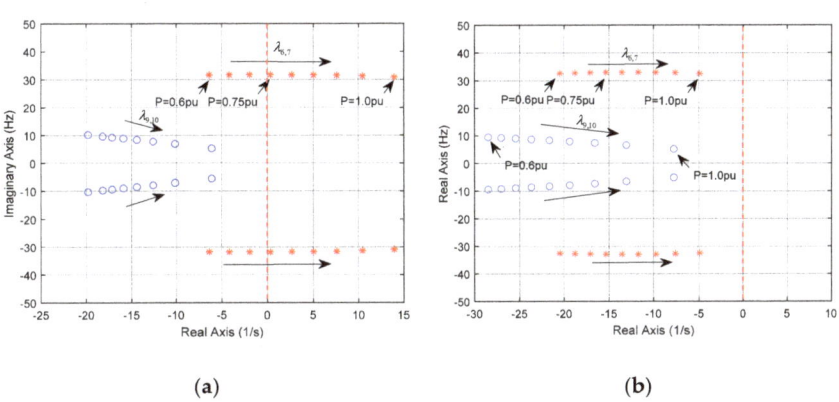

Figure 10. The impacts of the active power output on the SSO modes with the different control parameters. (**a**) Pre-set values. (**b**) Group 1.

When the control parameters of group 1 in Table 5 are used, the impact of the active power output on the eigenvalues of the SSO modes is shown in Figure 10b. Clearly, the eigenvalues are always in the LHP during the variations of active power output. The damping of the $\lambda_{6,7}$ mode is always positive, and the system remains stable. Therefore, it is proved that the system stability can be improved by decreasing K_{ipll} and increasing K_{pdc}.

In order to verify the above analysis, an electromagnetic transient simulation model of the grid-connected wind farm system in Figure 1 is built in MATLAB/Simulink. The studied system adopts the control parameters of the based-case and the group 1 control parameters, respectively. At $t = 3$ s, the active power output of the wind farm increases from 0.7 pu to 0.75 pu. Responses of active power output, DC voltage and phase-a voltage of the PCC are observed and analysed. The corresponding time-domain simulation results are presented in Figure 11. It can be seen that when the active power output increases from 0.7 pu to 0.75 pu, the system with the control parameters of the based-case oscillates and becomes unstable. As shown in Figure 11a, the wind power has 31Hz oscillation. This further confirms the conclusion in Section 3 that the active power output is negatively correlated to the damping of this $\lambda_{6,7}$ mode with a lager K_{ppll}. Furthermore, the system with the group 1 control parameters is able to keep stable after disturbance, indicating that the damping of the SSO mode increased after adjusting K_{ipll} and K_{pdc}. The simulation results are consistent with the analysis results above.

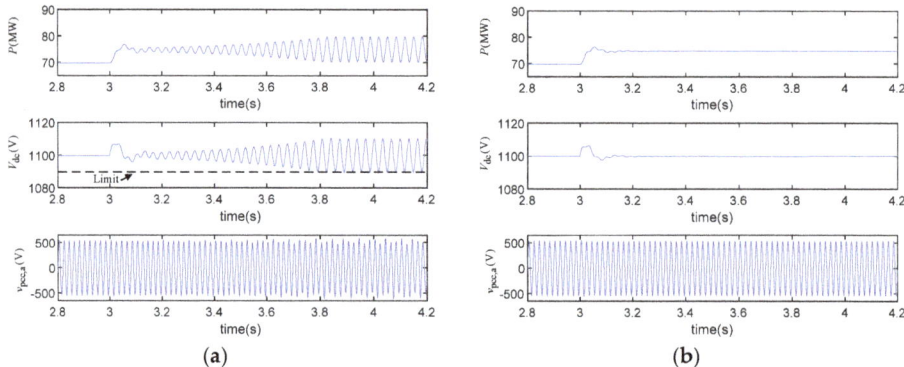

Figure 11. Simulation results of wind power increase with a large K_{ppll}. (**a**) Based-case. (**b**) Group 1.

4.2. Verification of the Positive Correlation when the PLL Proportional Gain Is Small

When the control parameters of group 2 in Table 5 are used (the smaller K_{ppll} is selected), the eigenvalue locus with varied active power output is depicted in Figure 12a. It can be seen that the eigenvalues $\lambda_{6,7}$ move to the LHP as the active power output increases. The damping of this SSO mode increases and the system stability is enhanced. Moreover, when the active power output of the wind farm is too small (less than 0.75 pu), the eigenvalue $\lambda_{6,7}$ will be in the RHP. The system will result in diverging oscillation and become unstable. That is, the active power output is positively correlated to the damping of the $\lambda_{6,7}$ mode. It is proved that when K_{ppll} is small, the active power output is positively correlated with the damping of this SSO mode.

Similarly, when the control parameters of group 3 in Table 5 are adopted, the impact of active power output on the eigenvalues of the SSO modes is shown in Figure 12b. In the process of active power output change, the eigenvalues are always in the LHP. The damping of the $\lambda_{6,7}$ mode is always positive, and the system remains stable. This analysis indicates again that the stability of the system can be enhanced by decreasing K_{ipll} and increasing K_{pdc}.

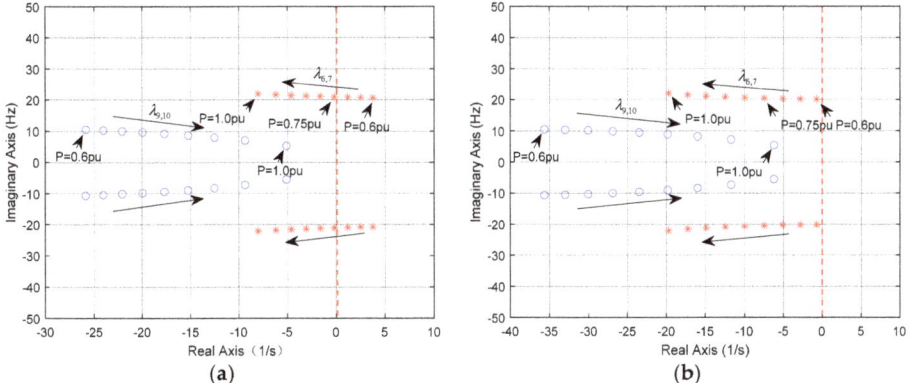

Figure 12. The impacts of the active power output on the SSO modes with the different control parameters. (**a**) Group 2. (**b**) Group 3.

To validate the above analysis, group 2 and group 3 were selected as the control parameters of the system respectively. At $t = 3$ s, the active power output of the wind farm decreases from 0.8 pu to 0.7 pu. Figure 13 presents the corresponding time-domain simulation results. It can be observed that when the active power output decreases from 0.8 pu to 0.7 pu, the system using group 2 of control parameters is unstable and the oscillation frequency of the wind power is 21 Hz. This result matches the conclusion in Section 3 well, which demonstrates that there is a positive correlation between the active power output and the damping of this $\lambda_{6,7}$ mode with a smaller K_{ppll}. Meanwhile, the system using the control parameters of group 3 can remain stable after disturbance. This indicates that the damping of the SSO mode increases after adjusting the parameters. The simulation results are in accordance with the analysis results above.

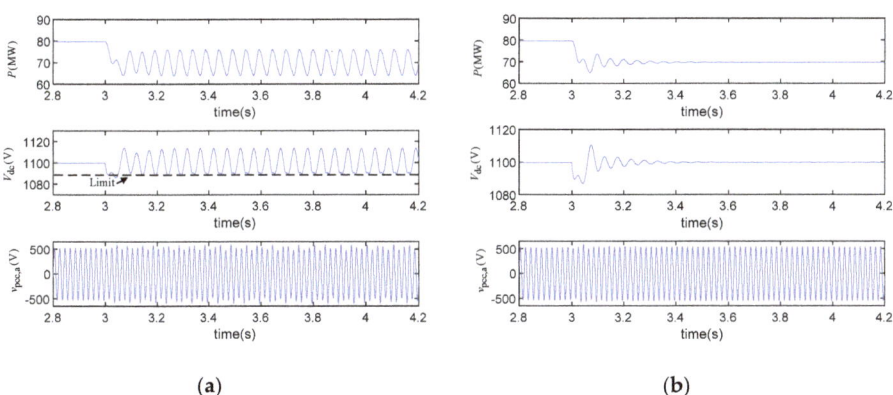

Figure 13. Simulation results of wind power decrease with a small K_{ppll}. (**a**) Group 2. (**b**) Group 3.

4.3. Simulation Verification for a Complex System

To verify the analyses, simulation has been carried out for a complex system with different wind farm ratings, grid configurations and grid voltage levels, as shown in Figure 14. The system parameters are given in Figure 14. The control parameters are given in Table 6. The simulation results are given in Figures 15 and 16.

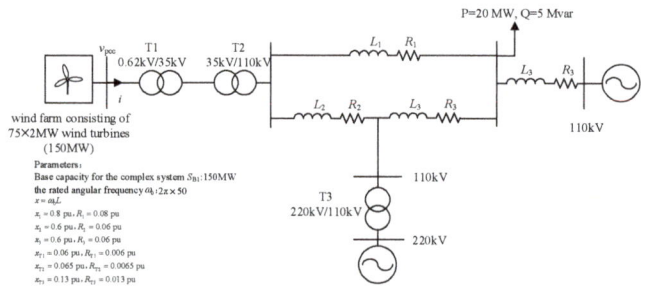

Figure 14. The diagram of a complex system.

Table 6. Four different control parameters.

Parameters		Group 4	Group 5	Group 6	Group 7
PLL	K_{ppll}	314	314	314 × 0.2	314 × 0.2
	K_{ipll}	24,700	24,700 × 0.8	24,700 × 0.8	24,700 × 0.7
DVC	K_{pdc}	1.1 × 1.2	1.1 × 1.8	1.1×1.8	1.1 × 2.0
	K_{idc}	137.5	137.5	137.5	137.5
Inner current control loop		$K_{pi} = 0.4758$		$K_{ii} = 3.28$	

In Figure 15, a large K_{ppll} is used. Figure 15a gives the simulation results when the Group 4 control parameters are used. When the wind power increases, system tends to be unstable. If the control parameters are adjusted properly, by reducing K_{ipll} and increasing K_{pdc}, as in Group 5, the system can be maintained stable, as shown in Figure 15b.

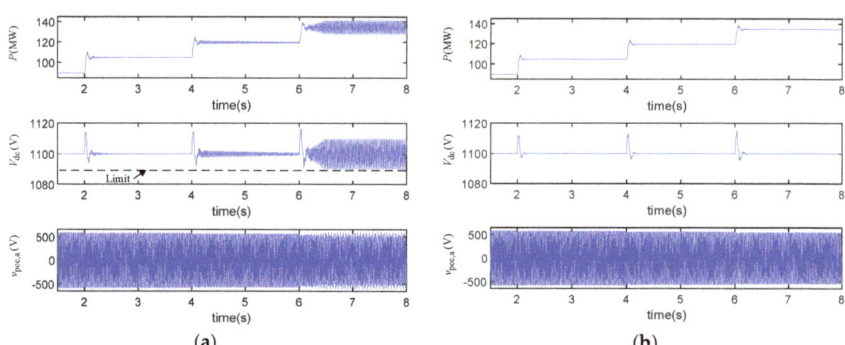

Figure 15. Simulation results of wind power increase with a large K_{ppll}. (a) Using the Group 1 parameter (b) Using the Group 2 parameter after adjustment.

In Figure 16, a small K_{ppll} is used. Figure 16a gives the simulation results when the Group 6 control parameters are used. When the wind power decreases, system tends to be unstable. If the control parameters are adjusted properly, by reducing K_{ipll} and increasing K_{pdc}, as in Group 7, the system can be maintained as stable, as shown in Figure 16b.

The simulation of the complex system further verifies the proposed theoretical analysis.

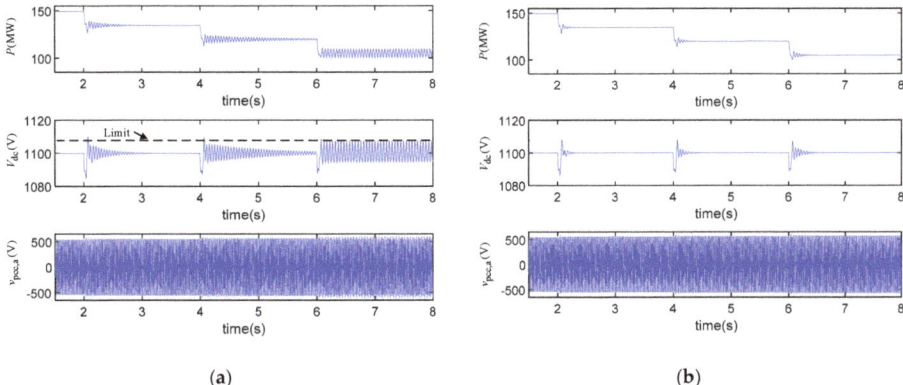

Figure 16. Simulation results of wind power decrease with a small K_{ppll}. (**a**) Using Group 3's parameters (**b**) Using Group 4's parameters after adjustment

5. Conclusions

This paper investigates the influence of active power output on subsynchronous oscillation characteristics in weak grids. Compared to the research in the literature, this is the first of its kind to investigate the distinctive correlations between active power output and the damping of the SSO mode. The reasons for the different correlation between active power output and SSO mode damping have been explained. The findings and contributions of the study include:

The change of active power output in one direction can either improve or reduce SSO mode damping. This work identifies that the correlation between active power variation and damping mainly depends on the proportional gain of the phase-locked loop (PLL).

- When the PLL proportional gain is large, the active power output is negatively correlated with the damping of the SSO mode. When the PLL proportional gain is small, the active power output is positively correlated with the damping of the SSO mode. This clarifies the confusions in the understanding of the correlation between active power output and SSO damping.
- The PLL integral gain and the DC voltage control proportional gain have little influence on the correlation between the active power output and SSO damping. However, the system stability can be improved by appropriately retuning the PLL integral gain and the DC voltage control proportional gain.
- There is a critical range for the PLL proportional gain, in which SSO damping is near consistent irrespective to the change of active power variation. The influence of active power output on the stability can be minimized by selecting proper the PLL proportional gain first when the damping variation is at the critical range. Then adjustment of other parameters will improve the stability. This is valuable for engineering applications in designing PLL parameters.

For full-converter wind power systems, the grid-connected dynamics mainly depend on the control of GSC and are not affected by the wind turbine types. The conclusions of this paper are applicable to full-converter wind farms with induction generators or permanent magnet synchronous generators. DFIG is not covered in the study, and the analysis of the DFIG-based wind farms and the auxiliary control design will be undertaken in future research.

Author Contributions: Conceptualization, Y.H., J.L. and G.W.; methodology, Y.H., J.L., K.W. and G.W.; validation, Y.H., J.L. and G.W.; formal analysis, Y.H. and J.L.; investigation, Y.H. and J.L.; writing—original draft preparation, Y.H., J.L., G.W. and R.S.; writing—review and editing, Y.H., J.L., T.J. and R.S.; supervision, J.L. and K.W.; funding acquisition, J.L. All authors have read and agreed to the published version of the manuscript.

Funding: This work is supported by National Natural Science Foundation of China under Grant 51507155 and State Key Laboratory Open Fund Project under Grant FXB51201901458.

Conflicts of Interest: The authors declare no conflict of interest.

Appendix A

The A matrix expression in Equation (8):

$$\begin{bmatrix}
\frac{K_1K_9+K_2K_{10}-R_{f1}}{L_{f1}} & \frac{K_1K_{10}-K_2K_9}{L_{f1}}-\omega_0 & 0 & 0 & \frac{K_1^2-1}{L_{f1}} & \frac{K_1K_2}{L_{f1}} & 0 & \frac{K_1K_{11}-K_2K_{12}+K_7}{L_{f1}} & \frac{2K_1K_{pdc}K_{pi}}{L_{f1}} & \frac{K_1K_{pi}}{L_{f1}} & \frac{K_1}{L_{f1}} & \frac{-K_2}{L_{f1}} \\
\frac{K_2K_9-K_1K_{10}}{L_{f1}}-\omega_0 & \frac{K_1K_9+K_2K_{10}-R_{f1}}{L_{f1}} & 0 & 0 & \frac{K_1K_2}{L_{f1}} & \frac{K_2^2-1}{L_{f1}} & 0 & \frac{K_2K_{11}+K_1K_{12}+K_8}{L_{f1}} & \frac{2K_2K_{pdc}K_{pi}}{L_{f1}} & \frac{K_2K_{pi}}{L_{f1}} & \frac{K_2}{L_{f1}} & \frac{K_1}{L_{f1}} \\
0 & 0 & \frac{-R_g}{L_g} & \omega_0 & \frac{1}{L_g} & 0 & 0 & 0 & 0 & 0 & 0 & 0 \\
0 & 0 & -\omega_0 & \frac{-R_g}{L_g} & 0 & \frac{1}{L_g} & 0 & 0 & 0 & 0 & 0 & 0 \\
\frac{1}{C_1} & 0 & \frac{-1}{C_1} & 0 & 0 & \omega_0 & 0 & 0 & 0 & 0 & 0 & 0 \\
0 & \frac{1}{C_1} & 0 & \frac{-1}{C_1} & -\omega_0 & 0 & 0 & 0 & 0 & 0 & 0 & 0 \\
0 & 0 & 0 & 0 & -K_2K_{ipll} & K_1K_{ipll} & 0 & K_6K_{ipll} & 0 & 0 & 0 & 0 \\
0 & 0 & 0 & 0 & -K_2K_{ppll} & K_1K_{ppll} & 1 & K_6K_{ppll} & 0 & 0 & 0 & 0 \\
\frac{-v_{pcc,d0}^g}{2\tau} & \frac{-v_{pcc,q0}^g}{2\tau} & 0 & 0 & \frac{-i_{1d0}^g}{2\tau} & \frac{-i_{1q0}^g}{2\tau} & 0 & 0 & 0 & 0 & 0 & 0 \\
0 & 0 & 0 & 0 & 0 & 0 & 0 & 0 & 2K_{idc} & 0 & 0 & 0 \\
-K_1K_{ii} & -K_2K_{ii} & 0 & 0 & 0 & 0 & 0 & -K_3K_{ii} & 2K_{ii}K_{pdc} & K_{ii} & 0 & 0 \\
K_2K_{ii} & -K_1K_{ii} & 0 & 0 & 0 & 0 & 0 & -K_4K_{ii} & 0 & 0 & 0 & 0
\end{bmatrix}$$

State variables:

$$x = [i^g_{1d}, i^g_{1q}, i^g_{2d}, i^g_{2q}, v^g_{pcc,d}, v^g_{pcc,q}, x_{pll}, \Delta\theta_{pll}, V_{dc}, x_1, x_2, x_3]$$

$$K_1 = \cos\theta_{pll0}, \quad K_2 = \sin\theta_{pll0}$$

$$K_3 = -\sin\theta_{pll0}i^g_{1d0} + \cos\theta_{pll0}i^g_{1q0}, \quad K_4 = -\cos\theta_{pll0}i^g_{1d0} - \sin\theta_{pll0}i^g_{1q0}$$

$$K_5 = -\sin\theta_{pll0}v^g_{pcc,d0} + \cos\theta_{pll0}v^g_{pcc,q0}, \quad K_6 = -\cos\theta_{pll0}v^g_{pcc,d0} - \sin\theta_{pll0}v^g_{pcc,q0}$$

$$K_7 = -\sin\theta_{pll0}v^c_{d0} - \cos\theta_{pll0}v^c_{q0}, \quad K_8 = \cos\theta_{pll0}v^c_{d0} - \sin\theta_{pll0}v^c_{q0}$$

$$K_9 = K_2\omega_0L_{f1} - K_1K_{pi}, \quad K_{10} = -K_1\omega_0L_{f1} - K_2K_{pi}$$

$$K_{11} = K_5 - K_4\omega_0L_{f1} - K_3K_{pi}, \quad K_{12} = K_3\omega_0L_{f1} - K_4K_{pi}$$

$$\tau = \frac{C_{dc}V_{dc}^2}{2P_{base}}$$

References

1. Global Wind Energy Council. *Global Wind Report 2019*; Global Wind Energy Council: Brussels, Belgium, 2019.
2. Leon, A.E.; Solsona, J.A. Sub-synchronous interaction damping control for DFIG wind turbines. *IEEE Trans. Power Syst.* **2015**, *30*, 419–428. [CrossRef]
3. Ren, W.; Larsen, E. A refined frequency scan approach to subsynchronous control interaction (SSCI) study of wind farms. *IEEE Trans. Power Syst.* **2016**, *31*, 3904–3912. [CrossRef]
4. Adams, J.; Pappu, V.A.; Dixit, A. ERCOT experience screening for sub-synchronous control interaction in the vicinity of series capacitor banks. In Proceedings of the 2012 IEEE Power and Energy Society General Meeting, San Diego, CA, USA, 22–26 July 2012; pp. 1–5.
5. Strachan, N.P.; Jovcic, D. Stability of a variable-speed permanent magnet wind generator with weak ac grids. *IEEE Trans. Power Del.* **2010**, *25*, 2779–2788. [CrossRef]
6. Liu, H.; Xie, X.; He, J.; Xu, T.; Yu, Z.; Wang, C.; Zhang, C. Subsynchronous interaction between direct-drive PMSG based wind farms and weak AC networks. *IEEE Trans. Power Syst.* **2017**, *32*, 4708–4720. [CrossRef]
7. Liu, H.; Xie, X.; Zhang, C.; Li, Y.; Liu, H.; Hu, Y. Quantitative SSR analysis of series-compensated DFIG-based wind farms using aggregated RLC circuit model. *IEEE Trans. Power Syst.* **2017**, *32*, 474–483. [CrossRef]

8. Wang, L.; Xie, X.; Jiang, Q.; Liu, H.; Li, Y.; Liu, H. Investigation of SSR in practical DFIG-based wind farms connected to a series compensated power system. *IEEE Trans. Power Syst.* **2015**, *30*, 2772–2779. [CrossRef]
9. Hu, J.; Yuan, H.; Yuan, X. Modeling of DFIG-based WTs for small signal stability analysis in DVC timescale in power electronized power systems. *IEEE Trans. Energy Convers.* **2017**, *32*, 1151–1165. [CrossRef]
10. Li, Y.; Fan, L.; Miao, Z. Wind in weak grids: Low-frequency oscillations, subsynchronous oscillations, and torsional interactions. *IEEE Trans. Power Syst.* **2020**, *35*, 109–118. [CrossRef]
11. Wang, D.; Liang, L.; Shi, L.; Hu, J.; Hou, Y. Analysis of modal resonance between PLL and DC-Link voltage control in weak-grid tied VSCs. *IEEE Trans. Power Syst.* **2019**, *34*, 1127–1138. [CrossRef]
12. Huang, Y.; Wang, D. Effect of control-loops interactions on power stability limits of VSC integrated to AC system. *IEEE Trans. Power Del.* **2018**, *33*, 301–310. [CrossRef]
13. Li, X.; Liang, J.; Li, G.; Joseph, T. Modeling and stability analysis of the sub-synchronous interactions in weak AC grids with wind power integration. In Proceedings of the 2018 International Universities Power Engineering Conference (UPEC), Glasgow, UK, 4–7 September 2018; pp. 1–6.
14. TAO, G.; Wang, Y.; WU, Y.; CHEN, Y. Subsynchronous interaction analysis of PMSG based wind farm with AC networks. In Proceedings of the 2019 International Conference on Electrical Machines and Systems (ICEMS), Harbin, China, 11–14 August 2019; pp. 1–5.
15. Alawasa, K.M.; Mohamed, Y.A.-R.I.; Xu, W. Modeling, analysis, and suppression of the impact of full-scale wind-power converters on subsynchronous damping. *IEEE Syst. J.* **2013**, *7*, 700–712. [CrossRef]
16. Cespedes, M.; Sun, J. Impedance Modeling and Analysis of Grid-Connected Voltage-Source Converters. *IEEE Trans. Power Electron.* **2014**, *29*, 1254–1261. [CrossRef]
17. Fan, L. Modeling type-4 wind in weak grids. *IEEE Trans. Sustain. Energy* **2019**, *10*, 853–864. [CrossRef]
18. Huang, Y.; Yuan, X.; Hu, J.; Zhou, P. Modeling of VSC connected to weak grid for stability analysis of DC-Link voltage control. *IEEE J. Emerg. Sel. Topics Power Electron.* **2015**, *3*, 1193–1204. [CrossRef]
19. Papangelis, L.; Debry, M.-S.; Prevost, T.; Panciatici, P.; Van Cutsem, T. Stability of a voltage source converter subject to decrease of short-circuit capacity: A case study. In Proceedings of the 2018 Power Systems Computation Conference (PSCC), Dublin, Ireland, 11–15 June 2018; pp. 1–7.
20. Huang, B.; Sun, H.; Liu, Y.; Wang, L.; Chen, Y. Study on subsynchronous oscillation in D-PMSGs-based wind farm integrated to power system. *IET Renew. Power Gener.* **2019**, *13*, 16–26. [CrossRef]
21. Lyu, J.; Cai, X.; Molinas, M. Optimal design of controller parameters for improving the stability of MMC-HVDC for wind farm integration. *IEEE J. Emerg. Sel. Topics Power Electron.* **2018**, *6*, 40–53. [CrossRef]
22. Jing, L.; Xu, C. Impact of controller parameters on stability of MMC-based HVDC systems for offshore wind farms. In Proceedings of the International Conference on Renewable Power Generation (RPG 2015), Beijing, China, 17–18 October 2015; pp. 1–6.
23. Pipelzadeh, Y.; Chaudhuri, N.R.; Chaudhuri, B.; Green, T.C. Coordinated Control of Offshore Wind Farm and Onshore HVDC Converter for Effective Power Oscillation Damping. *IEEE Trans. Power Syst.* **2017**, *32*, 1860–1872. [CrossRef]
24. Amin, M.; Molinas, M. Understanding the Origin of Oscillatory Phenomena Observed Between Wind Farms and HVdc Systems. *IEEE J. Emerg. Sel. Topics Power Electron.* **2017**, *5*, 378–392. [CrossRef]
25. Amin, M.; Rygg, A.; Molinas, M. Self-Synchronization of Wind Farm in an MMC-Based HVDC System: A Stability Investigation. *IEEE Trans. Energy Convers.* **2017**, *32*, 458–470. [CrossRef]
26. Fan, L. *Control and Dynamics in Power Systems and Microgrids*; CRC Press: Boca Raton, FL, USA, 2017.
27. Institute of Electrical and Electronics Engineers. *IEEE Guide for Planning DC Links Terminating at AC Locations Having Low Short-Circuit Capacities*; IEEE Standard 1204-1997; Institute of Electrical and Electronics Engineers: New York, NY, USA, 1997.
28. Joseph, T.; Ugalde-Loo, C.E.; Balasubramaniam, S.; Liang, J. Real-Time Estimation and Damping of SSR in a VSC-HVDC Connected Series-Compensated System. *IEEE Trans. Power Syst.* **2018**, *33*, 7052–7063. [CrossRef]

 © 2020 by the authors. Licensee MDPI, Basel, Switzerland. This article is an open access article distributed under the terms and conditions of the Creative Commons Attribution (CC BY) license (http://creativecommons.org/licenses/by/4.0/).

Article

Transition from Electromechanical Dynamics to Quasi-Electromechanical Dynamics Caused by Participation of Full Converter-Based Wind Power Generation

Jianqiang Luo [1,2], Siqi Bu [1,2,*] and Jiebei Zhu [3]

1. Shenzhen Research Institute, The Hong Kong Polytechnic University, Shenzhen 518057, China; jq.luo@connect.polyu.hk
2. Department of Electrical Engineering, The Hong Kong Polytechnic University, Kowloon, Hong Kong, China
3. School of Electrical Automation and Information Engineering, Tianjin University, Tianjin 300072, China; zhujiebei@hotmail.com
* Correspondence: siqi.bu@polyu.edu.hk

Received: 30 September 2020; Accepted: 23 November 2020; Published: 27 November 2020

Abstract: Previous studies generally consider that the full converter-based wind power generation (FCWG) is a "decoupled" power source from the grid, which hardly participates in electromechanical oscillations. However, it was found recently that strong interaction could be induced which might incur severe resonance incidents in the electromechanical dynamic timescale. In this paper, the participation of FCWG in electromechanical dynamics is extensively investigated, and particularly, an unusual transition of the electromechanical oscillation mode (EOM) is uncovered for the first time. The detailed mathematical models of the open-loop and closed-loop power systems are firstly established, and modal analysis is employed to quantify the FCWG participation in electromechanical dynamics, with two new mode identification criteria, i.e., FCWG dynamics correlation ratio (FDCR) and quasi-electromechanical loop correlation ratio (QELCR). On this basis, the impact of different wind penetration levels and controller parameter settings on the participation of FCWG is investigated. It is revealed that if an FCWG oscillation mode (FOM) has a similar oscillation frequency to the system EOMs, there is a high possibility to induce strong interactions between FCWG dynamics and system electromechanical dynamics of the external power systems. In this circumstance, an interesting phenomenon may occur that an EOM may be dominated by FCWG dynamics, and hence is transformed into a quasi-EOM, which actively involves the participation of FCWG quasi-electromechanical state variables.

Keywords: electromechanical dynamics; FCWG dynamics; strong interaction; electromechanical loop correlation ratio (ELCR); FCWG dynamic correlation ratio (FDCR); quasi- electromechanical loop correlation ratio (QELCR)

1. Introduction

The high penetration of renewables and power electronic domination are two important aspects of the future power system [1,2]. Converter interfaced generations (CIGs) such as wind power and photovoltaic (PV) generation have been increasingly integrated into the power system at an incredible scale and speed and play a pivotal role in rendering the power system more sustainable [3–5]. As one of the promising CIGs, full converter-based wind power generation (FCWG, e.g., permanent magnet synchronous generator (PMSG)), in which two full scale converters are employed to transfer wind power to the power system, has become prevalent in the wind market due to its concise physical structure and mature control techniques [6–9]. The ever-increasing share of wind generation and its replacement of conventional synchronous generators (SGs) involve profound challenges on the electromechanical dynamics and potential threats on the oscillatory stability of the power system [10–13]. Unlike conventional rotational power sources, the integration of FCWG may introduce complex interactions with the electromechanical dynamics, which is well worth investigating, whereas it has not been thoroughly studied.

Currently, many efforts have been endeavored as to how to utilize wind generation and employ various additional controllers to mitigate electromechanical oscillations. Despite the inertia-less characteristic under the maximum power point tracking (MPPT) control strategy, by emulating an inertia response, a double fed inductor generator (DFIG) is capable of damping electromechanical oscillations [14]. Both drivetrain torsional oscillations of a DFIG and electromechanical oscillations are further examined, and alternative dampers are designed to suppress these oscillations [15]. The potential of imposing inter-area oscillation damping control with wind power plants is studied in [16]. The effect of spatial correlation between wind speed of geographically close wind farms on the damping of electromechanical oscillations is examined in [17]. With the aid of the wide area measurement system, a wide area damping controller is designed for DIFGs to alleviate electromechanical oscillations [18]. A second-order sliding-mode based damping controller is proposed in [19,20] as a resort for inter-area oscillation mitigation. A reduced-order model based optimal oscillation damping controller is also designed in [21]. A residue-based evaluation method is implemented to provide an additional control design for the power oscillations [22]. In addition, modulation and coordination resorts such as active power modulation and reactive power modulation [23] and DC-link control [24] are also proposed to damp inter-area oscillations with wind generation.

Apart from mitigating electromechanical oscillations with wind generation, the dynamic interaction between wind generation and the electromechanical dynamics has also drawn attention and is defined as a converter-driven stability problem [3]. Model validation and reduction techniques for different types of wind power induction generators (i.e., a fixed-speed induction generator (FSIG), DFIG) are discussed in terms of oscillatory stability issues [25–27]. The dynamic interaction between wind generation and the electromechanical modes of the nearby synchronous generators (SGs) poses threats to the small signal stability with high penetration levels of wind power, which is verified with modal analysis techniques [28]. An electric torque analysis method is proposed in [29] to quantify the impact of wind generation integration on electromechanical oscillations. A novel modal superposition theory is proposed in [30] to classify the interaction categories between wind generation and the external power system. The impact of power electronic integrated wind generation considering increasing wind penetration and load conditions on the inter-area oscillation is investigated in [31]. An adaptive coordination strategy is proposed in [32] to eliminate the modal resonance between FCWG dynamics and electromechanical dynamics.

Although the above works validated the impact of wind generation on electromechanical dynamics and provided satisfactory solutions to tackle the electromechanical oscillations with various control resorts, the systematic modeling to deepen the understanding of FCWG participation in electromechanical dynamics is still worth further exploring. Especially, an interesting phenomenon is discovered that, in some circumstances, the electromechanical oscillation mode (EOM) may be dominated by the FCWG dynamics and become a quasi-EOM, which has not been studied before. Therefore, the main contributions of this paper are summarized: (1) the linearized open-loop and closed-loop power system models tailored for FCWG dynamics impact investigation are established; (2) together with the electromechanical loop correlation ratio (ELCR), the FCWG dynamics correlation ratio (FDCR) and the quasi-ELCR (QELCR) are proposed to quantify the participation of FCWG in electromechanical dynamics; (3) extensive case studies considering comprehensive wind penetration levels are thoroughly examined to uncover the unusual transition in electromechanical dynamics; and (4) useful findings and suggestions on how FCWG dynamics transform both local and inter-area modes are provided based on modal analysis and time domain simulations.

The remainder of this paper is organized as follows. Section 2 presents a typical configuration of FCWG. Section 3 proposes a method to investigate the participation of FCWG by constructing the open-loop linearized power system model and the closed-loop linearized power system model. In Section 4, the participation of FCWG is meticulously examined in a two-area test system, and important findings on the impact of the electromechanical dynamics are concluded. The main findings are summarized in Section 5, while conclusions are emphasized in Section 6.

2. Configuration of FCWG

The typical topology of an FCWG (e.g., permanent magnet synchronous generator (PMSG)) connected to the multi-machine power system is demonstrated in Figure 1.

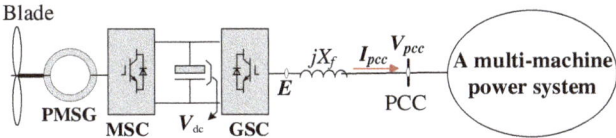

Figure 1. Physical configuration of a full converter-based wind power generation (FCWG) connected to a multi-machine power system.

The FCWG consists of three parts: (1) the PMSG, the machine side converter (MSC) and the associated control system (as demonstrated in Figure 2); (2) the DC-link, the grid side converter (GSC) and the associated control system (as shown in Figure 3); and (3) the synchronous reference frame phase-locked loop (SRF-PLL) (as presented in Figure 4), which is used to synchronize FCWG with the external power system.

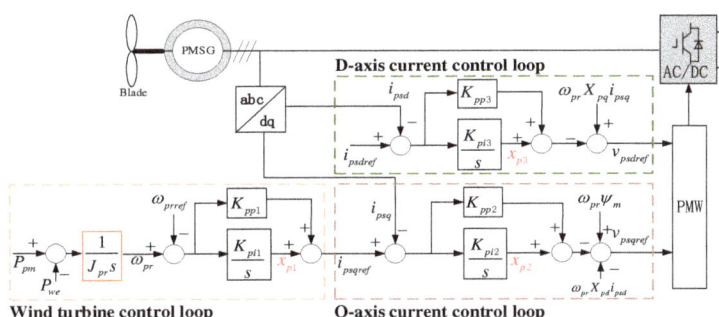

Figure 2. The control configuration of machine side converter (MSC).

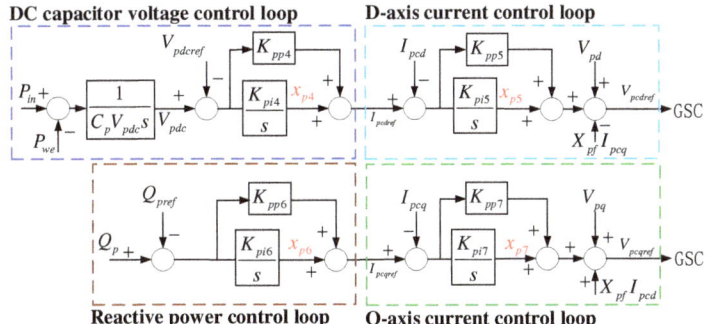

Figure 3. The control configuration of grid side converter (GSC).

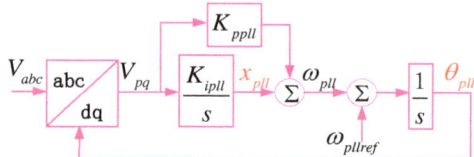

Figure 4. Block diagram of the synchronous reference frame phase-locked loop (SRF-PLL).

The electromechanical dynamics stem from the inertia sources of power systems. Regarding the large mass of physical rotors, SGs are the main inertia sources of conventional power systems, which actuate as a buffer under unintended disturbance contingencies and bolster the oscillatory stability. Owing to the AC-DC-AC configuration, the rotor inertia of wind turbine is decoupled from the multi-machine power system, and hence FCWG is normally regarded as a low-inertia source. Such low-inertia characteristic is significantly distinguished from conventional power sources. Therefore, the integration of FCWG is usually recognized to be inertia-less, which may not participate in electromechanical dynamics like SGs. Its impact on electromechanical dynamics is not taken into account meticulously.

3. Modal Analysis on Electromechanical Dynamics and FCWG Dynamics

Comprehensive modal analyses of the electromechanical oscillation modes (EOMs) are carried out to essentially reveal the participation mechanism of FCWG in electromechanical dynamics. To elaborate on the participation of FCWG, the power system that excludes the FCWG dynamics is denoted as the open-loop power system, while the entire system is the closed-loop power system. By comparing the EOMs of the open-loop and closed-loop power systems, the impact of FCWG is quantified.

3.1. State-Space Model of FCWG

The detailed modeling of FCWG can refer to [12,33]. The state-space model of FCWG is expressed as

$$\begin{cases} \frac{d}{dt}\Delta X_W = A_W \Delta X_W + B_W \Delta V_W \\ \Delta I_W = C_W \Delta X_W \end{cases} \quad (1)$$

where ΔX_W denotes all the state variables of FCWG (as illustrated in Figures 2–4); A_W, B_W, C_W are the state-space matrices after integrating all the linearized differential equations.

It is noteworthy that all the controller parameters of FCWG are included in Equation (1) and will be further integrated in the closed-loop power system in Section 3.3. Mathematically, this is how FCWG controller parameters affect the formation of the state matrix and thus influence the electromechanical dynamics of the external power system.

3.2. Open-Loop Power System

In the open-loop power system, FCWG is regarded as a constant power source, and thus its dynamics are excluded.

The state-space model of the ith SG in the power system can be expressed as

$$\begin{cases} \frac{d}{dt}\Delta X_{gi} = A_{gi}\Delta X_{gi} + B_{gi}\Delta V_{gi} \\ \Delta I_{gi} = C_{gi}\Delta X_{gi} + D_{gi}\Delta V_{gi} \end{cases} \quad (2)$$

where ΔX_{gi} is the state variables of SG i; A_{gi}, B_{gi}, C_{gi}, D_{gi} are the state-space matrices; ΔV_{gi} and ΔI_{gi} are voltage variation and current variation at the connecting bus of SG i, respectively.

The equation of the transmission network is expressed as

$$\begin{bmatrix} \Delta I_g \\ 0 \end{bmatrix} = \begin{bmatrix} Y_{gg} & Y_{gN} \\ Y_{Ng} & Y_{NN} \end{bmatrix} \begin{bmatrix} \Delta V_g \\ \Delta V_N \end{bmatrix} = Y_{open} \begin{bmatrix} \Delta V_g \\ \Delta V_N \end{bmatrix} \quad (3)$$

where ΔI_g denotes the current variation at the generator buses; ΔV_g and ΔV_N are the voltage variations at the generator buses and other buses, respectively; Y_{open} is the open-loop admittance matrix of the transmission network in which the FCWG is considered as a constant power source and modeled as a constant impedance. Assume that the total number of SGs is M, then

$$\begin{aligned}
\Delta I_g &= \begin{bmatrix} \Delta I_{g1} & \Delta I_{g2} & \cdots & \Delta I_{gM} \end{bmatrix}^T \\
\Delta V_g &= \begin{bmatrix} \Delta V_{g1} & \Delta V_{g2} & \cdots & \Delta V_{gM} \end{bmatrix}^T \\
\Delta X_g &= \begin{bmatrix} \Delta X_{g1} & \Delta X_{g2} & \cdots & \Delta X_{gM} \end{bmatrix}^T \\
A_g &= \text{diag}\begin{bmatrix} \Delta A_{g1} & \Delta A_{g2} & \cdots & \Delta A_{gM} \end{bmatrix}^T \\
B_g &= \text{diag}\begin{bmatrix} \Delta B_{g1} & \Delta B_{g2} & \cdots & \Delta B_{gM} \end{bmatrix}^T \\
C_g &= \text{diag}\begin{bmatrix} \Delta C_{g1} & \Delta C_{g2} & \cdots & \Delta C_{gM} \end{bmatrix}^T \\
D_g &= \text{diag}\begin{bmatrix} \Delta D_{g1} & \Delta D_{g2} & \cdots & \Delta D_{gM} \end{bmatrix}^T
\end{aligned} \quad (4)$$

where diag[] represents the diagonal matrix. By integrating all the SGs, the state-space model is expressed as

$$\begin{cases} \frac{d}{dt}\Delta X_g = A_g \Delta X_g + B_g \Delta V_g \\ \Delta I_g = C_g \Delta X_g + D_g \Delta V_g \end{cases} \quad (5)$$

From (3), the relationship between ΔI_g and ΔV_g can be expressed as

$$\Delta I_g = (Y_{gg} - \frac{Y_{gN} Y_{Ng}}{Y_{NN}}) \Delta V_g \quad (6)$$

Combine (5) and (6), and the state-space model of the open-loop power system is derived as

$$\frac{d}{dt}\Delta X_g = [A_g + \frac{B_g C_g}{Y_{gg} - \frac{Y_{gN} Y_{Ng}}{Y_{NN}} - D_g}] \Delta X_g = A_{open} \Delta X_g \quad (7)$$

where A_{open} is the state matrix of the open-loop power system.

3.3. Closed-Loop Power System

In the closed-loop power system, the dynamics of FCWG are included by injecting a current variation ΔI_W into the external power system. Accordingly, the network equation in Equation (3) should be modified as below

$$\begin{bmatrix} \Delta I_g \\ \Delta I_W \\ 0 \end{bmatrix} = \begin{bmatrix} Y_{gg} & Y_{gW} & Y_{gN} \\ Y_{Wg} & Y_{WW} & Y_{WN} \\ Y_{Ng} & Y_{NW} & Y_{NN} \end{bmatrix} \begin{bmatrix} \Delta V_g \\ \Delta V_W \\ \Delta V_N \end{bmatrix} = Y_{close} \begin{bmatrix} \Delta V_g \\ \Delta V_W \\ \Delta V_N \end{bmatrix} \quad (8)$$

where ΔI_g and ΔI_W are the current variations at generator buses and the FCWG bus, respectively; ΔV_g, ΔV_W and ΔV_N are the voltage variations at generator buses, FCWG bus, and other buses; and Y_{close} is the admittance matrix of the transmission network.

Eliminating the non-source buses, the network equation can be simplified as

$$\begin{bmatrix} \Delta I_g \\ \Delta I_W \end{bmatrix} = \begin{bmatrix} Y_{gg} - \frac{Y_{gN}Y_{Ng}}{Y_{NN}} & Y_{gW} - \frac{Y_{gN}Y_{NW}}{Y_{NN}} \\ Y_{Wg} - \frac{Y_{WN}Y_{Ng}}{Y_{NN}} & Y_{WW} - \frac{Y_{WN}Y_{NW}}{Y_{NN}} \end{bmatrix} \begin{bmatrix} \Delta V_g \\ \Delta V_W \end{bmatrix}$$
$$= \begin{bmatrix} Y_{11} & Y_{12} \\ Y_{21} & Y_{22} \end{bmatrix} \begin{bmatrix} \Delta V_g \\ \Delta V_W \end{bmatrix} \qquad (9)$$

From the second equation of (1), the second equation of (5), and (9), the relation between voltage variation and the state variables is expressed as

$$\begin{bmatrix} \Delta V_g \\ \Delta V_W \end{bmatrix} = \begin{bmatrix} Y_{11} - D_g & Y_{12} \\ Y_{21} & Y_{22} \end{bmatrix}^{-1} \begin{bmatrix} C_g & C_W \end{bmatrix} \begin{bmatrix} \Delta X_g \\ \Delta X_W \end{bmatrix} \qquad (10)$$

From the first equations of (1) and (5),

$$\frac{d}{dt}\begin{bmatrix} \Delta X_g \\ \Delta X_W \end{bmatrix} = \begin{bmatrix} A_g & 0 \\ 0 & A_W \end{bmatrix} \begin{bmatrix} \Delta X_g \\ \Delta X_W \end{bmatrix} + \begin{bmatrix} B_g & 0 \\ 0 & B_W \end{bmatrix} \begin{bmatrix} \Delta V_g \\ \Delta V_W \end{bmatrix} \qquad (11)$$

From (10) and (11), the closed-loop state-space model of the entire power system is derived as

$$\frac{d}{dt}\begin{bmatrix} \Delta X_g \\ \Delta X_W \end{bmatrix} = A_{closed} \begin{bmatrix} \Delta X_g \\ \Delta X_W \end{bmatrix} \qquad (12)$$

where A_{closed} is state matrix of the closed-loop power system considering the dynamics of FCWG and is defined as

$$A_{closed} = \begin{bmatrix} A_g & 0 \\ 0 & A_W \end{bmatrix} + \begin{bmatrix} B_g & 0 \\ 0 & B_W \end{bmatrix} \begin{bmatrix} Y_{11} - D_g & Y_{12} \\ Y_{21} & Y_{22} \end{bmatrix}^{-1} \begin{bmatrix} C_g & C_W \end{bmatrix} \qquad (13)$$

By performing modal analysis on the state matrices of the open-loop and closed-loop power systems and comparing all the essential information of the oscillation modes, the impact of FCWG dynamics can be revealed. One advantageous aspect of modal analysis is that it can give insight into the relationship between oscillation modes and physical components and reveal the interaction between FCWG dynamics and electromechanical dynamics. For example, by analyzing the participation factors of SGs, the local EOMs and the inter-area EOM can be identified [34,35]. Moreover, it is also easy to uncover those state variables and the corresponding controllers that are the most active in the interaction between electromechanical dynamics and FCWG dynamics via the same technique.

3.4. Impact of FCWG on Electromechanical Dynamics

The EOMs are identified with the electromechanical loop correlation ratio (ELCR), which is defined as

$$ELCR = \frac{PF_{rotor}}{PF_{total} - PF_{rotor}} \qquad (14)$$

where PF_{rotor} is the sum of participation factors (PFs) related to electromechanical oscillatory loop associated with state variables (i.e., the rotor speed ω and rotor angle δ) for all M SGs, and PF_{total} is the sum of PFs of all state variables.

For any oscillation mode, if its ELCR is larger than 1, the oscillation mode is identified to be an EOM. Similarly, the FCWG dynamic correlation ratio (FDCR) can also be proposed to distinguish FCWG oscillation modes, which is defined as

$$FDCR = \frac{PF_{FCWG}}{PF_{total} - PF_{FCWG}} \qquad (15)$$

where PF_{FCWG} is the sum of participation factors (PFs) related to all state variables of FCWG.

Since the MSC of FCWG is decoupled from the external power system, the most interactive part is the grid side converter (GSC) of FCWG, and it should be highlighted that the concept of FDCR can be extended to any other power source (e.g., voltage source converter (VSC)). Likewise, the methodology proposed in this paper can be applied to study the impact on electromechanical dynamics from any kind of converter-based power sources (such as PV, energy storage system (ESS)).

In an EOM, two state variables (i.e., the rotor speed ω and rotor angle δ) related to the SG rotor are recognized as the main contributors to electromechanical oscillatory dynamics. For FCWG, there are usually two state variables taking part in the EOM most actively when strong interaction happens. Normally, a pair of state variables, which is closely related to a controller of FCWG (e.g., PLL controller, or DC voltage controller), might participate actively in electromechanical dynamics, and thus can be regarded as quasi-electromechanical state variables. Though these state variables are not from any physical rotational storage, their participation in an EOM will inevitably affect the electromechanical dynamic responses and might incur unintended consequences if not properly tackled.

If strong interaction between FCWG and the external power system occurs, the quasi-electromechanical state variables may hold a considerable PF in an EOM, and thus ELCR in Equation (14) may fall below 1. As a result, ELCR is not suitable for identifying EOMs in such cases. To fill in this gap, a quasi-ELCR (QELCR) is proposed to account for the two quasi-electromechanical state variables and is defined as

$$QELCR = \frac{PF_{rotor} + PF_{QEWG}}{PF_{total} - PF_{rotor} - PF_{QEWG}} \quad (16)$$

where PF_{QEWG} is the sum of PFs of the two quasi-electromechanical state variables from FCWG.

For any oscillation mode with an FDCR larger than 1, it can be recognized as an FOM. By analyzing the ELCR and FDCR of the same EOM, it is possible to quantify the participation of FCWG. Normally, the dynamics of FCWG are not involved in the EOM, and thus it is straightforward to identify an EOM via ELCR. However, if FCWG dynamics are strongly coupled with the electromechanical dynamics, ELCR may be lower than 1. Hence, ELCR is no longer suitable for EOM identification. In such a situation, two possible results may emerge: 1) the electromechanical dynamics may mingle with the FCWG dynamics; an EOM may be dominated by FCWG dynamics instead of the rotors of SGs, and is no longer a typical EOM, and thus can be identified as a quasi-EOM; and 2) a new quasi-EOM may be introduced (which may also be dominated by FCWG) and imposed on the rotor swing movements of SGs. To be more specific, a very interesting phenomenon may appear, in which, with the increase of the FDCR and the decrease of the ELCR, a typical EOM will gradually turn into a quasi-EOM, and at the same time, the most interactive FOM may have an ELCR larger than 1, and can be considered as a new quasi-EOM. Such a transition from the electromechanical dynamics to the quasi electromechanical dynamics is rare but may occur if FCWG strongly interacts.

With the criteria of ELCR and FDCR, it is capable of distinguishing all the EOMs and FOMs, as presented in Figure 5. This mode identification criteria can be implemented to observe the unusual transition in electromechanical dynamics.

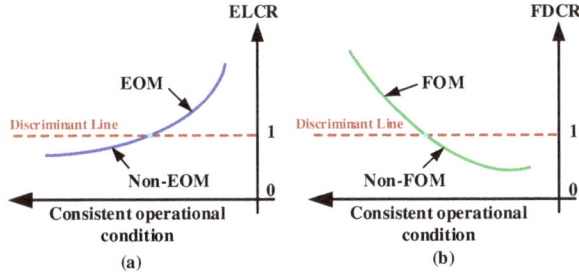

Figure 5. Criteria for electromechanical oscillation mode (EOM) and FCWG oscillation mode (FOM) identification: (**a**) EOM identification; (**b**) FOM identification.

4. Case Study

4.1. Introduction of Test System

An FCWG-integrated modified two-area power system is set up as a test system for investigation, as illustrated in Figure 6, in which the FCWG-based wind farm is connected at bus 12. Busbar 3 is the swing bus of the test system. To emulate the electromechanical dynamics, the simplified third-order model with a first order of the automatic voltage regulator (AVR) is adopted for each SG. No power system stabilizer (PSS) is equipped. All the parameters of SGs in [34] and the parameters of FCWG in [12] are used, and a detailed mathematical model can be found in [12,34].

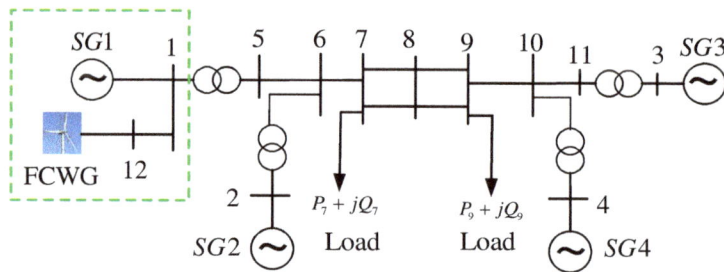

Figure 6. Configuration of two-area power system integrated with an FCWG wind farm.

To cover the participation level, the FCWG is used to replace the active power of SG1 step by step. The total active power output of FCWG and SG1 is 600 MW and kept constant. Other SGs and network and load parameters are the same throughout the following study.

The proportion of FCWG active power output increases from 0% to 100% with a change step of 2% (i.e., 12 MW). Meanwhile, the active power of SG1 decreases from 100% to 0% with the same amount of change step. To simplify the expression, "FCWG proportion" is used to represent the active power share of FCWG in the total active power of FCWG and SG1 (i.e., 600 MW). A higher FCWG proportion also indicates a higher wind power penetration level.

The impact of FCWG on all the EOMs are analyzed through modal analysis. Mathematically, the dynamic interaction between FCWG and the external power system can be seen as a modal coupling in which a major FOM interacts with the EOMs of the external power system. In other words, some state variables (usually two) of FCWG may participate in the EOMs, and state variables of rotor dynamics of SGs may also take part in the FOM that is determined by these FCWG state variables.

The original EOMs of the two-area power system (i.e., the output of FCWG is Pew = 0%) are identified and presented in Table 1.

Table 1. Electromechanical oscillation modes (EOMs) of two-area power system (Pew = 0.0%).

EOM No.	EOM1	EOM2	EOM3
Eigenvalue λ	$-0.0660 \pm 3.3891i$	$-0.3201 \pm 5.7346i$	$-0.2824 \pm 5.9767i$
Freq. (Hz)	0.5394	0.9127	0.9512
Damping Ratio	1.95%	5.57%	4.72%
Electromechanical loop correlation ratio (ELCR)	9.3952	23.7402	17.4803
FCWG dynamic correlation ratio (FDCR)	0	0	0
Major sources	SG3, SG4, SG1, SG2	SG2, SG1	SG4, SG3

The participation of power sources is also compared and demonstrated to clarify the relationship between the EOM and power sources, as shown in Figure 7. EOM1 is an inter-area oscillation mode that all SGs take part in, while EOM2 and EOM3 are two local oscillation modes that are dominated by

SG1, SG2, and SG3, SG4, respectively. Since FCWG is integrated into the left area, it is much more likely that FCWG will participate in two EOMs (i.e., EOM1 and EOM2) and will not affect EOM3.

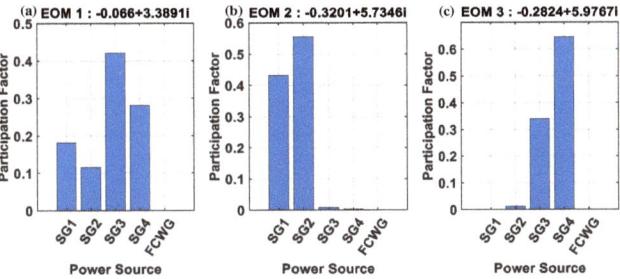

Figure 7. The participation of power sources in EOMs: (**a**) EOM1; (**b**) EOM2; (**c**) EOM3.

The participation of FCWG in the EOM is not only affected by the power injection level, but is also impacted by the parameters of the FCWG controllers. The interaction between FCWG and the external power system is strongly related to the relative locations of the FOM and the EOM. For a specific power system, EOMs normally do not vary too much and will stay at relatively stable frequencies. Meanwhile, the location of FOM is mainly determined by the controller parameters and operating conditions. The former is decisive as controller parameters can be designed with bandwidth to accommodate signals of various oscillation frequencies. While the latter also affects the FOM location with different power flows, such relation may not be decisive since it is mainly attributed to the variation of voltage and current, which are not strongly coupled in controller oscillation modes.

To give a thorough demonstration, two FOMs are selected to interact with the EOMs, i.e., the PLL-FOM which denotes the dynamics of the PLL controller, and the DC-FOM which represents the dynamics of the DC voltage controller. Different PI parameters are selected and denoted as different scenarios (under a 50% FCWG penetration level), as presented in Tables 2 and 3.

Table 2. Different scenarios with respect to PLL-FOM under 50% FCWG penetration level.

Scenario No.	Parameters of PLL Controller	PLL-FOM	EOM1	ELCR	FDCR
Scen. 1	Kipll = 6, Kppll = 1	−0.5243 ± 2.4481i	−0.1712 ± 3.4044i	9.8518	0.0260
Scen. 2	Kipll = 8, Kppll = 1	−0.5057 ± 2.8071i	−0.1851 ± 3.4137i	7.2631	0.0601
Scen. 3	Kipll = 10, Kppll = 1	−0.4627 ± 3.1275i	−0.2233 ± 3.4200i	3.6326	0.1793
Scen. 4	Kipll = 12, Kppll = 1	**−0.4412 ± 3.5208i**	**−0.2397 ± 3.3226i**	**2.3364**	**0.2905**
Scen. 5	Kipll = 14, Kppll = 1	−0.4870 ± 3.7901i	−0.1889 ± 3.3257i	4.4731	0.1148
Scen. 6	Kipll = 16, Kppll = 1	−0.4986 ± 4.0331i	−0.1724 ± 3.3359i	6.1709	0.0632
Scen. 7	Kipll = 100, Kppll = 1	−0.4364 ± 9.8689i	−0.1559 ± 3.3686i	11.7832	0.0023

Table 3. Different scenarios with respect to DC-FOM under the 50% FCWG penetration level.

Scenario No.	Parameters of DC Voltage Controller	DC-FOM	EOM1	ELCR	FDCR
Scen. 8	Kpi4 = 100, Kpp4 = 2	−0.0955 ± 1.8304i	−0.1614 ± 3.3743i	11.5318	0.0061
Scen. 9	Kpi4 = 200, Kpp4 = 2	−0.1558 ± 2.5672i	−0.1777 ± 3.3826i	8.8525	0.0340
Scen. 10	Kpi4 = 300, Kpp4 = 2	**−0.1651 ± 3.0997i**	**−0.2500 ± 3.4155i**	**2.1014**	**0.3827**
Scen. 11	Kpi4 = 400, Kpp4 = 2	−0.3863 ± 3.6959i	−0.1136 ± 3.2919i	3.0269	0.1943
Scen. 12	Kpi4 = 500, Kpp4 = 2	−0.4665 ± 4.0728i	−0.1201 ± 3.3287i	5.9994	0.0612
Scen. 13	Kpi4 = 600, Kpp4 = 2	−0.5485 ± 4.4338i	−0.1257 ± 3.3402i	7.7275	0.0316
Scen. 14	Kpi4 = 2000, Kpp4 = 2	−1.8423 ± 7.9388i	−0.1359 ± 3.3565i	10.5869	0.0053

For Scenarios 1–7, only the parameters of the PLL controller change, and all other parameters of FCWG stay at their original values. For Scenarios 8–14, only the parameters of the DC voltage controller vary, and all other parameters remain unchanged. It should be noted that the controller parameters are included in Equation (13), and the variation of these parameters will affect the state matrix A_{closed} and thus influence the eigenvalue of the EOMs. When controller parameters change, ELCR and FDCR also vary. Among all the scenarios, it is interesting that Scenario 4 and Scenario 10 have the lowest ELCR and the highest FDCR in the corresponding tables. This implies that in Scenario 4 and Scenario 10, FCWG dynamics are more active in interacting with the electromechanical dynamics. The decrease of ELCR and the increase of FDCR may lead to the transition in electromechanical dynamics. If the ELCR falls below 1, the unusual transition occurs, and thus an EOM will turn into a quasi-EOM, which will be demonstrated in Sections 4.2 and 4.3.

Comparing the FOMs in Tables 2 and 3, the oscillation frequencies of FOMs increase with the integral parameter of the controllers. When the oscillation frequencies of FOMs are close to the frequency of EOM1 (about 0.5 Hz), the participation of FCWG becomes very active. When the FOMs move away, the participation of FCWG becomes less active.

4.2. Interaction Between PLL-FOM and EOMs

Modal analyses are extensively implemented for every scenario, considering 50 operating conditions (0–100% penetration level of FCWG). For each group of controller parameters, eigenvalue analyses are implemented based on varying operating conditions and thus the eigenvalue loci of critical modes are drawn, which are significantly different from the parameter-based root locus. Therefore, the term "eigenvalue loci" is used in this paper to distinguish from "root locus".

The interactions between PLL-FOM and EOM1, EOM2 are demonstrated in Figures 8 and 9, respectively. It is worth mentioning that EOM3 is hardly influenced by the integration of FCWG since it is another local EOM which is closely related to SG3 and SG4, and thus is not presented due to space limit.

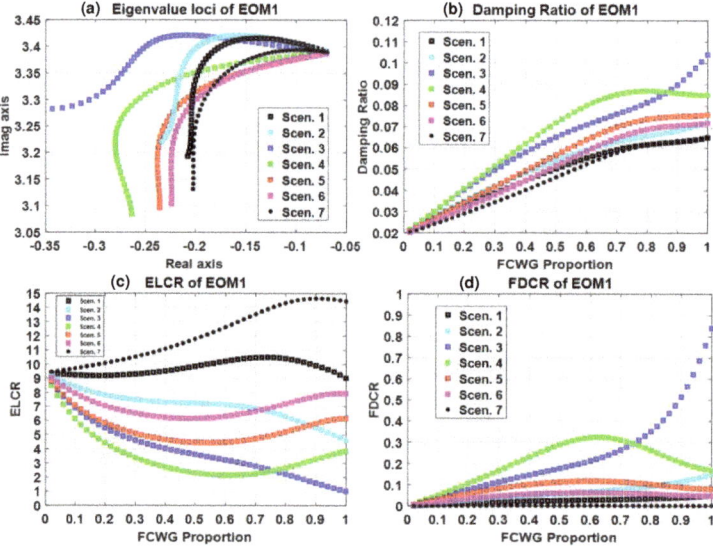

Figure 8. The interaction between PLL-FOM and EOM1 considering increasing FCWG proportion: (**a**) eigenvalue loci; (**b**) variation trend of damping ratio; (**c**) variation trend of ELCR; (**d**) variation trend of FDCR.

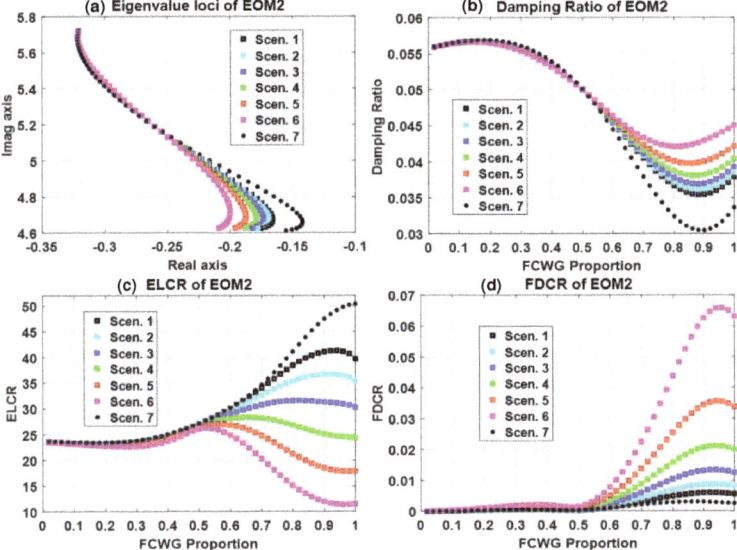

Figure 9. The interaction between PLL-FOM and EOM2 considering increasing FCWG proportion: (**a**) eigenvalue loci; (**b**) variation trend of damping ratio; (**c**) variation trend of ELCR; (**d**) variation trend of FDCR.

In Scenarios 1–7, the integration of FCWG is beneficial for EOM1, as it is shown that in Figure 8a, the eigenvalue of EOM1 turns to move towards the left in the complex plane, and Figure 8b further confirms that the damping ratio of EOM1 is enhanced. It is very interesting that, in Figure 8c, the ELCR may increase or decrease under different scenarios. Particularly, for Scenario 3, the ELCR continuously decreases to almost 1 when the FCWG proportion is near 100%, which indicates that EOM1 gradually interacts with FCWG dynamics while the electromechanical dynamics become less and less active. At the same time, the FDCR increases from 0 to almost 1. If ELCR falls below 1 and FDCR is above 1, the corresponding EOM will transform into a quasi-EOM.

Moreover, the FDCR in Figure 8d demonstrates that in most scenarios, the participation of FCWG in EOM1 stays at a low level (e.g., for Scenarios 1, 2, 5, 6, and 7, FDCR is less than 0.1), which suggests that the dynamic interaction is weak, and the damping enhancement is largely due to the power injection of FCWG. However, the FDCRs can increase drastically in Scenarios 3 and 4, and the damping ratio can also be raised a lot when FCWG participates actively. This indicates that dynamic interaction could be too strong to be ignored. It is surprising that such dynamic interaction is positive and may be utilized as a resort to enhance oscillatory stability.

From Figure 8, it is concluded that strong interaction between FCWG and the external power system is possible. In this case, the integration of FCWG is conducive for EOM1, and the damping ratio of the EOM is raised from 1.95% to over 10%, which is prominent from the perspective of low frequency oscillation suppression.

In Scenarios 1–7, the overall impact of FCWG integration is negative for EOM2, as it is shown that in Figure 9a, the eigenvalue of EOM1 tends to move towards the right in the complex plane, and Figure 9b further confirms that the damping ratio of EOM1 decreases.

The ELCRs in Figure 9c encounter some fluctuations under different scenarios, but stay above 10 all the time, which means EOM2 is always dominated by electromechanical dynamics, whereas the FDCRs in Figure 9d are always less than 0.1, which indicates the participation of FCWG dynamics is very limited or even can be ignored. This finding elucidates that the damping deterioration of EOM2 is mainly attributed to the power injection of FCWG and the power reduction of SG1.

From Figure 9, it is concluded that the FCWG dynamics may not always hold considerable participation in the EOMs, and the power injection may become the main influence. The damping ratio of EOM2 decreases from 5.57% to about 3%, which is a potential threat for this local EOM. Careful coordination for FCWG integration should be considered.

4.3. Interaction Between DC-FOM and EOMs

From the analyses in Section 4.2, the parameters of the PLL controller play an important role in the interaction between FCWG dynamics and electromechanical dynamics. With certain parameters in the PLL controller, FCWG can have very active participation in electromechanical dynamics (e.g., Scenarios 3 and 4).

To further validate the participation of FCWG, the parameters in the DC voltage controller are also examined to investigate the interaction of DC-FOM and the EOMs. Accordingly, Scenarios 8–14 are studied via modal analysis, considering 50 operating conditions (which covers from 0% to 100% penetration level of FCWG with a 2% step). The results of the interaction between DC-FOM and EOM1, EOM2 are depicted in Figure 10 and Figure 11, respectively.

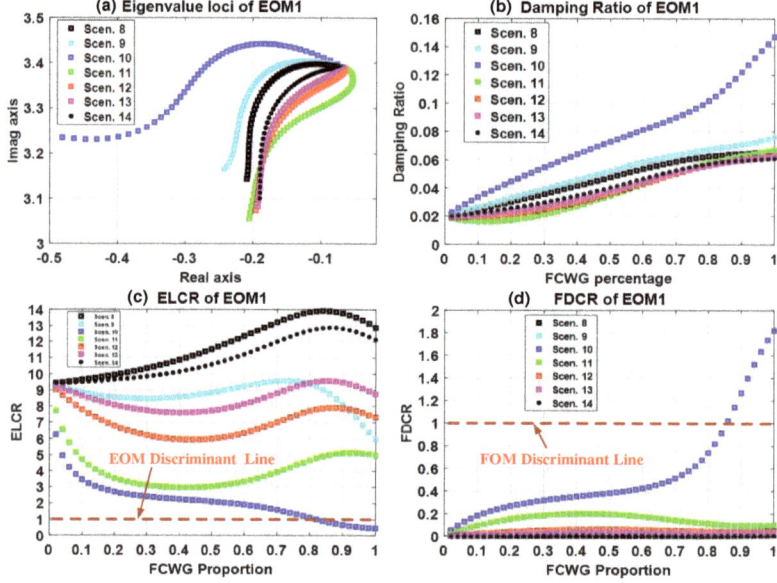

Figure 10. The interaction between DC-FOM and EOM1, considering increasing FCWG proportion: (**a**) eigenvalue loci; (**b**) variation trend of damping ratio; (**c**) variation trend of ELCR; (**d**) variation trend of FDCR.

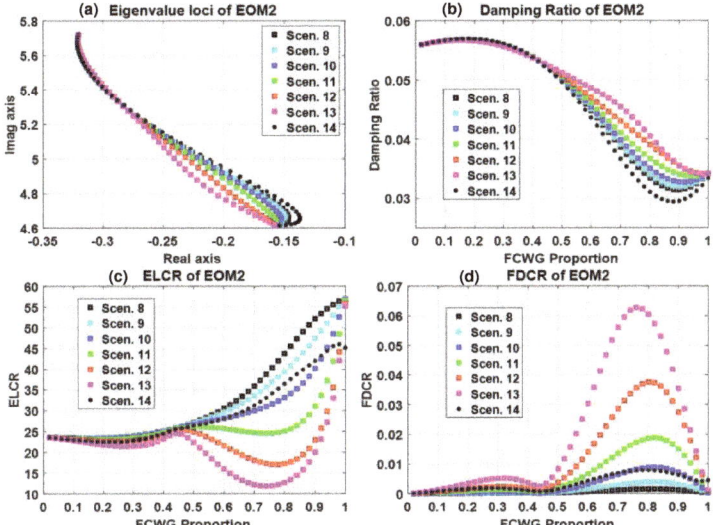

Figure 11. The interaction between DC-FOM and EOM2, considering increasing FCWG proportion: (**a**) eigenvalue loci; (**b**) variation trend of damping ratio; (**c**) variation trend of ELCR; (**d**) variation trend of FDCR.

In Scenarios 8–14, the overall impact of FCWG integration is also beneficial for EOM1, as it is shown that in Figure 10a, the eigenvalue of EOM1 tends to move towards the left in the complex plane. Figure 10b further ascertains that the damping ratio of EOM1 is enhanced.

In Figure 10c, the ELCR may also increase or decrease under different scenarios, which is similar to that of Figure 8c. However, when it comes to ELCRs and FDCRs, it should be highlighted that, for Scenario 10, the ELCR consistently decreases to below 1, which implies that EOM1 is no longer a typical EOM that is determined by the electromechanical dynamics, and the participation of FCWG becomes the primary domination. As also verified in Figure 10d, the FDCR in Scenario 10 can increase to above 1 at the 86% penetration level, which also indicates this oscillation mode (i.e., previously identified to be EOM1) now becomes a quasi-EOM.

It is worth pointing out that, although all scenarios can improve the damping ratio of EOM1, the contribution of damping enhancement due to FCWG integration are from two aspects: (1) the power flow impact (which refers to the low participation of FCWG, such as in Scenarios 8–9, 12–14); and (2) dynamic interaction impact (which could superpose dynamic impact on the electromechanical dynamics and even dominate the electromechanical oscillatory stability, such as in Scenario 10).

In Scenario 10, the integration of FCWG is conducive for EOM1, and the damping ratio of EOM1 can be raised from 1.95% to about 15%, which is quite impressive comparing with other scenarios (which can only reach about 8% damping ratio). This proves that dynamic interaction can be pronounced and should not be ignored, and the participation of FCWG is significant.

The interaction between DC-FOM and EOM2 is shown in Figure 11. The damping ratio of EOM2 decreases in all scenarios, which indicates that the integration of FCWG is negative for EOM2. Such influence is mainly attributed to the power flow impact of FCWG, since the FDCRs are at a very low level (less than 0.1 as illustrated in Figure 11d). Though the ELCRs in Figure 11c have encountered fluctuations, they stay at a very high level (over 10), and thus the electromechanical dynamics of EOM2 are hardly affected by FCWG dynamics.

Peculiarly, take Scenario 10 as an example, major modes related to electromechanical dynamics in both an open-loop power system and a closed-loop power system model are demonstrated in Table 4 (the FCWG proportion is 86%). Due to the strong interactions between FCWG dynamics and electromechanical dynamics, there are only two typical EOMs (i.e., EOM2 and EOM3) left in the closed-loop power system. The inter-area EOM 1 (i.e., 0.51 Hz) is now dominated by FCWG with an ELCR less than 1 and an FDCR larger than 1, and thus is a quasi-EOM. Local EOM2 is slightly affected while local EOM3 is hardly moved by comparing them with the closed-loop modes 3 and 4. The participation of power sources in four major oscillation modes is depicted in Figure 12. It is worthwhile mentioning that the active participation of FCWG not only dominates the inter-area mode (viz. Closed Mode 2) but also introduces a new quasi-EOM (i.e., Closed Mode 1), in which the electromechanical dynamics are involved.

Table 4. EOMs of two-area power system in Scenario 10 (Pew = 86%).

Mode No.	Eigenvalue λ	Freq. (Hz)	Damping Ratio	ELCR	FDCR
Open-Loop Power System					
Open EOM1	$-0.1994 \pm 3.2280i$	0.5138	6.17%	17.4809	0
Open EOM2	$-0.1677 \pm 4.7219i$	0.7515	3.55%	47.7340	0
Open EOM3	$-0.2839 \pm 5.9764i$	0.9512	4.75%	17.9009	0
DC-FOM	$-0.5333 \pm 3.1585i$	0.5027	16.65%	0	6.3587
Closed-Loop Power System					
Closed Mode 1	$-0.2777 \pm 3.0665i$	0.4880	9.02%	**0.8911**	**0.8601**
Closed Mode 2	$-0.3697 \pm 3.2492i$	0.5171	11.31%	**0.8194**	**1.0237**
Closed Mode 3	$-0.1563 \pm 4.6983i$	0.7478	3.33%	36.6454	0.0084
Closed Mode 4	$-0.2840 \pm 5.9762i$	0.9511	4.75%	17.8584	0

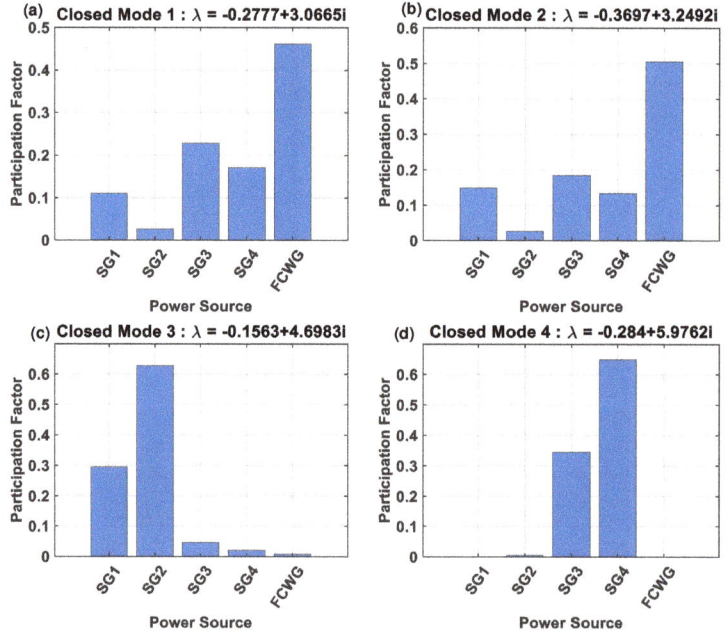

Figure 12. The participation of power sources in major oscillation modes of the closed-loop power system: (**a**) closed-loop Mode 1; (**b**) closed-loop Mode 2; (**c**) closed-loop Mode 3; (**d**) closed-loop Mode 4.

4.4. Time Domain Simulations for Verification of Frequency Domain Analysis

From the analysis above, in Scenario 4 and Scenario 10, FCWG has the most active participation. To verify the above analyses, time domain simulations are also performed. The simulation condition is set: a 5% increase of mechanical output occurs at SG2 at 0.2 s and then drops to the original value after 100 ms. The FCWG penetration level is set to be 50% (i.e., 300 MW), and all the parameters of the transmission network and generators are the same. To save space and maintain clarity, only Scenarios 4, 7, and 10 are selected to implement the small disturbance simulations.

The angular speed, bus voltage, and active power of SG3 are compared in Figure 13a–c. The reason why variables of SG3 are chosen for comparison is that, in time domain simulations, the variation of SG3 variables are formulated with the superposition of both local mode (EOM3) and inter-area mode (EOM1). The participation of FCWG affects both EOM1 and EOM2, whereas FCWG integration benefits EOM1 and deteriorates EOM2 at the same time, and these two EOMs will impose on the dynamic performances of SG1 and SG2 and may lead to misunderstanding. Therefore, by comparing variables of SG3 under different scenarios, the impact on EOM1 from the participation of FCWG can be clearly demonstrated.

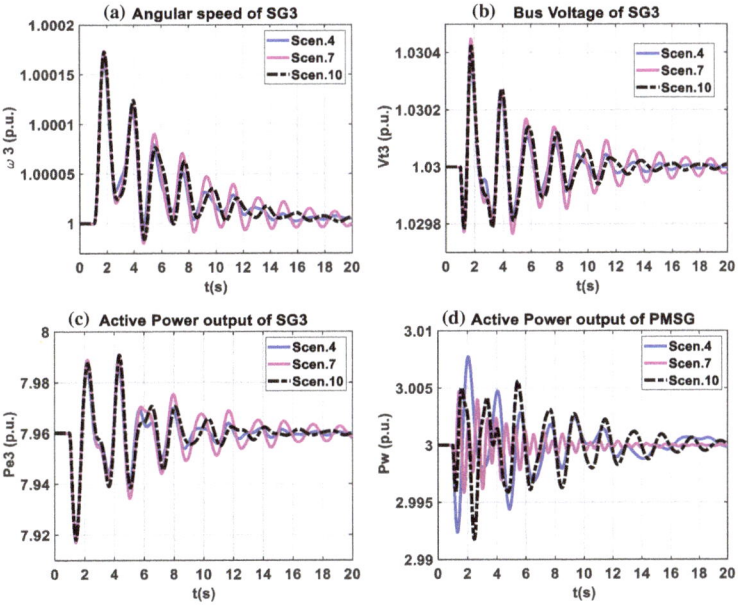

Figure 13. The dynamic responses of interaction between FCWG dynamics and electromechanical dynamics: (**a**) angular speed of SG3; (**b**) bus voltage of SG3; (**c**) active power of SG3; (**d**) active power of FCWG.

Scenarios 4 and 10 have better dynamic performances than that of Scenario 7 in terms of electromechanical dynamics. The only difference is whether FCWG actively participates or not. It is important to also mention that the participation of FCWG in electromechanical dynamics may introduce negative effects on its own dynamics. For example, the active powers of FCWG in Scenarios 4 and 10 have worse dynamic performances than that of Scenario 7, as demonstrated in Figure 13d. Therefore, the integration of FCWG may not only participate in the electromechanical dynamics and influence the oscillatory stability of the power system; additionally the side effects of its own dynamic performances should also be carefully considered. Appropriate coordination between FCWG dynamics and electromechanical dynamics is suggested when integrating FCWG into the power system.

4.5. Discussion on a Special Case: Replacement of SG with FCWG

The replacement of an SG with FCWG significantly affects the electromechanical dynamics. On one hand, with the removal of an SG from the grid, the rotor swing dynamics of this SG are now excluded from the inter-area EOM. On the other hand, a local EOM closely related to this SG may also disappear (e.g., EOM 2 of the two-area benchmark system in this paper). In a weak interaction case, FCWG hardly interacts with electromechanical dynamics, and hence the replacement of SG1 with FCWG will lead to the disappearance of the local EOM between SG1 and SG2, as confirmed in Figure 14. There are only two EOMs left in the power system, i.e., a local EOM associated with SG3 and SG4, and the inter-area EOM in which the remaining three SGs participate.

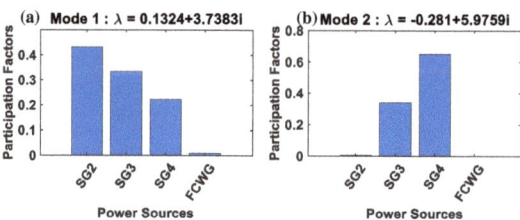

Figure 14. Weak interaction case: participation of power sources in EOMs when SG1 is replaced with FCWG: (**a**) closed-loop Mode 1; (**b**) closed-loop Mode 2.

However, if strong interaction between FCWG and the external power system occurs, two quasi-electromechanical state variables of FCWG may act as the electromechanical oscillatory loop associated state variables of the replaced SG, and hence introduce a new local quasi-EOM. In such circumstances, the integration of FCWG becomes vital in determining power system oscillatory stability. As demonstrated in Figure 15, the local EOM (Mode 4) and the inter-area EOM (Mode 3) still exist. FCWG could have a significant participation in the inter-area EOM (Mode 3) due to the active interaction between FCWG and electromechanical dynamics, which may pose threats to the power system oscillatory stability if not properly tackled. Moreover, two quasi-EOM are introduced (Mode 1 and Mode 2). Mode 1 can be regarded as a local quasi-EOM, since it is mainly dominated by SG2 and FCWG. Meanwhile, Mode 2 is largely a FOM, whereas all 3 SGs participate actively, and hence can also be recognized as an inter-area quasi-EOM.

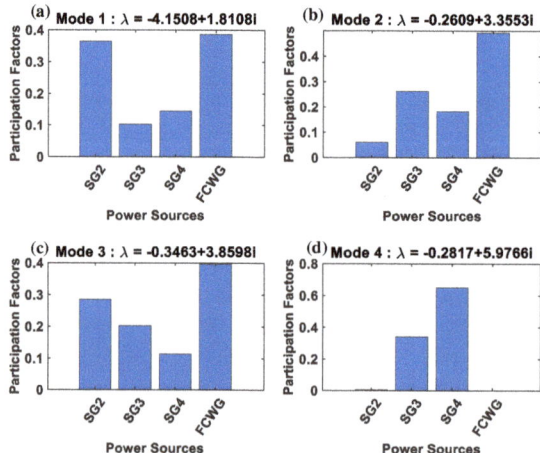

Figure 15. Strong interaction case: participation of power sources in EOMs when SG1 is replaced with FCWG: (**a**) closed-loop Mode 1; (**b**) closed-loop Mode 2; (**c**) closed-loop Mode 3; (**d**) closed-loop Mode 4.

Above all, the replacement of SG with FCWG should be carefully investigated. In the weak interaction case, it normally leads to the missing of a local EOM, while in the strong interaction case, new quasi-EOMs may be introduced. Such impact on the electromechanical dynamics may be critical for oscillatory stability and hence should be carefully tackled.

5. Discussion

Based on all the analyses above, some key findings with respect to the FCWG participation in electromechanical dynamics are summarized as below:

(1) FCWG dynamics might interact with both inter-area EOMs and local EOMs;
(2) The interaction can be either positive or negative, and may improve one EOM while deteriorate the other;
(3) Different FOMs with respect to different FCWG controllers may interact with the EOMs;
(4) For the same FCWG controller, the integral parameter plays a key role in determining the oscillation frequency of the relevant FOM and thus affects the participation of FCWG in the electromechanical dynamics;
(5) The degree of interaction is normally influenced by the penetration level of FCWG and the distance between the two affected modes. A strong interaction is more likely to occur at a high penetration level of FCWG, with the frequency of the FOM within the oscillation frequency range of the EOM (i.e., 0.2Hz–2.5Hz), especially when an FOM is close to an EOM;
(6) When a strong interaction occurs, if the FDCR of the EOM increases above 1, the ELCR of the EOM would drop below 1, which indicates that this EOM is no longer dominated by electromechanical dynamics, and thus is transformed into a quasi-EOM. The participation of FCWG may not only affect the system electromechanical dynamics, but also influence the FCWG dynamics; thus proper coordination of dynamic interaction is needed to avoid the negative effects;
(7) In the case of strong interaction, the integration of FCWG introduces a new quasi-EOM which relates to both system electromechanical dynamics and FCWG dynamics;
(8) The replacement of an SG with the FCWG significantly affects the system electromechanical dynamics. A local EOM may disappear in weak interaction cases, while new local quasi-EOMs may be introduced in strong interaction cases.

6. Conclusions

Due to the decoupling nature of FCWG, its dynamics can be normally neglected when studying the system electromechanical dynamics, whereas the exceptional case is a strong interaction between wind power generation and the grid. In this paper, we extensively investigated the participation of FCWG in the electromechanical dynamics of the power system and how it transforms the characteristics of the system's EOM. By using the mode identification criteria, the participation of FCWG in system electromechanical dynamics is quantified. It is found that in most scenarios when the FOMs have an oscillation frequency far from that of the EOMs, the participation of FCWG is quite limited, and the main impact of FCWG on the EOM is via the power flow injection. However, when an FOM has a similar frequency to that of an EOM, the participation of FCWG may become significantly active, or even dominate the EOM. In this condition, a transition from the traditional electromechanical dynamics to quasi-electromechanical dynamics was observed with the assistance of the proposed FDCR and QELCR.

Author Contributions: Completed the methodology, analysis, simulation, validation, and wrote the paper, J.L., supervised the research throughout, acquired funding, coordinated the project, and revised and edited the paper. S.B., reviewed and edited the paper, J.Z. All authors have read and agreed to the published version of the manuscript.

Funding: This research was funded by National Natural Science Foundation of China grant number (51807171), Guangdong Science and Technology Department grant number (2019A1515011226), Hong Kong Research Grant Council grant number (25203917), (15200418) and (15219619), and Department of Electrical Engineering,

The Hong Kong Polytechnic University grant number (1-ZE68). The APC was funded by TPS Scheme of Hong Kong Polytechnic University.

Acknowledgments: The authors would like to acknowledge the support from the National Natural Science Foundation of China for the Research Project (51807171), the Guangdong Science and Technology Department for the Research Project (2019A1515011226), the Hong Kong Research Grant Council for the Research Projects (25203917), (15200418) and (15219619), the Department of Electrical Engineering, and the Hong Kong Polytechnic University for the Start-up Fund Research Project (1-ZE68).

Conflicts of Interest: The authors declare no conflict of interest.

References

1. Milano, F.; Dörfler, F.; Hug, G.; Hill, D.J.; Verbič, G. Foundations and Challenges of Low-Inertia Systems. In Proceedings of the 2018 Power Systems Computation Conference (PSCC), Dublin, Ireland, 11–15 June 2018; pp. 1–25.
2. Sun, J.; Li, M.; Zhang, Z.; Xu, T.; He, J.; Wang, H.; Li, G. Renewable Energy Transmission by HVDC across the Continent: System Challenges and Opportunities. *CSEE J. Power Energy Syst.* **2017**, *3*, 353–364. [CrossRef]
3. Hatziargyriou, N.; Milanović, J.; Rahmann, C.; Ajjarapu, V.; Cañizares, C.; Erlich, I.; Hill, D.; Hiskens, I.; Kamwa, I.; Pal, B. *Stability Definitions and Characterization of Dynamic Behavior in Systems with High Penetration of Power Electronic Interfaced Technologies*; IEEE PES Technical Report PES-TR77; IEEE Power and Energy Society (PES) Resource Center: Atlanta, GA, USA, 2020.
4. Li, Y.; Gu, Y.; Zhu, Y.; Ferre, A.J.; Xiang, X.; Green, T.C. Impedance Circuit Model of Grid-Forming Inverter: Visualizing Control Algorithms as Circuit Elements. *IEEE Trans. Power Electron.* **2020**, *36*, 3377–3395. [CrossRef]
5. Gu, Y.; Li, Y.; Zhu, Y.; Green, T. Impedance-Based Whole-System Modeling for a Composite Grid via Embedding of Frame Dynamics. *IEEE Trans. Power Syst.* **2020**. [CrossRef]
6. Gautam, D.; Goel, L.; Ayyanar, R.; Vittal, V.; Harbour, T. Control Strategy to Mitigate the Impact of Reduced Inertia Due to Doubly Fed Induction Generators on Large Power Systems. *IEEE Trans. Power Syst.* **2011**, *26*, 214–224. [CrossRef]
7. Hedayati-Mehdiabadi, M.; Zhang, J.; Hedman, K.W. Wind Power Dispatch Margin for Flexible Energy and Reserve Scheduling with Increased Wind Generation. *IEEE Trans. Sustain. Energy* **2015**, *6*, 1543–1552. [CrossRef]
8. Garmroodi, M.; Hill, D.J.; Verbic, G.; Ma, J. Impact of Tie-Line Power on Inter-Area Modes with Increased Penetration of Wind Power. *IEEE Trans. Power Syst.* **2016**, *31*, 3051–3059. [CrossRef]
9. Bialasiewicz, J.T.; Muljadi, E. The wind farm aggregation impact on power quality. In Proceedings of the IECON 2006-32nd Annual Conference on IEEE Industrial Electronics, Paris, France, 6–10 November 2006; pp. 4195–4200.
10. Knüppel, T.; Nielsen, J.N.; Jensen, K.H.; Dixon, A.; Østergaard, J. Power Oscillation Damping Capabilities of Wind Power Plant with Full Converter Wind Turbines Considering Its Distributed and Modular Characteristics. *IET Renew. Power Gener.* **2013**, *7*, 431–442. [CrossRef]
11. Knüppel, T.; Nielsen, J.N.; Jensen, K.H.; Dixon, A.; Ostergaard, J. Small-signal stability of wind power system with full-load converter interfaced wind turbines. *IET Renew. Power Gener.* **2012**, *6*, 79–91. [CrossRef]
12. Luo, J.; Bu, S.; Zhu, J.; Chung, C.Y. Modal Shift Evaluation and Optimization for Resonance Mechanism Investigation and Mitigation of Power Systems Integrated with FCWG. *IEEE Trans. Power Syst.* **2020**, *35*, 4046–4055. [CrossRef]
13. Du, W.; Chen, X.; Wang, H. PLL-Induced Modal Resonance of Grid-Connected PMSGs with the Power System Electromechanical Oscillation Modes. *IEEE Trans. Sustain. Energy* **2017**, *8*, 1581–1591. [CrossRef]
14. Ying, J.; Yuan, X.; Hu, J.; He, W. Impact of Inertia Control of DFIG-Based WT on Electromechanical Oscillation Damping of SG. *IEEE Trans. Power Syst.* **2018**, *33*, 3450–3459. [CrossRef]
15. Sun, L.; Liu, K.; Hu, J.; Hou, Y. Analysis and Mitigation of Electromechanical Oscillations for DFIG Wind Turbines Involved in Fast Frequency Response. *IEEE Trans. Power Syst.* **2019**, *34*, 4547–4556. [CrossRef]
16. Singh, M.; Allen, A.J.; Muljadi, E.; Gevorgian, V.; Zhang, Y.; Santoso, S. Interarea Oscillation Damping Controls for Wind Power Plants. *IEEE Trans. Sustain. Energy* **2015**, *6*, 967–975. [CrossRef]
17. Silva-Saravia, H.; Pulgar-Painemal, H. Effect of Wind Farm Spatial Correlation on Oscillation Damping in the WECC System. In Proceedings of the 2019 North American Power Symposium (NAPS), Wichita, KS, USA, 13–15 October 2019; pp. 1–6.

18. Mokhtari, M.; Aminifar, F. Toward Wide-Area Oscillation Control through Doubly-Fed Induction Generator Wind Farms. *IEEE Trans. Power Syst.* **2014**, *29*, 2985–2992. [CrossRef]
19. Liao, K.; He, Z.; Xu, Y.; Chen, G.; Dong, Z.Y.; Wong, K.P. A Sliding Mode Based Damping Control of DFIG for Interarea Power Oscillations. *IEEE Trans. Sustain. Energy* **2017**, *8*, 258–267. [CrossRef]
20. Boubzizi, S.; Abid, H.; Chaabane, M. Comparative study of three types of controllers for DFIG in wind energy conversion system. *Prot. Control Mod. Power Syst.* **2018**, *3*, 21. [CrossRef]
21. Gurung, N.; Bhattarai, R.; Kamalasadan, S. Optimal Oscillation Damping Controller Design for Large-Scale Wind Integrated Power Grid. *IEEE Trans. Ind. Appl.* **2020**, *56*, 4225–4235. [CrossRef]
22. Morató, J.; Knüppel, T.; Østergaard, J. Residue-Based Evaluation of the Use of Wind Power Plants with Full Converter Wind Turbines for Power Oscillation Damping Control. *IEEE Trans. Sustain. Energy* **2014**, *5*, 82–89. [CrossRef]
23. Fan, L.; Yin, H.; Miao, Z. On Active/Reactive Power Modulation of DFIG-Based Wind Generation for Interarea Oscillation Damping. *IEEE Trans. Energy Convers.* **2011**, *26*, 513–521. [CrossRef]
24. Elhaji, E.M.; Hatziadoniu, C.J. Damping tie line oscillation using permanent magnet wind generators in the Libyan power system. In Proceedings of the 2014 North American Power Symposium (NAPS), Pullman, WA, USA, 7–9 September 2014; pp. 1–6.
25. Bu, S.; Du, W.; Wang, H. Model validation of DFIGs for power system oscillation stability analysis. *IET Renew. Power Gener.* **2017**, *11*, 858–866. [CrossRef]
26. Bu, S.Q.; Zhang, X.; Zhu, J.B.; Liu, X. Comparison analysis on damping mechanisms of power systems with induction generator based wind power generation. *Int. J. Electr. Power Energy Syst.* **2018**, *97*, 250–261. [CrossRef]
27. Xia, S.W.; Bu, S.Q.; Zhang, X.; Xu, Y.; Zhou, B.; Zhu, J.B. Model reduction strategy of doubly-fed induction generator-based wind farms for power system small-signal rotor angle stability analysis. *Appl. Energy* **2018**, *222*, 608–620. [CrossRef]
28. Jafarian, M.; Ranjbar, A.M. Interaction of the dynamics of doubly fed wind generators with power system electromechanical oscillations. *IET Renew. Power Gener.* **2013**, *7*, 89–97. [CrossRef]
29. Ge, Y.; Cai, H.; Cao, J.; Wang, H.F. Impact of large scale wind power penetration on power system oscillations based on electric torque analysis. In Proceedings of the International Conference on Sustainable Power Generation and Supply, Hangzhou, China, 8–9 September 2012; pp. 1–7.
30. Luo, J.; Bu, S.; Teng, F. An Optimal Modal Coordination Strategy based on Modal Superposition Theory to Mitigate Low Frequency Oscillation in FCWG Penetrated Power Systems. *Int. J. Electr. Power Energy Syst.* **2020**, *120*, 105975. [CrossRef]
31. Wilches-Bernal, F.; Chow, J.H.; Sanchez-Gasca, J.J. Impact of wind generation power electronic interface on power system inter-area oscillations. In Proceedings of the IEEE Power and Energy Society General Meeting (PESGM), Boston, MA, USA, 17–21 July 2016; pp. 1–5.
32. Luo, J.; Bu, S.; Zhu, J. A Novel PMU-based Adaptive Coordination Strategy to Mitigate Modal Resonance between Full Converter-based Wind Generation and Grids. *IEEE J. Emerg. Sel. Top. Power Electron.* **2020**. [CrossRef]
33. Du, W.; Bi, J.; Wang, H. Damping Degradation of Power System Low-Frequency Electromechanical Oscillations Caused by Open-Loop Modal Resonance. *IEEE Trans. Power Syst.* **2018**, *33*, 5072–5081. [CrossRef]
34. Kundur, P.; Balu, N.J.; Lauby, M.G. *Power System Stability and Control*; McGraw-Hill: New York, NY, USA, 1994; Volume 7.
35. Wang, X.-F.; Song, Y.; Irving, M. *Modern Power Systems Analysis*; Springer Science & Business Media: New York, NY, USA, 2010.

Publisher's Note: MDPI stays neutral with regard to jurisdictional claims in published maps and institutional affiliations.

© 2020 by the authors. Licensee MDPI, Basel, Switzerland. This article is an open access article distributed under the terms and conditions of the Creative Commons Attribution (CC BY) license (http://creativecommons.org/licenses/by/4.0/).

Article

A Novel Deep Learning Approach for Wind Power Forecasting Based on WD-LSTM Model

Bingchun Liu [1,*], Shijie Zhao [1], Xiaogang Yu [1], Lei Zhang [1] and Qingshan Wang [2]

1. Research Institute of Circular Economy, Tianjin University of Technology, Tianjin 300384, China; tjutzsj@163.com (S.Z.); 15935142263@163.com (X.Y.); tjutz_l@163.com (L.Z.)
2. School of Humanities, Tianjin Agricultural University, Tianjin 300384, China; tigermountain@yeah.net
* Correspondence: tjutlbc@tjut.edu.cn

Received: 9 August 2020; Accepted: 18 September 2020; Published: 22 September 2020

Abstract: Wind power generation is one of the renewable energy generation methods which maintains good momentum of development at present. However, its extremely intense intermittences and uncertainties bring great challenges to wind power integration and the stable operation of wind power grids. To achieve accurate prediction of wind power generation in China, a hybrid prediction model based on the combination of Wavelet Decomposition (WD) and Long Short-Term Memory neural network (LSTM) is constructed. Firstly, the nonstationary time series is decomposed into multidimensional components by WD, which can effectively reduce the volatility of the original time series and make them more stable and predictable. Then, the components of the original time series after WD are used as input variables of LSTM to predict the national wind power generation. Forty points were used, 80% as training samples and 20% as testing samples. The experimental results show that the MAPE of WD-LSTM is 5.831, performing better than other models in predicting wind power generation in China. In addition, the WD-LSTM model was used to predict the wind power generation in China under different development trends in the next two years.

Keywords: wind power generation; hybrid prediction model; wavelet decomposition; long short-term memory; scenario analysis

---

1. Introduction

Environmental pollution and serious shortage of energy have become the most pressing problems in the world today. With the increasing environmental pollution and the depletion of fossil energy, there is a strong demand for renewable energy generation [1]. Wind power generation is one of the main renewable energy generation methods, showing a good momentum of continuous growth. The Global Wind Energy Council (GWEC) emphasized in its 14th Global Wind Power Development Report that the value wind energy, as a new form of energy, brings to power systems and markets will contribute to the wind power integration and balance between supply and demand. Wind power generation can not only effectively relieve the pressure of energy crisis but as a kind of clean energy can also greatly reduce environmental pollution [2]. Wind power generation prediction is an effective measure to improve the acceptance capacity of wind power and ensure the stable operation of power grid. A high-precision wind power generation prediction model directly affects power quality, power grid stability and the balance between power grid processing load and power generation, which is of great practical significance for power grid security, stability and efficient operation [3]. Wind power generation is affected by wind speed fluctuation on three time scales: ultra-short-term fluctuation (a few minutes) influences the control of wind turbine to a certain extent, medium-term fluctuation (from a few hours to a few days) has a certain impact on wind power grid connection and power grid dispatch and long-term fluctuations (weeks or months) are related to maintenance plans for wind farms and power grids. Accurate long-term generation of wind power prediction is of great significance for improving power grid planning, optimizing power dispatching, management development and enhancement of power consumption. High-precision wind power generation prediction is also a key factor as well as an effective way to realize power mutual assistance and power generation complementary dispatching in the field of renewable energy [4].

To avoid the huge risks brought by randomness to wind power integration and improve the efficiency and safety of power grid operation, many scholars have conducted extensive and intensive studies on short-, medium- and long-term prediction of wind power generation. The research methods used in the wind power prediction are mainly divided into physical prediction method [5], traditional statistical prediction method [6], artificial intelligence prediction method [7,8] and mixed prediction method [9,10]. Physical prediction methods are based on digital weather prediction (NWP, numerical weather prediction), which uses many data from weather and environmental factors to calculate and predict wind power generation [11]. Because the models are relatively complex and have a large amount of calculation, they are generally used for medium- and long-term forecasting of wind power, showing lower prediction accuracy. Based on the historical data of wind power generation to predict the future power generation, the statistical prediction methods are relatively simple [12]. When there are obstacles to obtaining a large amount of data, the statistical method is suitable for prediction. Because the correlation of time series is fully considered, the accuracy of short-term prediction is improved. Due to the intermittent and fluctuating nature of wind power data, it exhibits extremely strong nonlinear characteristics. The introduction of artificial intelligence algorithms such as BP neural network and Recurrent Neural Network (RNN) more accurately fits the nonlinear relationship [13–15]. The time series variation of wind power signal is extremely complex. Although the single model has made a breakthrough in the prediction accuracy, it still cannot reach a satisfactory height. On this basis, some scholars use the method of preprocessing the wind power time series to reduce the impact of the non-stationarity of the original time series on the prediction accuracy. Wind power data are preprocessed by filtering, decomposition and other methods, and then the processed time series is input to the prediction model to obtain a more accurate prediction result [16].

Therefore, by eliminating the volatility of wind power data through preprocessing and speeding up the model convergence, the prediction accuracy of the model can be effectively improved. In this paper, macroeconomic indicators and related renewable energy generation are selected as input indicators of the prediction model, while wind power generation in China is taken as output indicators, which greatly reduces the randomness and uncertainty of input data and overcomes the limitation in previous studies of single wind farms being used as prediction objects. Besides, Wavelet decomposition is used to further reduce the volatility of input data and reflect data characteristics more clearly in data preprocessing. Finally, to avoid the problems of gradient disappearance and gradient explosion caused by the increase of time series length in the training process, the decomposed data are taken as the input data of the LSTM model, and the long-term correlation between the input samples and the output variables was fully learned, which improves the accuracy of the prediction model to some extent.

The contribution of this research consists mainly of two aspects: (1) establishing a national wind power generation forecasting model based on normalization and WD-LSTM; and (2) taking national macroeconomic indicators (gross domestic product, consumer price index, industrial added value and total imports and exports) and related renewable energy power generation (total power generation and hydropower generation) as input indicators, while the dimensionless data are realized through normalization and the data dimension is optimized by wavelet decomposition, which improves the convergence speed of the model and the prediction accuracy of the combined model.

2. Related Works

2.1. Data Preprocessing Models

Previous research on wind power prediction mainly took wind speed, wind direction and humidity as input variables of the model and preprocessed wind power signals by Empirical Mode Decomposition (EMD) [17], Ensemble Empirical Mode Decomposition (EEMD) [18], Complete Ensemble Empirical Mode Decomposition (CEEMD), Variational Mode Decomposition (VMD) [19] etc., which can more clearly reflect the characteristics of wind power signals. EMD was proposed by NordneE. Huang et al. to decompose signals into characteristic modes, which has the advantage that it does not use any defined function as a basis, but adaptively generates a natural mode based on the analyzed signal state function. With high signal-to-noise ratio and good time–frequency focus, it can be used to analyze nonlinear and non-stationary signal sequences. In the research of Jyotirmayee Naik et al., EMD was used as a data preprocessing method in short-term wind speed and wind power prediction. The original nonlinear non-stationary wind speed and wind time series data were decomposed by EMD. The accuracy of the proposed EMD-KRR and EMD-RVFL prediction models has been confirmed in experiments [20]. However, the Intrinsic Mode Function (IMF) after EMD will cause modal aliasing, while EEMD uses Noise-Assisted Signal Processing (NASP) to solve this problem effectively. As a preprocessing method for wind power prediction, the hybrid prediction model can improve the performance and prediction accuracy, and show good results in wind power signal processing [21]. As a preprocessing method for wind power time series in wind power prediction, the performance of the hybrid prediction model is improved, the prediction accuracy is improved and it shows good results in wind power signal processing. CEEMD has been further improved on the basis of EEMD, which makes up for the problem of EEMD's unclean noise removal in wind signal processing. To reduce the non-stationarity of the wind power time series, Wang et al. used CEEMD to decompose the wind power signal. The decomposed time series, as the input variables of the prediction model, can effectively improve the accuracy of short-term wind power prediction [22]. VMD is a completely non-recursive signal decomposition method based on the frequency domain, which to some extent overcomes many shortcomings of EMD. Li et al. used VMD to decompose wind power data into long-term modes, wave modes and random modes, which is more conducive for the prediction model to better understand the characteristics of the three constituent modes [23]. With the improvement of wind power prediction on the stability of sample data, data preprocessing has been improved on the original method.

2.2. Prediction Models

Through comparison, selection and improvement of the models, more accurate prediction models are obtained. The modeling methods mainly include Autoregressive models (AR) [24,25], Time Series Models [26,27], Support Vector Machine (SVM) [28,29], Artificial Neural Networks (ANN) [30,31], etc. The initial application of these prediction models in the field of wind power prediction has improved the accuracy of the prediction to a certain extent. However, these models do not fully consider the long-term correlation between the input samples, so the ability to improve the accuracy of wind power prediction models is also very limited. Li. et al. combined support vector machine (SVM) and improved dragonfly algorithm to forecast short-term wind power for a hybrid prediction model, and they found the proposed model suitable for short-term wind power prediction [32]. The SVM method can theoretically find a global optimal prediction. However, the calculation cost of SVM method will increase sharply, when the data volume is large. Under the circumstances, recursive neural network is introduced to improve the accuracy of wind power forecasts. RNN is a deep learning network, where there is a recursive link in the network structure. The relationship between the samples before and after the learning can be considered, which is especially suitable for processing time series signals. Aiming at the problems of gradient explosion and gradient disappearance, various improved methods have been studied. The emergence of LSTM neural network effectively solved the problems existing in previous models and achieved considerable results in the field of wind power prediction. At present, it is difficult for a single model to achieve good prediction effect, while the fusion method combining multiple models can improve the accuracy of prediction model more easily [33,34]. Erick Lopez et al. deeply integrated Long Short-Term Memory (LSTM) with Echo State Network (ESN) in their study, proposing an architecture similar to ESN. LSTM-ESN is superior to the WPPT model in all global indicators [35]. The wind power is predicted by the LSTM neural network algorithm, while the Gaussian Mixed Model (GMM) is used to analyze the error distribution characteristics of wind power short-term prediction. Both methods show better performance and evaluation [36]. On this basis, some scholars have made simple improvements to the structure of LSTM, reducing the influence of random components on prediction, effectively avoiding overfitting and making it more suitable for prediction [37]. Jyotirmayee Naik et al. used VMD to decompose the original nonlinear and non-stationary data and combined the VMD with 10 Multi-Kernel Regularized Pseudo Inverse Neural Network (MKPPINN), which showed the superiority of this model in wind power prediction [38]. Yu et al. proposed the Long Short-Term Memory and Enhanced Forget-Gate network model (LSTM-EFG), which can be used for wind power prediction. Based on correlation, the characteristic data of units within a certain distance are filtered, and the effect of wind power prediction is optimized by cluster analysis [39]. Lin. et al. integrated IF with deep learning and proposed a novel approach to perform power prediction using high-frequency SCADA data. Compared with the conventional predictive model used for outlier detection, the proposed deep learning prediction model shows superiority in wind power prediction [40].

3. Methodology

Wavelet Decomposition and Long Short-Term Memory neural network (WD-LSTM) is an intelligent network combining the advantages of WD and LSTM neural network. To better represent the data characteristics of the input index and facilitate the prediction of the neural network of LSTM, this paper adopts the loose WD and LSTM neural network, in which the WD is used as the preprocessing method of the prediction model of the LSTM neural network. According to the multi-fraction analysis function of WD, the original data are decomposed into time series with different frequency components to provide input vectors for LSTM neural network. S_{A1} is the approximate coefficient, while S_{D1}, S_{D2} and S_{D3} are the detail coefficients [41]. After the original data are decomposed by WD, the prediction is made by using the LSTM neural network, and the prediction results are obtained.

WD-LSTM prediction model combines the advantages of WD and LSTM neural network. This network can not only use the WD to analyze the subtle features of the original data but also can combine the self-learning and fault-tolerance capabilities of the neural network, which can improve both the accuracy of wind power generation prediction and the learning efficiency of the network. The steps of WD-LSTM neural network to predict wind power are shown in Figure 1.

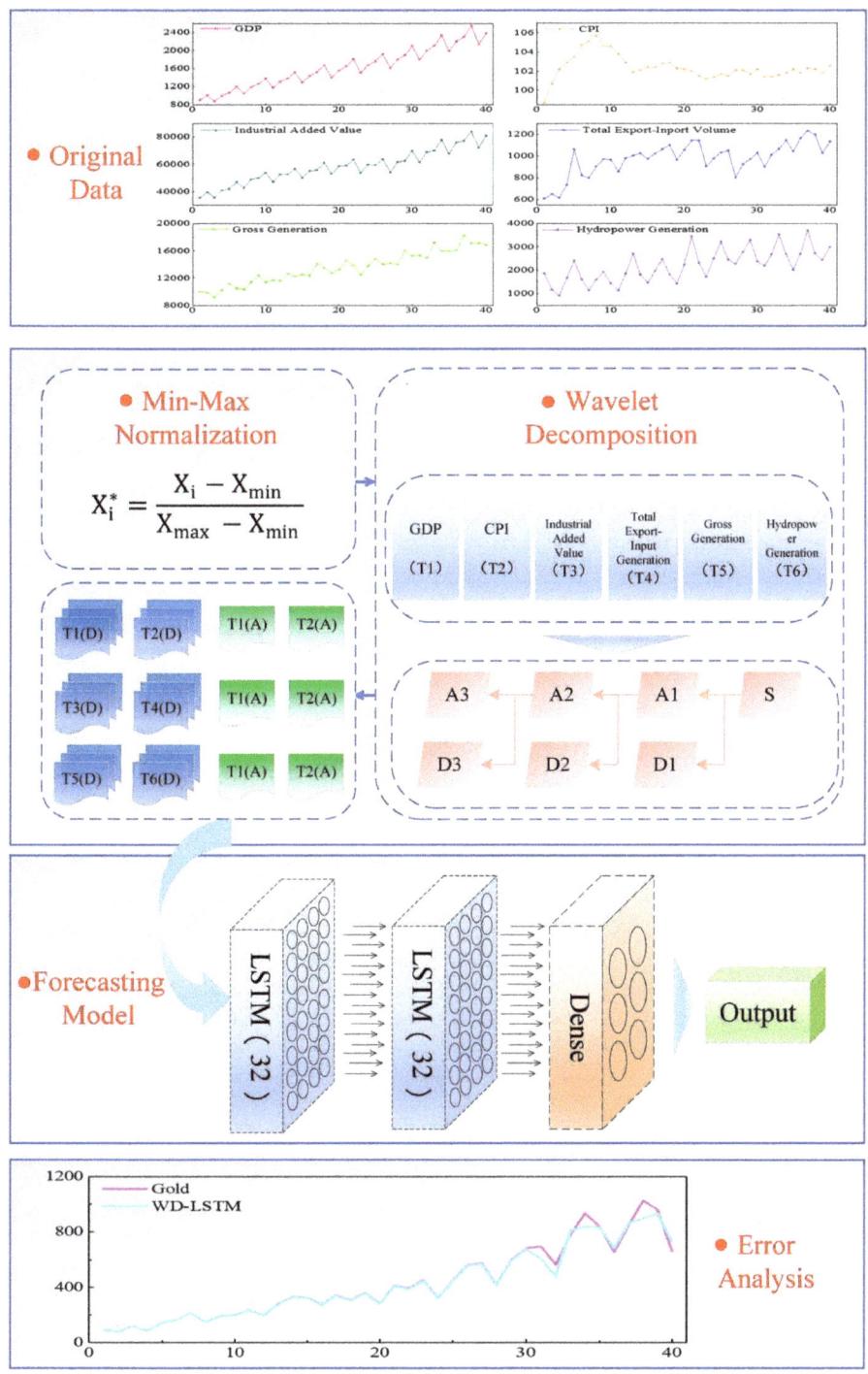

Figure 1. Calculation process of the WD-LSTM model.

3.1. Wavelet Decomposition

WD is an effective method to deal with non-stationary sequences. The multi-scale decomposition capability of WD can decompose the original time series into different frequency sequences according to different scales. WD is used to perform multi-scale analysis of various frequency components in the original signal, and noise frequency is screened out to obtain high-quality signals that can represent data characteristics, so as to improve the prediction accuracy of the model.

In the continuous wavelet transform, suppose $\varphi(t) \in L^2(R)$, $\varphi^*(w)$ as the results of Fourier transform $\varphi(t)$, and $\varphi^*(w)$ meet the conditions of Equation (1),

$$\int_{-\infty}^{+\infty} \frac{|\varphi^*(w)|^2}{|w|} dw < \infty \tag{1}$$

Then, $\varphi(t)$ can be considered as the parent wavelet function.

At the same time, $\varphi(t)$ can be obtained by stretching and shifting,

$$\varphi_{a,b}(t) = \frac{1}{\sqrt{|a|}} \varphi\left(\frac{t-b}{a}\right), a, b \in R; a \neq 0 \tag{2}$$

where a is the scaling variable and b is the translation variable.

For the square product function $f(t) \in L^2(R)$, the continuous wavelet transform is,

$$w_f(a,b) = \langle f, \varphi_{a,b} \rangle = \frac{1}{\sqrt{|a|}} \int_{-\infty}^{+\infty} f(t) \overline{\varphi\left(\frac{t-b}{a}\right)} dt \tag{3}$$

In Equation (3), a, b and t are continuous variables, while a is the expansion variable and b is the translation variable.

Continuous wavelets are usually sampled into discrete wavelets in practical applications, in order to facilitate calculation and analysis. Wavelet discretization is mainly for scaling variables a and shifting variables b. Then, the discrete wavelet function is Equation (4),

$$\varphi_{j,k}(t) = a_0^{-\frac{j}{2}} \varphi\left(\frac{t - k a_0^j b_0}{a_0^j}\right) = a_0^{-\frac{j}{2}} \varphi\left(a_0^{-j} t - k b_0\right) \tag{4}$$

In 1998, Mallet proposed wavelet multi-resolution analysis to perform J scale decomposition on the original sequence $s(t)$. In the first step, the original signal was first decomposed into low-frequency components a_1 and high-frequency components d_1. In the second step of decomposition, the high frequency part is retained and the low frequency component a_1 is further decomposed into a low frequency component a_2 and a high frequency component d_2. The low-frequency components obtained at each step are decomposed in turn to finally obtain the low-frequency components a_J and high-frequency components d_J in the J scale. Then, the original sequence can be expressed as Equation (5),

$$s(t) = a_J(t) + \sum_{r=1}^{J} d_r(t) \tag{5}$$

where J is the decomposition scale, $a_J(t)$ is the component approaching the original time series (low-frequency component) and $d_r(t)(r = 1, \ldots, J)$ is the detail signal component (high-frequency component).

The more important step in WD is to choose the wavelet function and the scale of WD to participate in the algorithm. The number of wavelet decompositions is small, and the approximate signal usually contains random interference signals, which cannot effectively reflect the change trend of the original wind speed sequence. If the number of decompositions is too large, there will be greater error accumulation and the training time will be longer. In this paper, Daubechies (DB) wavelet is used to decompose the original data, taking $J = 3$.

3.2. Basic Principles of LSTM

The traditional neural network model lacks the memory function of historical information, and the cyclic neural network (RNN) can apply the output information of previous neurons to the current task. However, conventional RNN has the problem of gradient disappearance or gradient explosion; in other words, when the time interval is large, the past learning results will disappear. To address these shortcomings, Hochreiter proposed the Long Short-Term Memory Neural Network (LSTM) in 1997. LSTM is a type of Recurrent Neural Network that can learn long-term dependent information. It not only has the memory function of historical information, but also overcomes the long-term dependence of the model and can selectively forget the invalid information and update the effective information, thus solving the problem of gradient dispersion to some extent. As shown in Figure 2, the LSTM network is composed of an input layer, an output layer and several recursive hiding layers between them. A recursive hiding layer is composed of multiple memory modules, each of which contains one or more self-connected memory units with three gates controlling the information flow: the input gate, the forgetting gate and the output gate. The state of LSTM cell is calculated as follows:

$$i_t = \sigma(W_i \times [h_{t-1}, x_t] + b_i) \tag{6}$$

$$f_t = \sigma(W_f \times [h_{t-1}, x_t] + b_f) \tag{7}$$

$$o_t = \sigma(W_o \times [h_{t-1}, x_t] + b_o) \tag{8}$$

$$\widetilde{c_t} = \tanh(W_c \times [h_{t-1}, x_t] + b_c) \tag{9}$$

$$c_t = f_t \times c_{t-1} + i_t \times \widetilde{c_t} \tag{10}$$

In Equations (6)–(8), i_t, f_t and o_t are, respectively, input gate, forgetting gate and output gate. In Equation (10), c_t is a new candidate for cell state. LSTM cells act as state information, updating the c_t of the old cell state c_{t-1} to the new cell state. W_i, W_f, W_o and W_c are, respectively, the weights of input, forgetting, output and current cell state. b_i, b_f, b_o and b_c are, respectively, the deviations of input, forgetting, output and current cell state.

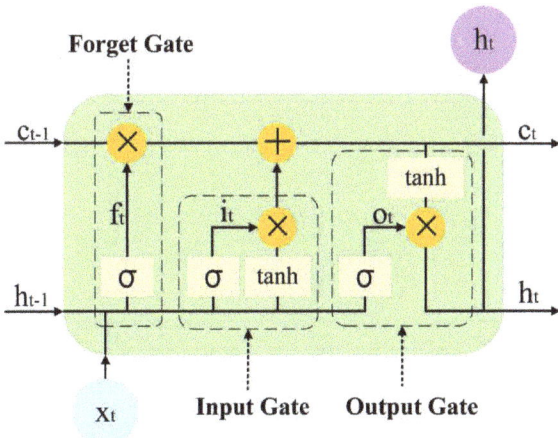

Figure 2. Schematic diagram of neurons.

4. Empirical Study

4.1. Data Description and Preprocessing

This paper selects four macroeconomic indicators of Gross Domestic Product (GDP), Consumer Price Index (CPI), Industrial Added Value (IAV) and Total Imports and Exports (TIE), as well as two related power generation indicators of National Total Power Generation (NTPG) and Hydropower Generation (HG), as input variables. To accurately evaluate the accuracy of wind power generation prediction model, this paper selects macroeconomic indicators and related power generation indicators from the National Bureau of Statistics of China. Macroeconomic indicators and related power generation data from the third quarter of 2009 to the second quarter of 2019 are selected, with a total of 40 data points. The original data samples are divided into two datasets: 80% of the original data (32 data points) are used as training samples and the remaining 20% (8 data points) are used as test samples to evaluate the predictive performance of the model.

Since the macroeconomic indicators and related power generation indicators have different dimensions and dimensional units, it is necessary to carry out data standardization processing for the original time series in order to eliminate the dimensional impact between indicators. According to Equation (11), each group of data is normalized into the interval 0–1 to solve the comparability between indicators, reduce the influence of outliers and noise and speed up the training speed of the model.

$$x_i^* = \frac{x_i - x_{\min}}{x_{\max} - x_{\min}} \tag{11}$$

The Matrix Laboratory (MATLAB) wavelet toolbox is used to decompose the normalized time series $s(t)$; then,

$$s(t) = a_J(t) + \sum_{r=1}^{3} d_r(t) \tag{12}$$

where J is the decomposition scale, $a_J(t)$ is the low-frequency component close to the original sequence, $d_r(t)$ is the detail signal component (high-frequency component) of the r-th decomposition and t is the discrete time.

4.2. Model Parameters

In this paper, the time series of multiple macroeconomic indicators and related power generation indicators after WD are taken as the input variables LSTM neural network, and the wind power generation of the whole country is taken as the output variable. The LSTM neural network contains four parameters that affect the prediction accuracy of the model, including the time step of each layer in the LSTM neural network, the number of hidden units in each layer in the model and the training times. In the process of training the model, the other parameters are the same each time, but the single parameter is different, so as to find the best prediction model. Each parameter setting in the model is shown in Table 1.

Table 1. Parameters for the LSTM network.

Dataset	Time Steps	Hidden Layers	Batch Size	Lr	Epoch
WD-LSTM	2	64	3	0.001	15,000

The LSTM model is a deep learning neural network, which has three layers: an input layer, a hidden layer and an output layer. The input is composed of six input variables: Gross Domestic Product (GDP), Consumer Price Index (CPI), Industrial Added Value (IAV), Total Imports and Exports (TIE), National Total Power Generation (NTPG) and Hydropower Generation (HG). The hidden layer consists of two LSTM units with time steps of 2, and each LSTM unit contains 64 cells. The output layer contains an output variable of wind power generation. The structure of the LSTM model is shown in Figure 3.

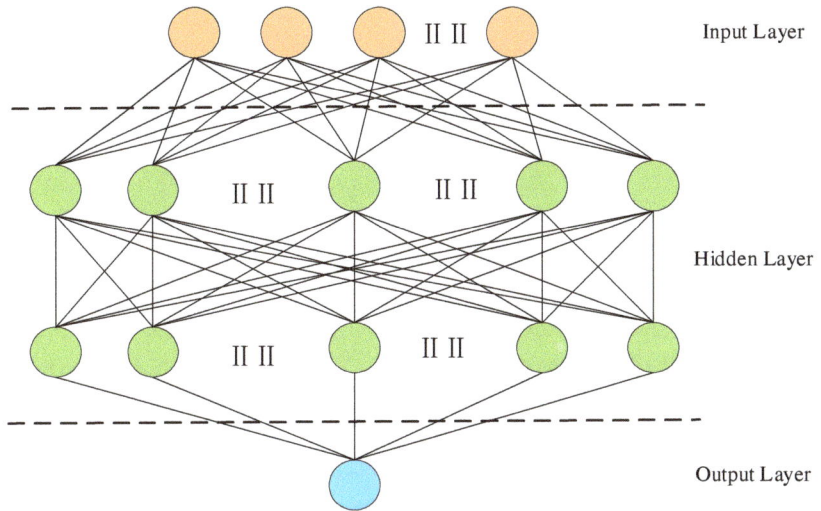

Figure 3. The structure of LSTM neural network.

4.3. Performance Indicators

To further verify the effectiveness and performance of the prediction method proposed for wind power prediction, three error analysis criteria are introduced to evaluate the proposed model, as given in Equations (13)–(15) (where y_real_i is the actual values and y_pred_i is predicted values). The mean absolute error (MAE), root mean square error (RMSE) and mean absolute percentage error (MAPE) are used to evaluate the performance of each method.

$$\text{MAE} = \frac{1}{N}\sum_{i=1}^{N}\left|y\_real_i - y\_pred_i\right| \tag{13}$$

$$\text{MAPE} = \frac{1}{N}\sum_{i=1}^{N}\left|\frac{y\_real_i - y\_pred_i}{y\_real_i}\right| * 100 \tag{14}$$

$$\text{RMSE} = \sqrt{\frac{\sum_{i=1}^{N}(y\_real_i - y\_pred_i)^2}{N}} \tag{15}$$

5. Results and Analysis

5.1. Model Accuracy

To evaluate the performance of WD-LSTM model in wind power prediction more effectively, other models are preliminarily selected for comparison in the paper. The physical or statistical models commonly used to predict by time series are selected. In addition, the models commonly used in machine learning and deep learning are selected. Bayesian Model Averaging and Ensemble Learning (BMA-EL) proposed by Wang et al. can accurately predict wind power generation under different meteorological conditions. Chen et al. built Multi-Resolution Multi-Learner Ensemble and Adaptive Model Selection (MRMLE-AMS), which presented high accuracy in wind power prediction by time series. Li et al. proposed the Support Vector Machine and Improved Dragonfly Algorithm model (SVR-IDA). In this paper, these models are selected as the comparison object for experimental verification. The results are shown in Table 2.

Table 2. Comparison of forecasting errors using different models.

Algorithms	BMA-EL	MRMLE-AMS	SVR-IDA	WD-LSTM
MAPE	22.328	20.624	15.679	5.831

As shown in Table 2, among the four models, the MAPE of WD-LSTM model is the lowest, reaching 5.831. The MAPE of SVR-IDA model comes in second at 15.679, more than 10. The errors of BMA-EL and MRMLE-AMS are relatively high, exceeding 20. In this experiment, the accuracy of machine learning and deep learning prediction models is generally better than that of physical or statistical prediction models. The presumed reason may be that machine learning and deep learning predictive models can fully learn the correlation between input and output variables, similar to human neural networks. In particular, the deep learning model can more fully learn the variation trend of data in time series, hence showing a higher prediction accuracy.

Based on previous comparisons, the prediction model proposed in this paper, which combines wavelet decomposition with long short-term memory neural network, has shown high prediction accuracy when predicting wind power generation in China. To effectively evaluate the performance of WD-LSTM in wind power prediction, traditional prediction methods of machine learning and deep learning are used in this paper as comparative experiments. Based on the same input time series, the learning situation of each model is tested, and its errors are compared and analyzed. During the experiment, Support Vector Regression (SVR), Gate Recurrent Unit (GRU), Wavelet Decomposition and Support Vector Regression (WD-SVR) and Wavelet Decomposition and Gated Recurrent Unit (WD-GRU) are used for time series prediction as comparative tests. In addition, the proportion of training set and test set is the same as that of WD-LSTM model and five comparative experiments are conducted. To objectively evaluate and describe the performance of the six prediction models, the prediction error values of each model are calculated according to the above formulas. The experimental results of MAE, MAPE and RMSE of the raw test set are shown in Table 3.

Table 3. Comparison of prediction performances using machine learning and deep learning models.

Algorithms	MAE	MAPE	RMSE	Computing Time (Minutes)
SVR	137.888	15.351	165.175	0.05
GRU	127.863	15.048	177.223	32
LSTM	101.511	13.715	169.644	32
WD-SVR	206.831	20.153	212.016	12.05
WD-GRU	144.321	18.034	226.302	44
WD-LSTM	49.896	5.831	63.991	44

Among all the experimental models, WD-SVR has the largest error, and its MAE, MAPE and RMSE are 206.831, 20.153 and 212.016, respectively. The MAE, MAPE and RMSE of WD-GRU are 144.321, 18.034 and 226.302, respectively. The MAPE of the three models of SVR, GRU and LSTM are similar, 15.351, 15.048 and 13.715, respectively. The error of WD-LSTM model is the smallest, and its MAPE is 5.831, which is significantly lower than the other five models. It can be seen from the data in Table 3 that WD-LSTM has a high accuracy in predicting wind power generation and is more effective than the traditional models and single models. Furthermore, Table 3 shows computing time cost of WD-LSTM and five other comparison models. In the machine learning models, the prediction using SVR model took 0.05 min while WD-LSTM took 12.05 min. In the deep learning models, GRU and LSTM cost the same time, 32 min, while WD-GRU and WD-LSTM cost the same time, 44 min. In general, compared with machine learning models, deep learning models take a longer time to predicate using time series. However, as for deep learning models, since the data samples are relatively small, there is no significant difference in the time spent on prediction.

Figure 4 shows the prediction results of WD-LSTM neural network and five other comparison models, which directly reflects the degree of fitting between the predicted values of the six models and the real values. Meanwhile, Figure 5 shows the predicted and original value based on WD-LSTM. As shown in Figure 4, the prediction curve of Support Vector Regression (SVR) is relatively stable and it is difficult to predict the dynamic change of data. When the data present a large fluctuation, the model presents a large error value. Gated Recurrent Unit (GRU) is a variant or simplification of the Long Short-Term Memory network (LSTM), which includes reset gate and update gate. From the forecast results, it can reflect the fluctuation of wind power generation, but the variation trend in a single quarter is opposite to the real value, leading to higher error value. The results show that the input indexes such as Gross Domestic Product (GDP), Consumer Price Index (CPI), Industrial Added Value (IAV), Total Imports and Exports (TIE), National Total Power Generation (NTPG) and Hydropower Generation (HG) can be used as the input data of wind power generation. WD-LSTM can accurately predict the fluctuation of wind power generation, and the error value is lower than other models.

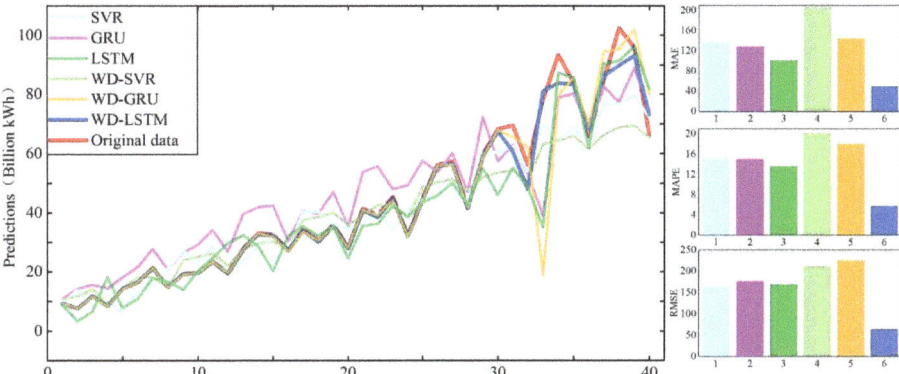

Figure 4. Forecasting performance of the WD-LSTM model and five other models.

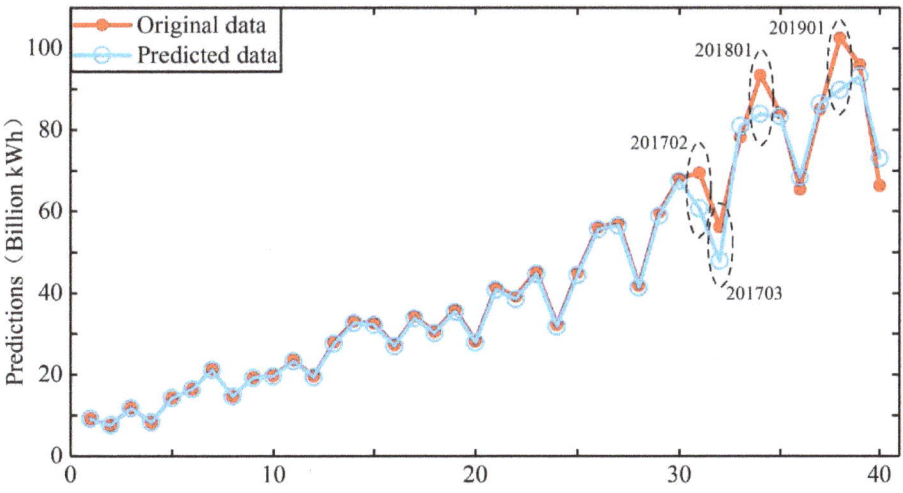

Figure 5. Forecasting results of wind power generation based on WD-LSTM model.

On the basis of the above research, the paper further studies the influence of different types of input indicators on the accuracy of wind power generation prediction. The six input indicators are divided into two categories: (1) macroeconomic indicators, including GDP, CPI, IAV and TIE; and (2) power generation indicators, including NTPG and HG. The two kinds of indicators are, respectively, taken as input variables, and the WD-LSTM model is used to predict wind power generation on the condition that the model parameters are kept consistent. When macroeconomic indicators are taken as input variables, the experimental result of MAPE is 19.732. When the related power generation index is used as the input variables, the MAPE is 16.298. The results show that wind power forecast achieves the best prediction accuracy when six indicators are used as input variables.

5.2. Sensitivity Analysis

Sensitivity analysis is a common method to study and analyze the effect of parameter changes on system behavior. The sensitivity of variables to test parameters can be calculated as follows:

$$S_t = \left| \frac{(Y'_t - Y_t)/Y_t}{(X'_t - X_t)/X_t} \right| \tag{16}$$

where S_t is the sensitivity of variables to test parameters at time t, setting the third quarter of 2009 as $t = 1$; Y_t and Y'_t are the value of output variable before and after change at time t; and X_t and X'_t are the value of input variables before and after change at time t. The maximum sensitivity of wind power generation during 2006–2017 is:

$$s = max(S_t), 1 \leq t \leq 40 \tag{17}$$

The sensitivity of the wind power generation variable on the six main input variable in the proposed WD-LSTM model is studied and analyzed by changing the corresponding input variables by −5%, −3%, −1%, 1%, 3% and 5%, and the maximum sensitivity of wind power generation from the third quarter of 2009 to the second quarter of 2019 with respect to the six input variables change is shown in Table 4.

Table 4. The maximum sensitivity of wind power generation with respect to the six variables.

Input Variables	Change Rate					
	−5%	−3%	−1%	1%	3%	5%
Gross Domestic Product (GDP)	0.09635	0.09475	0.09520	0.09520	0.09455	0.09635
Consumer Price Index (CPI)	0.00615	0.00605	0.00675	0.00675	0.00595	0.00615
Industrial Value Added (IVA)	0.07087	0.06913	0.07000	0.07000	0.06913	0.07087
Total Imports and Exports (TIE)	0.04830	0.04885	0.04715	0.04715	0.04885	0.04830
Total Power Generation (TPG)	0.05910	0.06875	0.06045	0.06045	0.06370	0.05910
Hydroelectricity Generation (HG)	0.00523	0.00467	0.00480	0.00480	0.00467	0.00523

It is found that the maximum sensitivity of wind power generation in the proposed model with respect to the six input variables from the third quarter of 2009 to the second quarter of 2019 is less than 0.10, which means the maximum sensitivity of wind power generation in the proposed WD-LSTM model is less sensitive. Therefore, the proposed model is stable and does not cause abnormal fluctuations in the output variable data due to the small changes of input variables.

5.3. Scenarios Setting

Different scenarios for forecasting are set in this paper, in which different scenarios match different input data to explore the changing trend of national wind power generation under different development situations and reduce the uncertainty of forecasting. Taking historical data rates and national economic, energy and social macro-development plans into account to make more realistic predictions and analyze the future trends of each characteristic value, this paper sets the following three scenarios to predict the wind power generation in China under different development trends in the next two years. Scenario 1 is a low-growth scenario, which keeps the country's recent development trend sustainable and calculates the minimum growth rate (non-negative) of macroeconomic indicators and related power generation indicators based on the growth rate of historical data to predict the national wind power generation. Scenario 2 is the base scenario, in which the development trend of each indicator is predicted more accurately and in line with the actual development trend from the fourth quarter of 2019 to the fourth quarter of 2022, and the average year-on-year growth rate of the data from the third quarter of 2009 to the second quarter of 2019 is calculated. Scenario 3 is a high-growth scenario, which maintains a high growth rate according to the historical development trend. According to the data from the third quarter of 2009 to the second quarter of 2019, the average year-on-year growth rate of each quarter is calculated and increased by 1.2 times on the basis of the average growth rate of each quarter. The specific growth rates under each scenario are shown in Table 5.

Table 5. Growth rates in different scenarios.

Different Scenarios	Gross Domestic Product (GDP)	Consumer Price Index (CPI)	Industrial Value Added (IVA)	Total Imports and Exports (TIE)	Total Power Generation (TPG)	Hydroelectricity Generation (HG)
Scenario 1	2.622	0.097	0.774	0.053	0.306	5.739
Scenario 2	3.146	0.101	2.645	2.345	1.603	6.455
Scenario 3	3.775	0.122	3.174	2.815	1.924	7.746

5.4. Future Prediction Results

By comparing the error values of each single model and the hybrid model in the prediction of wind power generation across the country through testing, we obtained that the prediction accuracy of WD-LSTM is relatively high, and predicted the wind power generation of China in the next two years by this model (from the fourth quarter of 2019 to the fourth quarter of 2021). The index data of different growth trends are substituted into WD-LSTM for the prediction of wind power generation in China based on the above scenario settings, and the prediction results under three different scenarios are compared, as shown in Figure 6. Wind power generation is projected to grow at a slower pace in Scenario 1, from 66.3 billion kWh in the third quarter of 2019 to 81.6 billion kWh in the fourth quarter of 2021, an increase of 15.3 billion kWh over nine quarters. Wind power generation forecasting show an increase in Scenario 2 from 66.3 billion kWh in the third quarter of 2019 to 95.6 billion kWh in the fourth quarter of 2021, an increase of 29.4 billion kWh in nine quarters. Scenario 3 is of high growth, from 66.3 billion kWh in the third quarter of 2019 to 105.6 billion kWh in the fourth quarter of 2021, an increase of 39.4 billion kWh in nine quarters. In summary, the forecast results of the three scenarios show that the total national wind power output will fluctuate between 283.1 and 300.4 billion kWh in 2020, and the total national wind power output will be between 303.1 and 363.9 billion kWh in 2021, floating from time to time.

From the overall trend, the country's wind power generation will continue to increase. Under the three scenarios, the national wind power generation will decline slightly in the first quarter of 2020, and the growth rate will peak in the fourth quarter of 2021. In 2017, the State Grid pointed out at a press conference that, by 2020, the problem of new energy consumption will be completely solved, and the rate of abandoned wind and light will be controlled within 5%. According to the 13th Five-Year Plan for Wind Power Development, by 2020, non-fossil energy will account for 15% of primary energy consumption, and the country's annual wind power generation will need to reach 42 billion kWh and 6% of the total power generation. The sustained growth of wind power generation in China in the future may be affected by the following factors: (1) The sustained and steady development of China's economy. At the present stage, China's economic development model is changing from high-speed development to high-quality development. The steady high-quality economic development has laid a solid foundation for the development of the wind power industry in China, thus realizing the sustainable growth of the country's wind power generation. (2) Environmental protection brings development opportunities for renewable energy such as wind energy. From the overall perspectives, the development of renewable energy is a common goal of mankind and an important support for the global response to future climate, environmental and economic changes. The development of wind power as an energy source can become more affordable than traditional coal power, and the parity of wind power will release new market space, which is also an important reason for the continuous growth of wind power generation across the country.

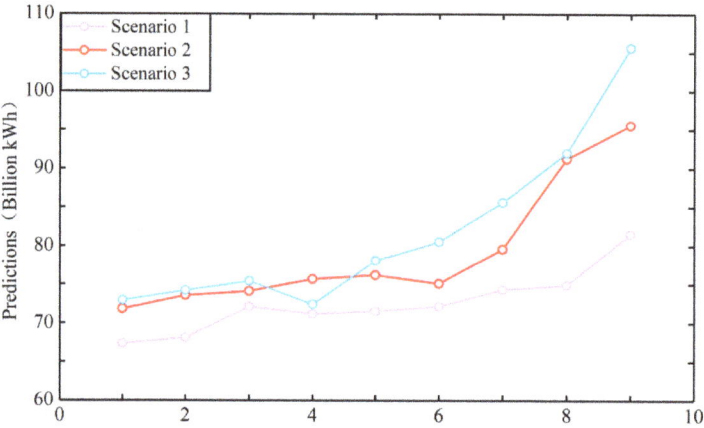

Figure 6. Predictions for wind power generation in different scenarios.

Wind power in China shows a trend of rapid development. Wind curtailment and power limiting has become the focus of society and a major problem that needs to be solved urgently in power grid planning and dispatching operation. In 2017, the National Energy Administration's Guidance on the Implementation of the 13th Five-Year Plan for the Development of Renewable Energy was released. At the same time, the target of consumption and utilization is also proposed to effectively solve the problem of wind power curtailment by 2020. From 2011 to 2016, the wind curtailment rate showed a trend of first decreasing and then increasing, reaching the highest value of 17.1% in 2016. According to the Clean Energy Consumption Action Plan (2018–2020) jointly issued by the National Development and Reform Commission and the National Energy Administration, the wind curtailment rate will be kept at a reasonable level (aiming at around 5%) by 2020. Based on previous predictions, the total national wind power output will fluctuate between 283.1 and 300.4 billion kWh in 2020. Consequently, wind curtailment power will keep between 14.1 and 15.0 billion kWh in 2020.

6. Conclusions

As a kind of renewable energy, wind power generation plays a crucial role in China's electric energy production. Therefore, accurate prediction of wind power generation is helpful to optimize the power grid dispatching, reduce the reserve capacity of the system and reduce the operating cost of the power system. In this paper, a hybrid LSTM model for predicting wind power generation in China is constructed based on six index factors: gross domestic product, consumer price index, industrial added value, total imports and exports, total power generation and hydropower generation. Based on wavelet decomposition and long short-term memory neural network methods, a hybrid WD-LSTM model for predicting national wind power generation is constructed. The following conclusions can be reached through experiments:

(1) Wind power generation is related to GDP, CPI, IAV, TIE, TPG and HG. The selection of these six input indexes can, to a certain extent, predict the wind power generation of the country.
(2) The time series of macroeconomic indicators and related power generation indicators are decomposed into low-frequency components and high-frequency components through wavelet decomposition, which increases the data dimension of the input variables of the prediction model to some extent. The time series data of macroeconomic and related power generation indexes of different frequencies are used as input variables to effectively improve the accuracy of the prediction model.
(3) In this paper, the WD-LSTM hybrid prediction model is selected to predict the wind power generation in China. The experimental results show that the MAPE of the mixed prediction model is 5.831. Compared with machine learning and a single prediction model, the model can predict wind power generation more accurately across the country.
(4) In addition, the prediction of national wind power generation in this paper still needs to be improved and deepened. Due to the difficulty in obtaining some index data and the inconsistency of some data in scale, the paper has the limitation in the selection of input indices. The limitations of the samples themselves will lead to a certain range of errors in the process of data processing and prediction. Therefore, other possible influencing factors can be considered as input variables.
(5) The next step of the study will consider whether the time series with different scales can be used as the input index of the same model. At the same time, Information Gain (IG) will also be used to sort and filter input indicators by correlation, and then make prediction using WD-LSTM model. The application of the proposed model in primary energy consumption or renewable energy consumption will also be considered.

Author Contributions: Data curation, S.Z., X.Y. and L.Z.; Formal analysis, S.Z.; Investigation, X.Y. and L.Z.; Methodology, B.L.; Software, Q.W.; Supervision, B.L.; Writing—original draft, S.Z.; and Writing—review and editing, B.L. and Q.W. All authors have read and agreed to the published version of the manuscript.

Funding: This research was funded by the National Natural Science Foundation of China (grant number 71503180) and Humanities and Social Sciences Fund of the Ministry of Education (grant number 20YJA630042).

Conflicts of Interest: The authors declare no conflict of interest.

References

1. Wuyong, Q.; Jue, W. An improved seasonal GM (1,1) model based on the HP filter for forecasting wind power generation in China. *Energy* **2020**, *209*, 118499.
2. Xu, X.; Niu, D.; Xiao, B.; Guo, X.; Zhang, L.; Wang, K. Policy analysis for grid parity of wind power generation in China. *Energy Policy* **2020**, *138*, 111225. [CrossRef]
3. Messner, J.W.; Pinson, P. Online adaptive lasso estimation in vector autoregressive models for high dimensional wind power forecasting. *Int. J. Forecast.* **2019**, *35*, 1485–1498. [CrossRef]
4. Han, S.; Qiao, Y.-H.; Yan, J.; Liu, Y.-Q.; Li, L.; Wang, Z. Mid-to-long term wind and photovoltaic power generation prediction based on copula function and long short term memory network. *Appl. Energy* **2019**, *239*, 181–191. [CrossRef]
5. Wang, Y.; Liu, Y.; Li, L.; Infield, D.; Han, S. Short-Term Wind Power Forecasting Based on Clustering Pre-Calculated CFD Method. *Energies* **2018**, *11*, 854. [CrossRef]
6. Natapol, K.; Thananchai, L. Robust short-term prediction of wind power generation under uncertainty via statistical interpretation of multiple forecasting models. *Energy* **2019**, *180*, 387–397.
7. Shao, H.; Deng, X.; Jiang, Y. A novel deep learning approach for short-term wind power forecasting based on infinite feature selection and recurrent neural network. *J. Renew. Sustain. Energy* **2018**, *10*, 043303. [CrossRef]
8. Wang, H.; Lei, Z.; Wu, Q.H.; Peng, J.; Liu, J. Echo state network based ensemble approach for wind power forecasting. *Energy Convers. Manag.* **2019**, *201*, 112188. [CrossRef]
9. Wang, G.; Jia, R.; Liu, J.; Zhang, H. A hybrid wind power forecasting approach based on Bayesian model averaging and ensemble learning. *Renew. Energy* **2020**, *145*, 2426–2434. [CrossRef]

10. Wang, C.; Zhang, H.; Ma, P. Wind power forecasting based on singular spectrum analysis and a new hybrid Laguerre neural network. *Appl. Energy* **2020**, *259*, 114139. [CrossRef]
11. Wang, H.; Han, S.; Liu, Y.; Yan, J.; Li, L. Sequence transfer correction algorithm for numerical weather prediction wind speed and its application in a wind power forecasting system. *Appl. Energy* **2019**, *237*, 1–10. [CrossRef]
12. Pearre, N.S.; Swan, L.G. Statistical approach for improved wind speed forecasting for wind power production. *Sustain. Energy Technol. Assess.* **2018**, *27*, 180–191. [CrossRef]
13. Leng, H.; Li, X.; Zhu, J.; Tang, H.; Zhang, Z.; Ghadimi, N. A new wind power prediction method based on ridgelet transforms, hybrid feature selection and closed-loop forecasting. *Adv. Eng. Inform.* **2018**, *36*, 20–30. [CrossRef]
14. Wang, K.; Qi, X.; Liu, H.; Song, J. Deep belief network based k-means cluster approach for short-term wind power forecasting. *Energy* **2018**, *165*, 840–852. [CrossRef]
15. Semero, Y.K.; Zhang, J.; Zheng, D.; Wei, D. A GA-PSO Hybrid Algorithm Based Neural Network Modeling Technique for Short-term Wind Power Forecasting. *Distrib. Gener. Altern. Energy J.* **2018**, *33*, 26–43. [CrossRef]
16. Hong, D.; Ji, T.; Li, M.; Wu, Q. Ultra-short-term forecast of wind speed and wind power based on morphological high frequency filter and double similarity search algorithm. *Int. J. Electr. Power Energy Syst.* **2019**, *104*, 868–879. [CrossRef]
17. Du, P.; Wang, J.; Yang, W.; Niu, T. A novel hybrid model for short-term wind power forecasting. *Appl. Soft Comput.* **2019**, *80*, 93–106. [CrossRef]
18. Lu, P.; Ye, L.; Sun, B.; Zhang, C.; Zhao, Y.; Teng, J. A New Hybrid Prediction Method of Ultra-Short-Term Wind Power Forecasting Based on EEMD-PE and LSSVM Optimized by the GSA. *Energies* **2018**, *11*, 697. [CrossRef]
19. Yagang, Z.; Yuan, Z.; Chunhui, K.; Bing, C. A new prediction method based on VMD-PRBF-ARMA-E model considering wind speed characteristic. *Energy Convers. Manag.* **2020**, *203*, 112254.
20. Naik, J.; Satapathy, P.; Dash, P. Short-term wind speed and wind power prediction using hybrid empirical mode decomposition and kernel ridge regression. *Appl. Soft Comput.* **2018**, *70*, 1167–1188. [CrossRef]
21. Wang, C.; Zhang, H.; Fan, W.; Ma, P. A new chaotic time series hybrid prediction method of wind power based on EEMD-SE and full-parameters continued fraction. *Energy* **2017**, *138*, 977–990. [CrossRef]
22. Wang, K.; Niu, D.; Sun, L.; Zhen, H.; Liu, J.; De, G.; Xu, X. Wind Power Short-Term Forecasting Hybrid Model Based on CEEMD-SE Method. *Processes* **2019**, *7*, 843. [CrossRef]
23. Han, L.; Zhang, R.; Wang, X.; Bao, A.; Jing, H. Multi-step wind power forecast based on VMD-LSTM. *IET Renew. Power Gener.* **2019**, *13*, 1690–1700. [CrossRef]
24. Cavalcante, L.; Bessa, R.; Reis, M.; Browell, J. LASSO vector autoregression structures for very short-term wind power forecasting. *Wind. Energy* **2016**, *20*, 657–675. [CrossRef]
25. Wang, Y.; Hu, Q.; Meng, D.; Zhu, P. Deterministic and probabilistic wind power forecasting using a variational Bayesian-based adaptive robust multi-kernel regression model. *Appl. Energy* **2017**, *208*, 1097–1112. [CrossRef]
26. Chen, C.; Liu, H. Medium-term wind power forecasting based on multi-resolution multi-learner ensemble and adaptive model selection. *Energy Convers. Manag.* **2020**, *206*, 112492. [CrossRef]
27. Ouyang, T.; Huang, H.; He, Y.; Tang, Z. Chaotic wind power time series prediction via switching data-driven modes. *Renew. Energy* **2020**, *145*, 270–281. [CrossRef]
28. Liu, M.; Cao, Z.; Zhang, J.; Wang, L.; Huang, C.; Luo, X. Short-term wind speed forecasting based on the Jaya-SVM model. *Int. Jelec Power* **2020**, *121*, 106056. [CrossRef]
29. Zhang, Y.; Le, J.; Liao, X.; Zheng, F.; Li, Y. A novel combination forecasting model for wind power integrating least square support vector machine, deep belief network, singular spectrum analysis and locality-sensitive hashing. *Energy* **2019**, *168*, 558–572. [CrossRef]
30. Nielson, J.; Bhaganagar, K.; Meka, R.; Alaeddini, A. Using atmospheric inputs for Artificial Neural Networks to improve wind turbine power prediction. *Energy* **2020**, *190*, 116273. [CrossRef]
31. Qin, Y.; Li, K.; Liang, Z.; Lee, B.; Zhang, F.; Gu, Y.; Zhang, L.; Wu, F.; Rodriguez, D. Hybrid forecasting model based on long short term memory network and deep learning neural network for wind signal. *Appl. Energy* **2019**, *236*, 262–272. [CrossRef]
32. Li, L.; Zhao, X.; Tseng, M.-L.; Tan, R. Short-term wind power forecasting based on support vector machine with improved dragonfly algorithm. *J. Clean. Prod.* **2020**, *242*, 118447. [CrossRef]

33. Yuan, X.; Chen, C.; Jiang, M.; Yuan, Y. Prediction interval of wind power using parameter optimized Beta distribution based LSTM model. *Appl. Soft Comput.* **2019**, *82*, 105550. [CrossRef]
34. Lu, K.; Sun, W.X.; Wang, X.; Meng, X.R.; Zhai, Y.; Li, H.H.; Zhang, R.G. Short-term Wind Power Prediction Model Based on Encoder-Decoder LSTM. *IOP Conf. Series: Earth Environ. Sci.* **2018**, *186*, 012020. [CrossRef]
35. López, E.; Valle, C.; Allende, H.; Gil, E.; Madsen, H. Wind Power Forecasting Based on Echo State Networks and Long Short-Term Memory. *Energies* **2018**, *11*, 526. [CrossRef]
36. Jinhua, Z.; Yan, J.; Infield, D.; Liu, Y.; Lien, F.-S. Short-term forecasting and uncertainty analysis of wind turbine power based on long short-term memory network and Gaussian mixture model. *Appl. Energy* **2019**, *241*, 229–244. [CrossRef]
37. Han, L.; Jing, H.; Zhang, R.; Gao, Z. Wind power forecast based on improved Long Short Term Memory network. *Energy* **2019**, *189*, 116300. [CrossRef]
38. Naik, J.; Dash, S.; Dash, P.; Bisoi, R. Short term wind power forecasting using hybrid variational mode decomposition and multi-kernel regularized pseudo inverse neural network. *Renew. Energy* **2018**, *118*, 180–212. [CrossRef]
39. Yu, R.; Gao, J.; Yu, M.; Lu, W.; Xu, T.; Zhao, M.; Zhang, J.; Zhang, R.; Zhang, Z. LSTM-EFG for wind power forecasting based on sequential correlation features. *Futur. Gener. Comput. Syst.* **2019**, *93*, 33–42. [CrossRef]
40. Lin, Z.; Liu, X.; Collu, M. Wind power prediction based on high-frequency SCADA data along with isolation forest and deep learning neural networks. *Int. J. Electr. Power Energy Syst.* **2020**, *118*, 105835. [CrossRef]
41. Akira, R.; Hiroka, R. Application of multi-dimensional wavelet transform to fluid mechanics. *Theor. Appl. Mech. Lett.* **2020**, *10*, 98–115.

© 2020 by the authors. Licensee MDPI, Basel, Switzerland. This article is an open access article distributed under the terms and conditions of the Creative Commons Attribution (CC BY) license (http://creativecommons.org/licenses/by/4.0/).

Article

Ultra-Short-Term Prediction of Wind Power Based on Error Following Forget Gate-Based Long Short-Term Memory

Pei Zhang [1,2,*], Chunping Li [1], Chunhua Peng [1] and Jiangang Tian [1]

1. School of Electrical and Automation Engineering, East China Jiaotong University, Nanchang 330013, China; 2018028085207014@ecjtu.edu.cn (C.L.); chpeng@ecjtu.edu.cn (C.P.); 2018028085207009@ecjtu.edu.cn (J.T.)
2. School of Electrical Engineering, Beijing Jiaotong University, Beijing 100044, China
* Correspondence: 8618@ecjtu.edu.cn

Received: 6 June 2020; Accepted: 2 October 2020; Published: 16 October 2020

Abstract: To improve the accuracy of ultra-short-term wind power prediction, this paper proposed a model using modified long short-term memory (LSTM) to predict ultra-short-term wind power. Because the forget gate of standard LSTM cannot reflect the correction effect of prediction errors on model prediction in ultra-short-term, this paper develops the error following forget gate (EFFG)-based LSTM model for ultra-short-term wind power prediction. The proposed EFFG-based LSTM model updates the output of the forget gate using the difference between the predicted value and the actual value, thereby reducing the impact of the prediction error at the previous moment on the prediction accuracy of wind power at this time, and improving the rolling prediction accuracy of wind power. A case study is performed using historical wind power data and numerical prediction meteorological data of an actual wind farm. Study results indicate that the root mean square error of the wind power prediction model based on EFFG-based LSTM is less than 3%, while the accuracy rate and qualified rate are more than 90%. The EFFG-based LSTM model provides better performance than the support vector machine (SVM) and standard LSTM model.

Keywords: error following forget gate-based long short-term memory; long short-term memory; ultra-short-term prediction; wind power

1. Introduction

Renewable energy is increasingly being discussed to phase out fossil fuel power generation to address changes in conditions, such as the new climate system and serious air pollution. Wind power, as one of major source of renewable energies, varies through time and space due to various factors, such as wind speed, wind direction, and temperature. When large-scale wind power is integrated into the grid, fluctuations of wind power bring challenges to the system operations of power systems [1]. Accurate wind power prediction enables secure and economic operation, as well as better utilization of wind power [2]. According to the forecast time scale, wind power forecast is divided into ultra-short-term forecast within 4 h ahead, short term forecast within 3 days, medium-term forecast ranges from 1 week to 1 month, and long-term forecast within 1 year [3]. Ultra-short-term wind power prediction refers to forecast wind power in the next 15 min to 4 h. The prediction interval is 15 min, and the predicted wind power is rolled in next time [4]. Existing wind power forecasting methods can be divided into two categories: physical methods and data-driven methods (statistical and artificial intelligence methods) [5].

The physical models first compute the wind speed at the hub height of wind turbines based on terrain, wind farm layout, numerical weather prediction (NWP) data, and environmental characteristics (topography, land use, etc.). The physical models then calculate the wind power according to the wind speed power curve or meteorological power parameter model. This method can better reflect the physical essence of meteorological factors. References [6,7] used weather (wind, temperature, lightning density, humidity, barometric pressure, etc.) data to predict wind power. However, physical models rely heavily on meteorological data. However, the meteorological data typically provided once or twice a day, the physical model has to wait for hours between two iterations, which limit their application for ultra-short term forecasting.

To solve the problems of the physical model, researchers have shifted to a data-driven model that can deal with a large number of multivariate data. Data-driven methods usually use historical data of wind power and weather for prediction. The auto-regression integrated moving average (ARIMA), artificial neural network (ANN) method, and support vector machine (SVM) method are commonly used at present [8,9]. Some researchers decompose the wind power and then establish a prediction method based on the time series model [10,11]. However, the decomposition error will be generated by decomposing the wind speed and power sequence. This error will be transferred to the prediction model, which will reduce the prediction accuracy. Reference [12] used wind speed, wind direction, air temperature, and air pressure as input variables, and established an SVM model for short-term wind power forecasting. Its prediction accuracy is better than the artificial neural network. Wind power has time series characteristics. However, the wind power at the predicted time is related, not only to the state at the last moment, but also to the past moments. The existing time series models (such as ARMA) and commonly used neural networks (ANN, SVM, etc.) cannot learn the correlation between wind power and wind speed, wind direction, etc. Therefore, these prediction models are difficult to further improve the prediction accuracy.

To solve the problems of the traditional neural network model, researchers shifted to the based on deep learning model which can deal with time series data. Long short-term memory (LSTM) is a neural network based on deep learning. Compared with the traditional neural network, LSTM has obvious advantages in dealing with a large number of samples and nonlinear data. LSTM has recently received researchers' attention and has been used in the field of power system prediction [13,14]. Some researchers have applied it to the field of wind power prediction [15–17]. For example, there are prediction methods based on traditional LSTM or LSTM combined with optimization algorithm [18,19]. It shows that the prediction accuracy of LSTM is better than ANN, support vector regression (SVR), back propagation (BP), and Bayesian network. Because the LSTM network has short-term memory capability, it can model the influence of the wind power at the previous time on the wind power at the current time. Reference [20] used empirical mode decomposition and vibration mode decomposition to decompose the wind power sequence, then applied the LSTM model to predict wind power. However, the method using multiple decompositions will increase the prediction error. Because decomposition of the original sequence will produce decomposition errors, these errors will be transferred to the prediction model. Reference [21] used principal component analysis to select input variables and established a short-term wind power forecasting model based on the LSTM network. The input data is NWP data without considering the impact of historical data on the prediction moment [22]. Most of these methods do not improve LSTM for wind power prediction. Because the wind power and meteorological data are dynamic time series with large randomness.

To solve the above problems of the traditional LSTM model for wind power prediction, this paper proposes a modified LSTM, called error following forget gate (EFFG)-based LSTM, which updates the output of forget gate using the difference between the predicted value and the actual value. The input data of this paper is the integrated data of NWP and historical wind power. First, Spearman rank correlation analysis is performed between wind power and NWPs to select influential weather factors on wind power. Second, Spearman rank correlation analysis is performed among historical wind powers to determine the timestep of the prediction network. Finally, the EFFG-based LSTM is developed for ultra-short-term wind power prediction. Therefore, this paper provides a kind of one-step ahead forecasting on a 15-min resolution.

2. Correlation Analysis to Determine Input Variables for LSTM

There are many factors affecting wind power. Redundant information will be introduced if all meteorological factors are used as input variables. Nevertheless, few factors will result in insufficient information. Wind power has time series characteristics. The wind power at the present moment is related to historical wind power. The time step in the LSTM model determines how long historical wind power data needs to be used as input variables. If the time step is too small, it will cause a lack of prediction information. Otherwise, it will decrease model performance.

Because the probabilistic distribution of NWP data and wind power data is not a normal distribution, Spearman rank correlation coefficient analysis is used to select meteorological factors and determine the time step. The Spearman correlation coefficient is calculated as follows

$$\rho_s = 1 - \frac{6\sum_{i=1}^{n} d_i^2}{n(n^2-1)} \tag{1}$$

$$d_i = rg(X_i) - rg(Y_i)$$

where $rg(X_i)$ and $rg(Y_i)$ are the ranks of each sequence of X and Y, and n is the number of samples. X_i and Y_i are two sequences of wind power.

Figure 1 shows that the correlation analysis between wind power and NWP data and the correlation analysis among wind power time series are carried out using the Spearman rank coefficient to determine the input variables of LSTM. A time series, including real wind power (RWP) at the historical time and NWP at forecast time is reconstructed, which is used as the input of each step of prediction for wind power prediction.

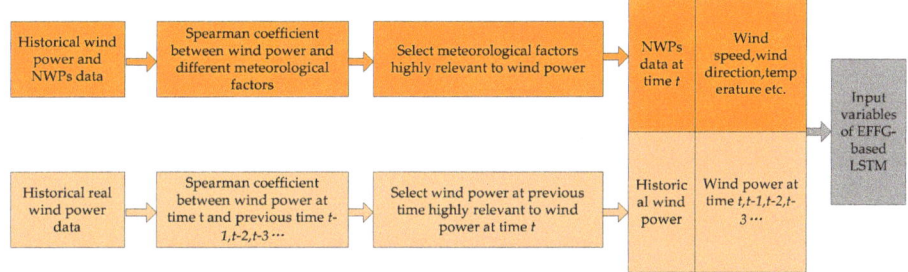

Figure 1. Correlation analysis to determine input variables for long short-term memory (LSTM).

3. Error Following Forget Gate-Based LSTM

LSTM has three gates: input gate, output gate, and forget gate, to protect and control the state of LSTM cell [23]. The inputs of the LSTM cell at time t include the inputs variables at time t, the outputs at time $t-1$, and the state variable at time $t-1$. The outputs of the LSTM cell include the output value at time t and LSTM cell state at time t. Figure 2 shows the LSTM model and its internal structure.

The forget gate update of traditional LSTM uses the output of the previous moment and the input of the prediction moment. According to the output at time $t-1$ and the input at time t, the forget gate f (t) determines how much cell information is saved to t time cell state. The mathematical expression of the forget gate is:

$$f_t = S(w_f \cdot [h_{t-1}, x_t] + b_f) \tag{2}$$

where $S(\cdot)$ is the sigmoid activation function; W_f is the weight matrix of the forgetting gate, and b_f is the bias of the forgetting gate.

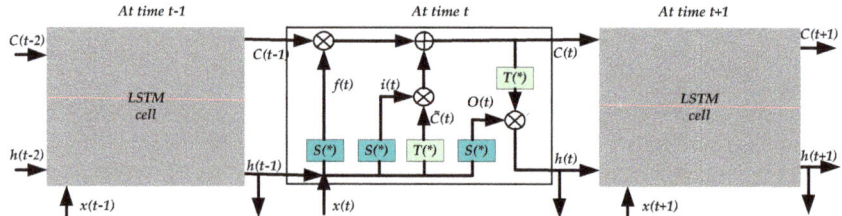

Figure 2. LSTM timing prediction model and its internal structure.

The ultra-short-term wind power prediction is to predict the wind power in the next 15 min. Therefore, when the prediction at time t is carried out, the actual power value at the t time and the predicted value are obtained from the model output. The error between the actual value and the predicted value determines how much historical information is forgotten. From Equation (2), it can be seen that the traditional LSTM forgetting gate cannot update with the error between the actual value and the predicted value. The forget gate of traditional LSTM cannot take into account the adjustment effect of the deviation between the predicted value and the actual value at the next predicted moment.

In the ultra-short-term prediction of wind power, the deviation between the actual and the predicted value at the previous moment cannot only reflect the prediction ability of the model but also reflect the positive effect of historical information on the output. If the deviation is large, the influence of the previous moment should have little impact on the prediction moment. The deviation between the predicted value and the actual value at the previous moment is proposed in this paper to update the input for the forget gate. Therefore, the newly proposed forget gate is called error following the forget gate. The calculation formula of the new forget gate is as follows

$$f_t = S(W_f k + b_f) \tag{3}$$

where $k = h'_{t-1} - h_{t-1}$, $h'_{(t-1)}$ is the actual value at time $t-1$; $h_{(t-1)}$ is the predicted value at time $t-1$. W_f and b_f are the weight matrix and bias of the forget gate. The weight W_f and bias b_f will be optimized by the optimization algorithm in the model training stage, such as Adam.

In comparison with the traditional LSTM network, the forget gate input is enhanced to take into account of the deviation between the predicted value and the actual value at the previous moment. The input gate and output gate remain the same as the ones in traditional LSTM. Figure 3 shows the new structure of the EFFG-based LSTM network at time t.

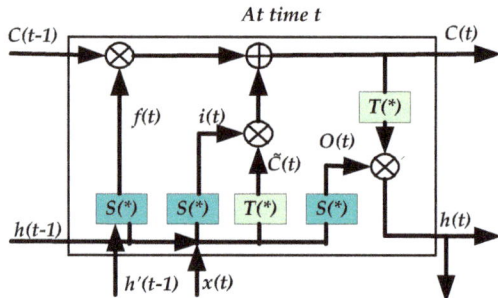

Figure 3. Error following forget gate (EFFG)-based LSTM network internal structure.

The update formula of the input gate is as Formula (4).

$$i_t = s(w_i \cdot [h_{t-1}, x_t] + b_i) \tag{4}$$

where $S(\cdot)$ is the sigmoid activation function; W_i is the weight matrix of the input gate, and b_i is the offset term of the input gate.

The update formula of the current time memory is as Formula (5)

$$\widetilde{c_t} = T(w_c \cdot [h_{t-1}, x_t] + b_c) \tag{5}$$

where $T(\cdot)$ is the tanh activation function; W_c is the weight matrix of the input gate, and b_c is the offset term of the input gate.

The update formula of the new cell state is as Formula (6).

$$c_t = f_t \cdot c_t + i_t \cdot \widetilde{c_t} \tag{6}$$

The update formula of the output gate is as Formula (7).

$$o_t = s(w_o \cdot [h_{t-1}, x_t] + b_o) \tag{7}$$

where $S(\cdot)$ is the sigmoid activation function; W_o is the weight matrix of the input gate, and b_o is the offset term of the input gate.

The output of the EFFG-based LSTM network is shown in Formula (8).

$$h_t = o_t \cdot \tanh(c_t) \tag{8}$$

4. Ultra-Short-Term Wind Power Prediction Model Based on EFFG-Based LSTM

The prediction model of wind power based on EFFG-based LSTM is shown in Figure 4. Assume a wind farm, according to Spearman correlation analysis, the RWP takes power at the first three moments of the predicted moment, and NWP takes the wind speed and direction at the prediction moment. The deviation between the predicted value and the actual value at the last moment is used as the input of EFFG-based LSTM model to update the forget gate.

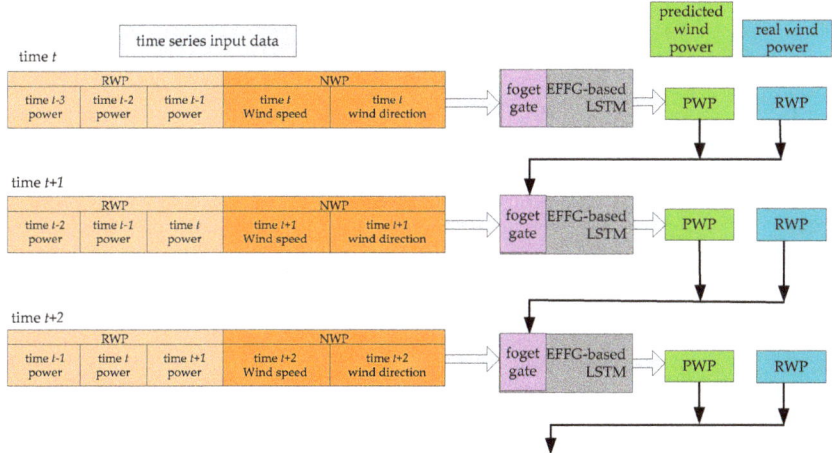

Figure 4. The dynamic structure of wind power forecasting model based on EFFG-based LSTM.

The structure of EFFG-based LSTM is composed of one input layer, one hidden layer, and one output layer. The static network structure is shown in Figure 5.

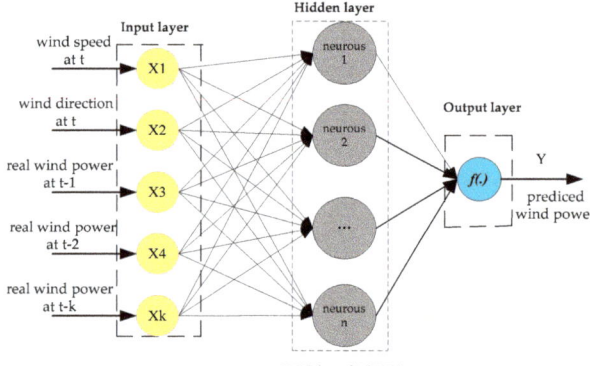

Figure 5. The static structure of wind power prediction model based on EFFG-based LSTM.

Input layer: X is the historical wind power of the past time and the wind speed and direction at the predicted time. The input layer normalizes the input data according to Equation (9).

$$X' = \frac{x - (x_{max} - x_{min})/2}{(x_{max} - x_{min})/2} \qquad (9)$$

where x_{max} and x_{min} are the maximum and minimum values of the variable, respectively.

Hidden layer: The hidden layer is the EFFG-based LSTM network. One hidden layer can ensure the faster prediction calculation speed and avoid overfitting phenomenon caused by too many hidden layers.

Output layer: $f(\cdot)$ is the activation function of the output layer. Y is the predicted value of wind power at the next time. The output layer weights and biases the output of the hidden layer, and outputs one-dimensional predicted wind power. The inverse normalization is calculated according to Equation (10). Finally, the predicted value of wind power is obtained.

$$Y = \frac{1}{2}[y'(y_{max} - y_{min}) + (y_{max} + y_{min})] \qquad (10)$$

5. Case Study

5.1. Data Description and Test Design

The data (NWP and historical real wind power) is from a practical wind farm in northwest China. The period is from January to December 2017. The sampling time interval of wind power is 15 min. the time interval of NWP data is 15 min. NWP data includes wind speed and direction of 170 m, 100 m, and 30 m, temperature, pressure, humidity. The NWP dada is every 15 min as a rapid refresh model. The model is a method of weather forecast based on the mathematical model of atmospheric movement and using the current weather conditions as input data. The forecast is computed for the next 15 min time step.

Firstly, the correlation coefficient between the wind power and NWP features, RWP historical moments is calculated by using the Spearman correlation coefficient, and then the input variables of the prediction model are selected.

It can be seen from Table 1 that the correlation coefficient average value of the wind speed and wind direction at 100 m is 0.57 and 0.52. The correlation coefficient average value of the wind speed at 30 m is 0.51. The two variables are strongly correlated when the correlation coefficient is greater than 0.5 [24]. When forecasting the wind power at time t, the wind speed and wind direction at 100 m, wind speed 30 m at time t are put into the newly constructed input time series.

Table 1. Spearman correlation coefficient between wind power and numerical weather prediction (NWP) features. RWP = real wind power.

NWPs \ RWP	January to March	April to June	July to September	October to December	Average Value
wind speed at 170 m	0.45	0.38	0.40	0.48	0.43
wind speed at 100 m	0.53	0.57	0.61	0.55	0.57
wind speed at 30 m	0.56	0.46	0.54	0.49	0.51
wind direction at 170 m	0.33	0.45	0.38	0.41	0.39
wind direction at 100 m	0.48	0.55	0.58	0.49	0.52
wind direction at 30 m	0.26	0.36	0.33	0.34	0.32
temperature	0.15	0.09	0.05	0.17	0.12
pressure	0.42	0.37	0.28	0.36	0.36
humidity	0.03	0.05	0.11	0.08	0.07

It can be seen from Table 2 that the correlation coefficient of the power at time $t-1$, $t-2$, and $t-3$ are greater than 0.5. Therefore, the current wind power at time t has a strong correlation with the power at time $t-1$, $t-2$, and $t-3$. When forecasting the wind power at time t, the wind power at the past three times is put into the newly constructed input time series.

Table 2. Spearman correlation coefficient between power at t time and at time t before.

RWP at Time t \ RWP at Time t before	15 min Interval ($t-1$)	30 min Interval ($t-2$)	45 min Interval ($t-3$)	60 min Interval ($t-4$)	75 min Interval ($t-5$)
RWP at 8:00	0.83	0.72	0.51	0.37	0.21
RWP at 10:15	0.78	0.70	0.48	0.23	0.11
RWP at 13:30	0.81	0.67	0.50	0.18	0.07
RWP at 16:00	0.68	0.60	0.52	0.22	0.03
RWP at 21:45	0.73	0.66	0.49	0.15	0.06

EFFG-based LSTM network parameter set timestep is set to 4, which is the three historical moments of RWP and at t time of NWP. The number of hidden layer neurons (EFFG-based LSTM cell) is not linearly related to the forecast accuracy (such as root mean square error (*RMSE*)). The number of neurons in the hidden layer should be determined according to the number of input features and the model training accuracy. The number of neurons in the hidden layer is set to 12 to achieve the best prediction accuracy. The EFFG-based LSTM gate activation function remains at the default value.

Finally, use the historical data of March, June, September, and December 2017 of the wind farm as the training data set of the model. Then, the four groups of prediction tests were conducted as follows:

(1) Group 1: to predict the wind power within 24 h of February 22.
(2) Group 2: to predict the wind power within 24 h of May 13.
(3) Group 3: to predict the wind power within 24 h of September 30.
(4) Group 4: to predict the wind power within 24 h of November 17.

5.2. Forecast Error Computation

Because the actual wind power value has zero value, the mean absolute percentage error (MAPE) in the forecast effect evaluation index will be meaningless [25]. Therefore, root mean square error (*RMSE*), accuracy *R1*, and qualification *R2*, are used to evaluate the prediction results.

The calculation formula of *RMSE* is:

$$RMSE = \sqrt{\frac{1}{n}\sum_{i=1}^{n}(P_i - P'_i)^2} \quad (11)$$

where n is the number of forecast samples; P is the actual value of wind power and P' is predicted value of wind power; i is the serial number of the actual value and the predicted value.

The calculation formula of the accuracy rate R_1 is:

$$R_1 = [1 - \sqrt{\frac{1}{N}\sum_{K=1}^{N}\left(\frac{P_{MK} - P_{PK}}{P_{cap}}\right)^2}] \times 100\% \quad (12)$$

In the formula, P_{MK} is the average value of the actual power in the K period, P_{PK} is the average value of the predicted power in the K period, P_{cap} is the starting capacity of the wind farm in the corresponding period, and N is the total number of predicted periods.

The calculation formula for the pass rate R_2 is:

$$R_2 = \frac{1}{N}\sum_{K=1}^{N} B_K \times 100\% \quad (13)$$

In the formula, $\left(1 - \frac{P_{MK}-P_{PK}}{P_{cap}}\right) \times 100\% \geq 75\%$, $B_K = 1$; $\left(1 - \frac{P_{MK}-P_{PK}}{P_{cap}}\right) \times 100\% < 75\%$, $B_K = 0$.

5.3. Comparison of Prediction Results

To verify that EFFG-based LSTM model has higher prediction accuracy, SVM and standard LSTM wind power prediction model are used for prediction and comparative analysis. Therefore, the same input data for all three methods for the prediction test. The predicted wind power and prediction error curve of Group 1 as shown in Figures 6 and 7. The predicted wind power and prediction error curve of Group 2 as shown in Figures 8 and 9. The predicted wind power and prediction error curve of Group 3 as shown in Figures 10 and 11. The predicted wind power and prediction error curve of Group 4 as shown in Figures 12 and 13. The time in the x-axis is the local time in all figure.

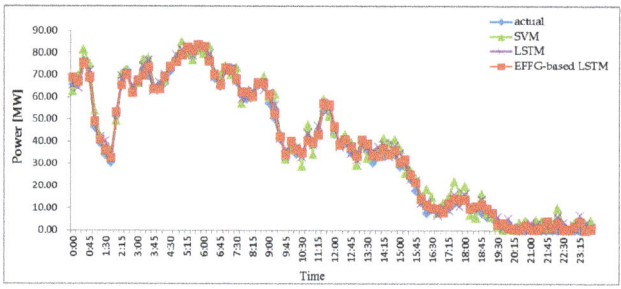

Figure 6. Each model predicts wind power of 24 h of Group 1.

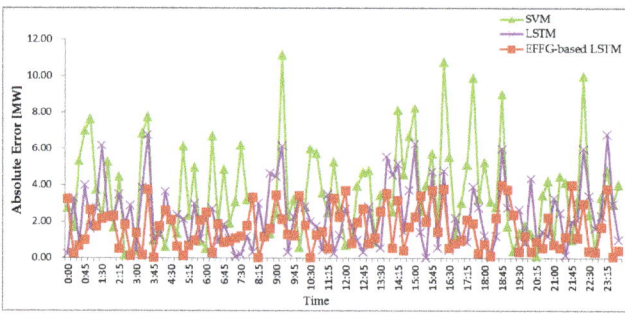

Figure 7. Each model prediction error of 24 h of Group 1.

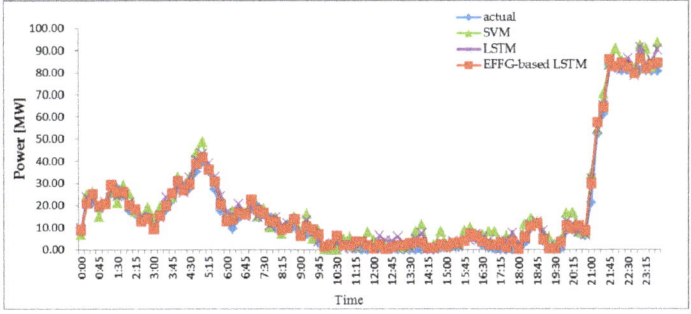

Figure 8. Each model predicts wind power of 24 h of Group 2.

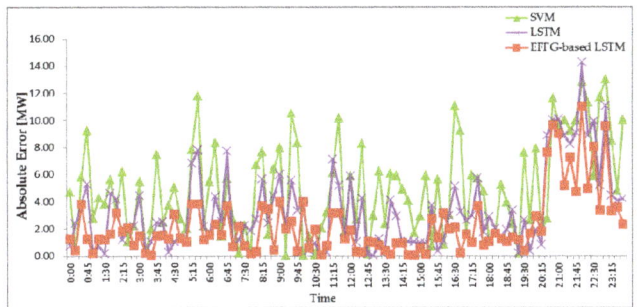

Figure 9. Each model prediction error of 24 h of Group 2.

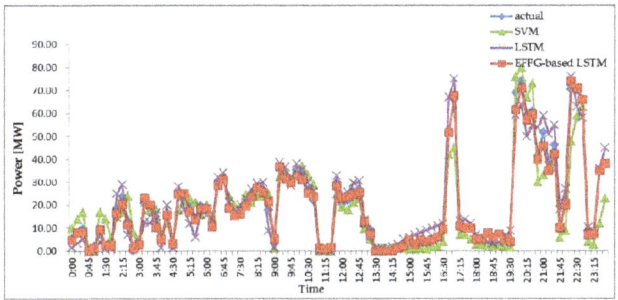

Figure 10. Each model predicts wind power of 24 h in Group 3.

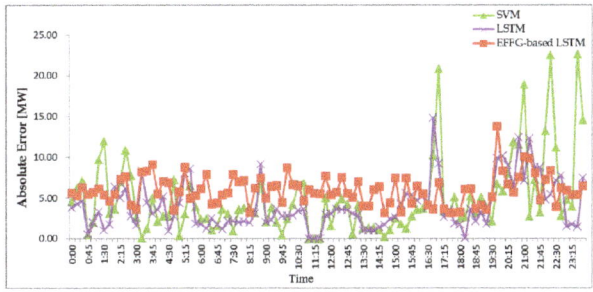

Figure 11. Each model prediction error of 24 h of Group 3.

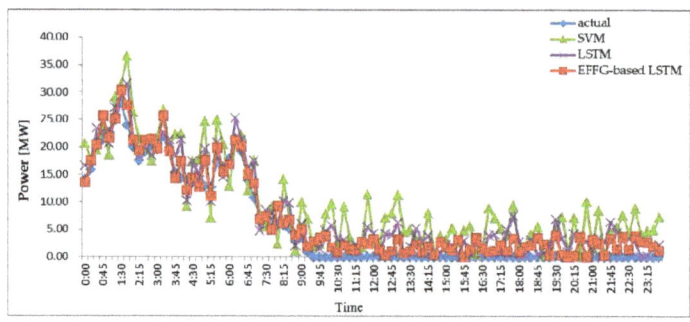

Figure 12. Each model predicts wind power of 24 h of Group 4.

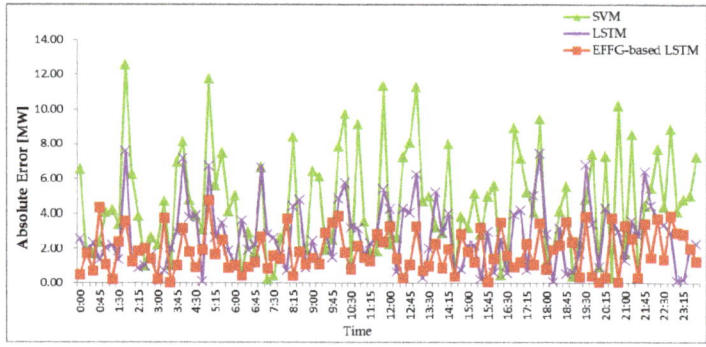

Figure 13. Each model prediction error of 24 h of Group 4.

It can be seen from Figures 6–13 that the SVM prediction model is the worst of the three models. The reason is that SVM is not a prediction model suitable for time series and cannot process time series information. LSTM model and EFFG-based LSTM model not only use deep learning technology to optimize network parameters, but also deal with the correlation information and time correlation between wind power time series as a time series model. When the wind power fluctuates, the accuracy of EFFG- based LSTM model is better than traditional LSTM model. The reason is that the traditional LSTM forget gate is updated by the last time output and input data, which cannot reflect the influence of the error between the predicted value and the actual power value on the forget gate. The EFFG-based LSTM forget gate is updated by the error between the predicted value and the actual power value. When the error is large, the forgetting coefficient of the model is large. Then the historical output value will be forgotten more, and the effect of historical value on the model will be smaller. When the wind power suddenly changes, there is no correlation between the next moment's wind power and the previous moment's power, it is necessary to reduce the role of historical value. Therefore, EFFG- based LSTM prediction model has the highest prediction accuracy.

It is difficult to predict the wind power slope, this paper uses the historical wind power slope data to train the prediction model. However, the following Figure 14 shows the up ramp period of wind power in Figure 8 of our paper. It can be seen from the figure that there is a prediction delay and the prediction error is relatively large. The maximum prediction error of the better model (EFFG-based LSTM) is over 10 MW. The same problem appears in Figure 10 of the up ramp period. This is one of the tasks we need to further study.

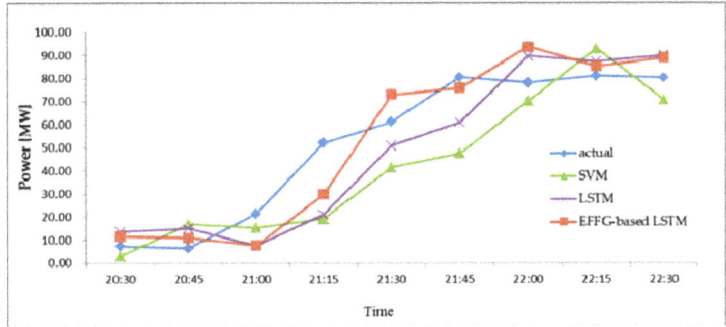

Figure 14. Detail of the ramp in Figure 8.

Tables 3–6 show *RMSE*, *R*1, and *R*2 of 4 h and 24 h forecast results of Group 1, Group 2, Group 3, and Group 4.

Table 3. Root mean square error (*RMSE*), accuracy rate (*R*1), pass rate (*R*2) of prediction results of Group 1 of each model.

	SVM	LSTM	EFFG-Based LSTM
RMSE	6.71	3.75	2.57
R1	73.19	83.54	90.81
R2	74.07	82.10	90.62

Table 4. Root mean square error (*RMSE*), accuracy rate (*R*1), pass rate (*R*2) of prediction results of Group 2 of each model.

	SVM	LSTM	EFFG-Based LSTM
RMSE	7.81	3.23	2.97
R1	73.78	82.44	90.78
R2	75.04	82.22	90.82

Table 5. Root mean square error (*RMSE*), accuracy rate (*R*1), pass rate (*R*2) of prediction results of Group 3 of each model.

	SVM	LSTM	EFFG-Based LSTM
RMSE	7.81	3.88	2.87
R1	73.78	81.02	90.11
R2	74.54	80.47	90.65

Table 6. Root mean square error (*RMSE*), accuracy rate (*R*1), pass rate (*R*2) of prediction results of Group 4 of each model.

	SVM	LSTM	EFFG-Based LSTM
RMSE	7.01	3.13	2.44
R1	74.68	81.40	90.06
R2	73.04	81.77	90.20

It can be seen from Tables 3–6 that the *RMSE*, *R*1, and *R*2 of the EFFG-based LSTM prediction model are optimal compared with the SVM and LSTM prediction models. The *RMSE* of the EFFG-based LSTM model changed small from Group 1 to Group 4 prediction, and the *R*1 and *R*2 were above 90% in 24 h prediction. The accuracy and qualified rate of SVM and LSTM models decreased from Group 1 to Group 4 prediction. SVM and traditional LSTM model can't deal with the period when wind power changes dramatically. This problem is solved by an improved forgetting gate, so the performance of the EFFG-based LSTM prediction model is the best.

6. Conclusions

This paper proposed the EFFG-based LSTM model for an ultra-short-term wind power forecasting method. The experimental outcomes are as follows: (1) Spearman correlation coefficient method can better find the relationship between predictive factors affecting wind power because the probabilistic distribution of NWP data and wind power data are not known distributions. (2) The input data is the integrated data of NWP and historical wind power. Compared with the monotype input data, it can better reflect the support effect of the wind power prediction. (3) The proposed method can realize an ultra-short-term wind power prediction, considering the influence of the error between the predicted value and the actual value on the prediction model. The forecast accuracy of the EFFG-based LSTM model is better than SVM and the traditional LSTM model.

Author Contributions: Conceptualization P.Z. and C.P.; software, C.L. and R.Y.; validation, C.L. and P.Z.; formal analysis, C.L. and R.Y.; writing—original draft preparation, J.T. and C.L.; writing—review and editing, P.Z. and C.P.; visualization, J.T. and C.L.; supervision, P.Z. and M.S. All authors have read and agreed to the published version of the manuscript.

Funding: The authors gratefully acknowledge the financial support of the research and application of new energy power prediction and evaluation based on micrometeorological big data (19214310D), National Natural Science Foundation of China (51867008), Natural Science Foundation of Jiangxi Province of China (20192ACBL20007).

Acknowledgments: Thanks to Ningxia Helanshan No.3 wind farm for providing us with wind farm data.

Conflicts of Interest: The authors declare no conflict of interest. The founding sponsors had no role in the design of the study, in the collection, analyses, or interpretation of data, in the writing of the manuscript, and in the decision to publish the results.

References

1. Lujano-Rojas, J.M.; Osório, G.J.; Matias, J.C.O. A heuristic methodology to economic dispatch problem incorporating renewable power forecasting error and system reliability. *J. Renew. Energy* **2016**, *87*, 731–743. [CrossRef]
2. Carta, J.A.; Velazquez, S.; Cabrera, P. A review of measure-correlate-prediction (MCP) methods used to estimate long-term wind characteristics at a target site. *J. Renew. Sustain. Energy Rev.* **2013**, *4*, 362–400. [CrossRef]
3. Tascikaraoglu, A.; Uzunoglu, M. A review of combined approaches for prediction of short-term wind speed and power. *J. Renew. Sustain. Energy Rev.* **2014**, *34*, 785–792. [CrossRef]
4. Tan, L.; Han, J.; Zhang, H. Ultra-short-term wind power prediction by salp swarm algorithm-based optimizing extreme learning machine. *IEEE Access* **2015**, *8*, 44470–44484. [CrossRef]
5. Okumus, I.; Dinler, A. Current status of wind energy forecasting and a hybrid method for hourly predictions. *Energy Convers. Manag.* **2016**, *123*, 362–371. [CrossRef]
6. Sarwat, A.; Amini, M.; Domijan, A.; Damnjanovic, A.; Kaleem, F. Weather-based interruption prediction in the smart grid utilizing chronological data. *J. Mod. Power Syst. Clean. Energy* **2016**, *4*, 308–315. [CrossRef]
7. Zhao, W.; Weisheng, W.; Bo, W. Regional wind power forecasting model with NWP grid data optimized. *Front. Energy* **2017**, *11*, 175–183.
8. Giorgi, M.D.; Campilongo, S.; Ficarella, A.; Congedo, P.M. Comparison between wind power prediction models based on wavelet decomposition with least-squares support vector machine (LS-SVM) and artificial neural network (ANN). *Energies* **2014**, *7*, 5251–5272. [CrossRef]
9. Liu, H.; Xiwei, M.; Yanfei, L. Wind speed forecasting method based on deep learning strategy using empirical wavelet transform long short term memory neural network and Elman neural network. *J. Energy Convers. Manag.* **2018**, *156*, 498–514. [CrossRef]
10. Renani, E.; Elias, M.; Rahim, N.A. Using Data-driven approach for wind power prediction: A comparative study. *J. Energy Convers. Manag.* **2016**, *118*, 193–203. [CrossRef]

11. Liu, X.J.; Mi, Z.Q.; Lu, B. A Novel Approach for Wind Speed Forecasting Based on EMD and Time-Series Analysis. In Proceedings of the Asia-pacific Power & Energy Engineering Conference, Denver, CO, USA, 18–20 September 2009.
12. Zeng, J.; Qiao, W. Short-term wind power prediction using a wavelet support vector machine. *IEEE Trans. Sustain. Energy* **2012**, *3*, 255–264. [CrossRef]
13. Zhang, S.; Wang, Y.; Liu, M.; Bao, Z. Data-based line trip fault prediction in power systems using LSTM networks and SVM. *IEEE Access* **2018**, *6*, 7675–7686. [CrossRef]
14. Zheng, C. A novel equivalent model of active distribution networks based on LSTM. *IEEE Trans. Neural Netw. Learn. Syst.* **2019**, *30*, 2611–2624. [CrossRef] [PubMed]
15. Xu, G.; Xia, L. Short-Term Prediction of Wind Power Based on Adaptive LSTM. In Proceedings of the 2018 2nd IEEE Conference on Energy Internet and Energy System Integration (EI2), Beijing, China, 20–22 October 2018.
16. Qiaomu, Z.; Hongyi, L.; Ziqi, W.; Jinfu, C.; Bo, W. Short-term wind power forecasting based on LSTM. *J. Power Syst. Technol.* **2017**, *41*, 3797–3802.
17. Shahid, F.; Zameer, A.; Mehmood, A.; Raja, M.A.Z. A novel wavelet's long short-term memory paradigm for wind power prediction. *Appl. Energy* **2020**, *269*, 1509–1517. [CrossRef]
18. Xiaoyun, Q.; Xiaoning, K.; Chao, Z.; Shuai, J.; Xiuda, M. Short-Term Prediction of Wind Power Based on Deep Long Short-Term Memory. In Proceedings of the 2016 IEEE PES Asia-Pacific Power and Energy Engineering Conference (APPEEC), Xi'an, China, 25–28 October 2016; pp. 1148–1152.
19. Shi, X.; Lei, X.; Huang, Q.; Huang, S.; Ren, K.; Hu, Y. Hourly day-ahead wind power prediction using the hybrid model of vibrational model decomposition and long short-term memory. *Energies* **2018**, *11*, 3227. [CrossRef]
20. Hao, Y.; Zuhong, O.; Shengquan, H.; Anbo, M. A cascaded deep learning wind power prediction approach based on a two-layer of mode decomposition. *Energy* **2019**, *189*, 116316. [CrossRef]
21. Qu, X.Y.; Kang, X.N.; Zhang, C. Short-Term Prediction of Wind Power Based on Deep Long Short-Term Memory. In Proceedings of the Asia-Pacific Power & Energy Engineering Conference, Suzhou, China, 15–17 April 2016.
22. Giorgi, M.G.D.; Ficarella, A.; Tarantino, M. Assessment of the benefits of numerical weather predictions in wind power forecasting based on statistical methods. *Energy* **2018**, *36*, 3968–3978. [CrossRef]
23. Sepp, H.; Jü, S. Long short-term memory. *J. Neural Comput.* **1997**, *8*, 782–790.
24. Piantadosi, J.; Howlett, P.; Boland, J. Matching the grade correlation coefficient using a copula with maximum disorder. *J. Ind. Manag. Optim.* **2007**, *3*, 305–312. [CrossRef]
25. Sreelakshmi, K.; Ramakanth, K.P. Performance evaluation of short term wind speed prediction techniques. *IJCSNS Int. J. Comput. Sci. Netw. Secur.* **2008**, *8*, 572–581.

Publisher's Note: MDPI stays neutral with regard to jurisdictional claims in published maps and institutional affiliations.

© 2020 by the authors. Licensee MDPI, Basel, Switzerland. This article is an open access article distributed under the terms and conditions of the Creative Commons Attribution (CC BY) license (http://creativecommons.org/licenses/by/4.0/).

Article

Capacity Planning of Distributed Wind Power Based on a Variable-Structure Copula Involving Energy Storage Systems

Yurong Wang [1,*], Ruolin Yang [1], Sixuan Xu [2] and Yi Tang [1]

1. School of Electrical Engineering, Southeast University, Nanjing 210096, China; yang_ruolin@163.com (R.Y.); tangyi@seu.edu.cn (Y.T.)
2. State Grid Jiangsu Electric Power Design Consulting Co., Ltd. State Grid Jiangsu Electric Power Co., Ltd. Economic Research Institution, Nanjing 210018, China; xusixuan111@163.com
* Correspondence: wangyurong@seu.edu.cn

Received: 10 June 2020; Accepted: 11 July 2020; Published: 13 July 2020

Abstract: Distributed wind power (DWP) needs to be consumed locally under a 110 kV network without reverse power flow in China. To maximize the use of DWP, this paper proposes a novel method for capacity planning of DWP with participation of the energy storage system (ESS) in multiple scenarios by means of a variable-structure copula and optimization theory. First, wind power and local load are predicted at the planning stage by an autoregressive moving average (ARMA) model, then, variable-structure copula models are established based on different time segment strategies to depict the correlation of DWP and load, and the joint typical scenarios of DWP and load are generated by clustering, and a capacity planning model of DWP is proposed considering investment and operation cost, and environmental benefit and line loss cost under typical scenario conditions. Moreover, a collaborative capacity planning model for DWP and ESS is prospectively proposed. Based on the modified IEEE-33 bus system, the results of the case study show that the DWP capacity result is more reasonable after considering the correlation of wind and load by using a variable-structure copula. With consideration of the collaborative planning of DWP and load, the consumption of DWP is further improved, the annual cost of the system is more economical, and the quality of voltage is effectively improved. The study results validate the proposed method and provide effective reference for the planning strategy of DWP.

Keywords: collaborative capacity planning; distributed wind power (DWP); energy storage system (ESS); optimization; variable-structure copula

1. Introduction

There are generally two typical integration forms of wind power into power systems: centralized and distributed. Distributed wind farms do not transport wind power over large-scale long-distance transmission lines, they are directly provided to the load center of the power system [1,2], and the generated electricity is consumed locally. Distributed wind power (DWP) has become an effective solution for improving China's environmental issues, and it will be an important form of wind power integration into the power grid.

For the development and construction of DWP projects, China's 2018 document [3] presented that the DWP needs to be locally consumed through a 110 kV network with no power delivery at higher voltage levels, and the installed DWP capacity limit should be based on the lowest consumption of load. There is no doubt that this capacity planning principle for DWP will reduce the use rate of wind power greatly, lead to waste of wind resources, and decrease the revenue of wind power industries, further hindering the development of DWP.

At present, many effective models and algorithms for wind power planning in distribution networks have been explored [4–6]. A multi-objective DWP planning model was proposed in [7] to meet the operation of unbalanced distribution systems, and the decision framework provided in [8] could optimize DWP planning through technology selection. These studies provide a good reference for further development of wind power in the distribution system.

To comply with the regulations under the premise of local consumption in the distribution system without reverse power flow, it is necessary to further investigate the correlation between DWP and load. At present, Nataf inverse transformation [9,10] and the correlation coefficient matrix method [11] are usually employed for multivariable correlation analysis. However, the correlation feature or correlation matrix between random variables must be determined in advance. When the correlation between variables is complex or the features are not obvious, the fitting effect with the above commonly used models is usually not good. In addition, it is also necessary to take into consideration different scenarios of DWP and load because of their stochastic characteristics [12–14]. In [15], the correlation among historical wind, photovoltaic power and electricity demand and the random moments is captured by generating a scenario matrix, but the variable structure is not sufficiently considered. In view of the above issues, copula theory [16] is employed in this work to better describe the correlation between DWP and load, and at different time segments, a variable-structure copula model is established to construct the correlation between DWP and load.

To follow the principle of no reverse power flow to higher voltage level and make the best consumption of renewable resources, one effective way is to bring in the energy storage system (ESS) at the planning stage [17–19]. A joint optimization in [17] was proposed to plan the capacity and location of ESS, and distributed generating units in a stand-alone micro-grid were presented. These studies mainly implement collaborative planning from the perspective of economics and pricing-based demand response [20], thus providing a good reference for this paper. Still, the consideration of construction investment, system line loss cost and the maintenance cost of ESS as part of the model's objective requires further improvement. Based on the above research, this paper takes the network line loss, the investment operation cost and environmental income, and the time-of-day tariff into the objective functions of the planning strategy, and prospectively proposes a feasible collaborative capacity planning model of DWP and ESS. This paper contributes as follows:

(1) To describe the tail correlation and the change of correlation between DWP and load, a variable-structure copula model is employed. In this paper, the variable-structure copula models are constructed using two different time division methods (monthly and quarterly), and strategies are evaluated by constructing an empirical copula model.

(2) Based on the correlation model of the variable-structure copula, the typical scenarios containing correlation information are obtained through the clustering method, and then the DWP capacity planning model is constructed under the typical scenarios. Furthermore, in order to increase the consumption of DWP, a collaborative planning model is established for wind storage, and the consumption of wind power and the quality of voltage level are analyzed based on a typical schedule day.

The structure of this paper is as follows. In Section 2, the autoregressive moving average model (ARMA) model for data prediction of DWP and load is introduced, the correlation of DWP and load based on the variable-structure copula is investigated, and joint typical scenarios are generated using the clustering method. In Section 3, a novel capacity planning model of DWP is proposed under typical scenarios, and a collaborative capacity planning model is further established for DWP and ESS to increase the consumption of wind power. A case study is then carried out in Section 4 based on practical data for wind farm and load. Section 5 summarizes the findings.

2. Correlation Analysis between DWP and Load Based on a Variable-Structure Copula

The wind power output of a distributed wind farm should be consumed by its local load. Since DWP has the characteristics of intermittency and inverse peak shaving, the correlation between

DWP and load consumption should be carefully investigated. In this section, a variable-structure copula model is employed to describe the joint density of DWP and local load. This method can well capture the nonlinear, asymmetry and tail correlation characteristics among variables, it can analyze the marginal distribution of each random variable individually, and can also illustrate the varied correlated structure between variables.

2.1. Data Preparation Based on an ARMA Model

To carry out the correlation study between DWP and load, the predicted data of DWP and local load in the planning stage are first obtained by ARMA model. The ARMA short-term prediction model includes the autoregressive part and the moving average part, and its formula is:

$$Y_t = \sum_{i=1}^{q} \alpha_i \varepsilon_{t-i} + \sum_{i=1}^{p} \beta_i Y_{t-i} + \varepsilon_t, \tag{1}$$

where Y_t is the value of DWP or load at point t of series; ε_t and ε_{t-i} are the prediction error term at t and i time points ahead of t, respectively; α is the correlation coefficient, which reflects the dependence of the prediction error at different segments; Y_{t-i} is the value with i time points ahead of t; β is the correlation coefficient; p is the order of autoregressive process; and q is the order of moving average process.

The order of the ARMA model can be determined by calculating the Akaike Information Criterion (AIC) value of the ARMA with different (p, q) pairs. The optimal ARMA (p, q) model is selected when the AIC value is the smallest.

For cases where there is no historical data of DWP in the local area, the centralized wind power data near the area can be used as a reference, since they have a similar wind source, and the data can be converted proportionally into the DWP capacity for prediction and planning analysis.

2.2. Theory of Copula Function

Based on the predicted time series of DWP and load at the planning stage, in this subsection, this paper proposes a variable-structure copula to depict the correlation between DWP and load.

2.2.1. The Definition and Properties of the Copula Function

Because DWP and load have the characteristics of fluctuation, and DWP also has the characteristic of inverse peak shaving, the correlation between wind power and load is very complicated, and correlation under extreme conditions (tail correlation) cannot be ignored. Therefore, the copula is a useful tool for characterizing nonlinear correlation and tail correlation [21,22] between DWP and load.

Copula theory states that there must exist a copula function that satisfies $F(x,y) = C(F_1(x), F_2(y))$ [23], where $F(\cdot)$ is a 2-dimensional cumulative distribution function, $C(\cdot)$ is a distribution function of two-element copula function, x and y are the samples of DWP and load (MW), respectively, and $F_1(\cdot), F_2(\cdot)$ are the marginal probability density functions of DWP and load, respectively. To simplify, let $u = F_1(x)$, $v = F_2(y)$, and vector u_i and v_i are the values of $F_1(x)$ and $F_2(y)$ at point i.

As a result, to construct a copula model, the first step is to estimate the marginal distribution of DWP and load. Next, a copula function should be carefully selected to fit the correlation between marginal distributions based on some evaluation indices.

2.2.2. Evaluation Indices of Copula Function

After estimating the marginal distribution function of DWP and load, respectively, this paper estimates the parameters based on maximum likelihood estimation (MLE), and the evaluation indices can subsequently be calculated.

The parameter estimation results of Gaussian copula and t copula are the same as Pearson coefficient, which can reflect variables' linear correlation

$$\rho_p = \frac{\sum_{i=1}^{n}(u_i - \overline{u})(v_i - \overline{v})}{\sqrt{\sum_{i=1}^{n}(u_i - \overline{u})^2} \sqrt{\sum_{i=1}^{n}(v_i - \overline{v})^2}}, \quad (2)$$

where $\overline{u}$ and $\overline{v}$ are the expected values of the vectors u_i and v_i, respectively, and ρ_p is the Pearson coefficient of the vector u_i and v_i.

The evaluation indices also include Kendall coefficient, Spearman coefficient and Euclidean index [24].

(1) Kendall coefficient ρ_k can reflect the nonlinear correlation of the change trend of the vectors u_i and v_i

$$\rho_k = \frac{4r}{n(n-1)} - 1, \quad (3)$$

where r is the number of the vectors u_i and v_i, whose two attribute values have the same size relationship.

(2) Spearman coefficient ρ_z can reflect the correlation of the rank of the variables

$$\rho_z = 1 - \frac{6\sum d_i^2}{n(n^2 - 1)}, \quad (4)$$

where d_i is the rank differences between two vectors u_i and v_i.

(3) Euclidean index can reflect the distance between the model and the empirical copula model [25].

$$d_x^2 = \sum_{i=1}^{n} |C_m(u_i, v_i) - C_x(u_i, v_i)|^2, \quad (5)$$

where $C_m(u_i, v_i)$ is the empirical copula function of DWP and load, and $C_x(u_i, v_i)$ is the basic copula functions, where the smaller the Euclidean distance is, the more accurate the model is.

Copula functions used in this paper include Gaussian copula, t copula, Frank copula, Gumbel copula and Clayton copula. To carry out the model evaluation of copula functions, the evaluation indices should be calculated and compared with the empirical copula, as shown in (6) [25].

$$C_m(u_i, v_i) = \frac{1}{n} \sum_{i=1}^{n} I_{(F_m(x_i) \leq u_i)} I_{(G_m(y_i) \leq v_i)}, \quad (6)$$

where $F_m(x_i)$ and $G_m(y_i)$ are the empirical distribution functions of DWP and load, respectively, $I(\cdot)$ represents explanatory function, and u and v follow 0-1 distribution satisfying $F(x_i) \leq u_i$, $I_{(F(x_i) \leq u_i)} = 1$.

2.2.3. Variable-Structure Copula

According to the stochastic characteristic of DWP and load, their joint distribution can exhibit varied correlation features at different periods; under these conditions, a unique copula function cannot sufficiently describe the change. The variable-structure copula provides the most suitable copula model for the description of correlation at different stages according to the varied structural features of DWP and load, and is able to capture the changes of related structures between them more flexibly [26].

In general, the variable-structure copula can be divided into three types [27,28]:

(1) Only the marginal distribution of a single variable has a variable structure;
(2) The copula function part with a definite marginal distribution possesses a variable structure;
(3) Both the marginal distribution of a single variable and the copula function possess variable structures.

In this work, both the marginal distribution of a single variable and the joint copula function are modeled with variable structures. Based on different time division strategies, the main steps of constructing the variable-structure copula model are as follows:

(1) Divide the time series of DWP and load into multiple time segments;
(2) Apply non-parametric estimation to determine the marginal distribution of each variable at each time segment;
(3) Construct the copula model at each time segment;
(4) Perform parameter estimation, evaluate the candidate models and select the optimal copula for each time segment;
(5) Compare the results based on different time division methods based on (6), and choose the most appropriate division strategy.

For each phased copula function, a binary frequency histogram between variables can be used intuitively as a first estimate of the joint density function selection of DWP and load. By means of the MLE method, the parameters of each basic copula model can be calculated [29]. Based on the evaluation indices of candidate copulas and empirical copula, the two-stage filtration method [30] is used to choose the optimal copula model.

After the modeling of the variable-structure copula, typical scenarios can be generated for further DWP planning.

2.3. Typical Scenario Generation

Based on the continuous variable-structure copula function, it is necessary to discretize it to obtain discrete DWP and load data pairs, so as to provide typical scenarios for the capacity planning of a distributed wind farm.

This paper uses K-means clustering to classify typical scenarios. The specific steps are as follows:

(1) Discretize each phased copula function to generate two-dimensional discrete data pairs.
(2) Set the number of typical scenarios and select the initial condensation point.
(3) Calculate the distance from the discrete points to each condensation point, selecting the minimum distance, and divide them into each class.
(4) Update the location of the condensate points for each class, and re-calculate step (3) to obtain a new clustering result until the set number of cycles is reached.
(5) Choose the best clustering result and find the corresponding original quantile by inverting the probability distribution function.

The joint typical scenarios of DWP and load can reflect the volatility of them with different conditions, and provide a feasible reference for the rational capacity planning of DWP.

3. Capacity Planning Model for Regional Distributed Wind Farms

Based on the established typical scenarios between DWP and load, this section firstly proposes an optimal capacity planning model for DWP. Then, a collaborative planning method with ESS is further proposed in order to improve the consumption of DWP.

3.1. Capacity Planning Model of DWP

In this subsection, the capacity planning model for DWP is set up based on the typical scenarios of DWP and load.

3.1.1. Optimization Function

The investment cost of the distributed wind farm, the environmental income provided by the government, and the cost of line loss are included in the objective function:

$$F_c = \min(f_1 + f_2), \tag{7}$$

First, f_1 is the annual investment cost of the distributed wind farm and the annual environmental income provided by the government [31].

$$f_1 = P_{DWG}\left[C_{wt} - C_{sr}\frac{r_0(1+r_0)}{(1+r_0)^T - 1}\right], \tag{8}$$

where P_{DWG} is the planning capacity of DWP in MW, C_{wt} is the annual initial investment cost of the distributed wind farm (RMB/kW), C_{sr} is the environmental income per unit capacity (RMB/kW), r_0 is the discount rate, and T is the operating life of the distributed wind farm (year).

Second, f_2 is the annual line loss cost of the power system:

$$f_2 = 8760\sum_{i=1}^{N}\sum_{\substack{j=1 \\ j \neq i}}^{N} \frac{\Delta U_{ij}^2}{|Z_{ij}|}\cos(\varphi)C_d, \tag{9}$$

where N is the number of system buses, ΔU_{ij} is the voltage difference between bus i and bus j of the system (kV), Z_{ij} is the impedance of the branch i-j (Ω), C_d is the electricity price (RMB/kWh), and φ is the power factor angle (rad).

3.1.2. Constraints

Considering the system power balance, the capacity limit of the generator, the constraints of bus voltage and phase angle, and the constraints are listed as follows:

$$P_{Gi} + P_{DWGi} - P_{Di} - U_i\sum_{j=1}^{N} U_j(G_{ij}\cos\theta_{ij} + B_{ij}\sin\theta_{ij}) = 0, \tag{10}$$

$$Q_{Gi} - Q_{Di} + U_i\sum_{j=1}^{N} U_j(G_{ij}\sin\theta_{ij} - B_{ij}\cos\theta_{ij}) = 0, \tag{11}$$

$$0 \leq P_{Gi} \leq P_{Gimax}, \tag{12}$$

$$0 \leq P_{DWGi} \leq P_{DWGmax}, \tag{13}$$

$$U_{imin} \leq U_i \leq U_{imax}, \tag{14}$$

$$|\theta_{ij}| \leq |\theta_{ij}|_{max}, \tag{15}$$

where P_{Gi} and Q_{Gi} are the active and reactive power from the reference bus (MW,Mvar), respectively, P_{Di}, Q_{Di} are the active and reactive power of nodal load(MW,Mvar), respectively, and P_{DWGi} is the DWP capacity to be optimized at bus i (MW). G_{ij}, B_{ij} are the conductance and susceptance of the branch i-j (S), respectively. U_i is the voltage magnitude at bus i (kV), θ_i is the voltage phase angle of bus i (rad), $\theta_{ij} = \theta_i - \theta_j$. The subscript min, max indicate the lower and upper limits of the variable (p.u.), respectively. $P_{DGWmax} = 0.5$ p.u., $U_{imin} = 0.95$ p.u., $U_{imax} = 1.05$ p.u. To make sure power flow from distributed wind farm does not transform to a higher voltage level, the active power at the reference bus is strictly non-negative.

3.2. Collaborative Capacity Planning of DWP and ESS

In Figure 1 the daily curve of DWP and local load demand is shown based on the actual historical data from a city in eastern China. It can be found that wind power is sometimes higher than load, an appropriate capacity of ESS installation in a distributed wind farm could help absorb the extra wind power and then satisfy the load demand when the wind power output is lower than load. In this section, a collaborative capacity planning model of DWP and ESS are prospectively proposed.

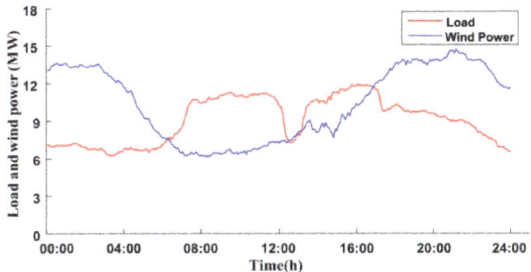

Figure 1. The daily curve of DWP and local load demand.

3.2.1. Objective Function

Considering the investment cost of DWP, the environmental income contributed by the government, the initial investment cost of the ESS, and the arbitrage gains of ESS into the capacity planning model, the objective function for the collaborative planning of DWP and ESS is

$$G_c = \min (g_1 + g_2 + g_3 + g_4), \quad (16)$$

First, the annual storage cost of ESS g_1 is [32]:

$$g_1 = (C_p P_e + C_e E_e) \frac{(1+r_0)^T r}{(1+r_0)^T - 1} + C_{yw} P_e, \quad (17)$$

where C_p is the power cost of ESS (RMB/kW); C_e is the capacity cost of ESS (RMB/kVA); C_{yw} is the annual operation and maintenance costs (RMB/kW); P_e is the active power of ESS (MW); E_e is the capacity of ESS(MVA); and T' is the operating life of energy storage (year) [33].

Second, the distribution line loss g_2 is

$$g_2 = 8760 \sum_{i=1}^{N} \sum_{\substack{j=1 \\ j \neq i}}^{N} \frac{\Delta U_{ij}^2}{|Z_{ij}|} \cos(\varphi) C_d, \quad (18)$$

Third, the investment cost of the distributed wind farm and the environmental income provided by the government g_3 are

$$g_3 = P_{DWG} \left[C_{wt} - C_{sr} \frac{r_0(1+r_0)}{(1+r_0)^T - 1} \right], \quad (19)$$

Fourth, the arbitrage gains of ESS g_4 are

$$g_4 = 365 E_{BESS} \sigma(t), \quad (20)$$

E_{BESS} is the energy absorbed by ESS (MVA), and $\sigma(t)$ is the time-of-day tariff (RMB/kWh), which satisfies

$$\sigma(t) = \begin{cases} 0.32 & 0 \leq t \leq 7, 22 < t \leq 24 \\ 0.72 & 7 < t \leq 9, 21 < t \leq 22 \\ 1.12 & 9 < t \leq 21 \end{cases}, \quad (21)$$

3.2.2. Constraints

Optimization constraints include (10)–(15) and the ESS operational constraints in (22)–(27).

$$SOC = SOC_{-1} - \frac{\eta P_{e,-1} \Delta t}{E_e}, \quad (22)$$

$$\eta = \begin{cases} \eta_{out}, P_{e,-1} > 0 & \eta_{out} = 95\% \\ \eta_{in}, P_{e,-1} < 0 & \eta_{in} = 90\% \end{cases} \quad (23)$$

$$p_e^{min} \leq p_e \leq p_e^{max}, \quad (24)$$

$$E_e^{min} \leq E_e \leq E_e^{max}, \quad (25)$$

$$SOC^{min} \leq SOC \leq SOC^{max}, \quad (26)$$

$$\sqrt{P_e^2 + Q_e^2} = E_e, \quad (27)$$

where SOC is the state of charging/discharging of ESS. η is the charge and discharge efficiency (%); Q_e are the reactive power of ESS (kvar), respectively; the superscripts min, max represent the lower and upper limits of the variables, respectively. The subscript "−1" represents the value of the previous moment. To prevent the ESS from overcharging or discharging, the range of SOC is generally 0.1~0.9 [34].

A system diagram is illustrated in Figure 2 to convey the main process of planning.

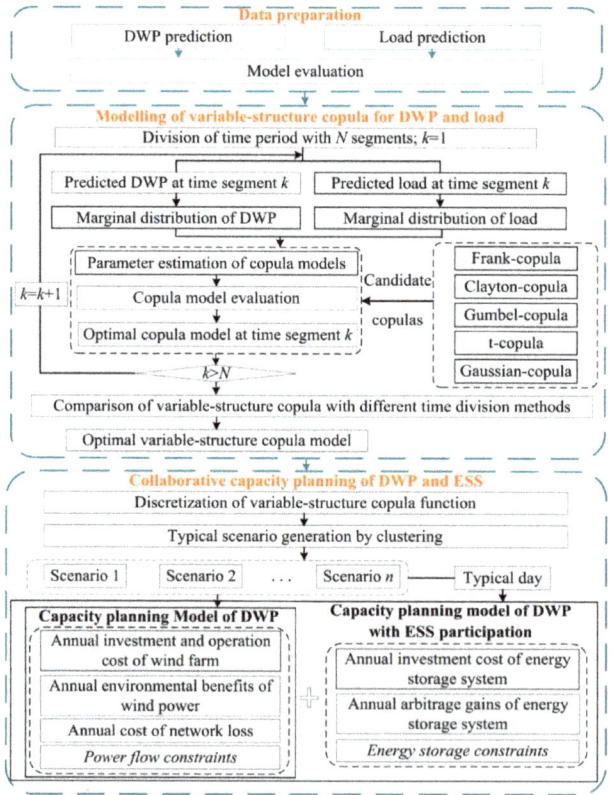

Figure 2. The system diagram of the DWP planning process.

Based on the sequence of DWP planning, a case study is carried out in the following section.

4. Case Study

A case study is applied to the modified IEEE 33-bus test system in Figure 3. The distributed wind farm and ESS are integrated at bus 6, the system-based capacity is 100 MVA, and Matlab^{TM} is used for analysis.

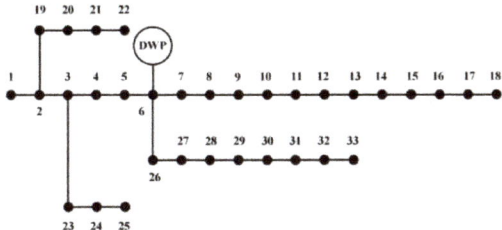

Figure 3. The modified IEEE 33-bus system.

In this study, with the assumption that the distributed wind farm has the same/similar wind source as that of the centralized wind farm, the centralized wind power data are used, and the wind power data are proportionally converted into DWP. Both wind power and local load data below 110 kV level are practical operation data from an economically developed area in Xuzhou, a city in eastern China. According to the distribution of the load at each bus in the test system, the practical load data in Xuzhou are allocated in the modified IEEE33-bus system. The time series include the 5-min data pairs of DWP and load from 1 January 2016 to 31 December 2018.

4.1. Data Preprocessing

The centralized wind power data is first proportionally converted into distributed wind power. Based on the historical wind power output and load data, the orders of ARMA model for both time series are shown in Table 1, where p, q is the order of the autoregressive, moving average process respectively, and AIC is the value determined by Akaike Information Criterion. Based on the ARMA model, the DWP and load are predicted at the planning stage.

Table 1. Order selection for ARMA models.

Data Type	p	q	AIC
Load	7	10	2.3870
DWP	9	8	1.4892

In this work, the Support vector machine (SVM) prediction method is used to evaluate the accuracy of the ARMA model, this paper employs root mean squared error (RMSE), mean absolute percentage error (MAPE), R^2 and mean absolute error (MAE) as indices to evaluate ARMA and SVM.

The smaller the RMSE, MAPE and MAE are, the more accurate the model is. The larger the R^2 is, the more credible the model is.

The evaluation indices are calculated for DWP prediction by using ARMA and SVM model. The comparison is listed in Table 2.

Table 2. The evaluation indices of ARMA and SVM for wind power prediction.

Model	RMSE	MAPE (%)	MAE	R^2
SVM	7.6806	31.7063	4.8614	0.5651
ARMA	4.6309	13.4228	2.2388	0.8418

The performance of load prediction by ARMA and SVM are further compared based on the four evaluation indices in Table 3.

Table 3. The evaluation indices of ARMA and SVM for load prediction.

Model	RMSE	MAPE (%)	MAE	R^2
SVM	1.0677	25.0545	0.8097	0.7244
ARMA	0.3064	3.1339	0.1391	0.9773

Based on the calculation of evaluation indices, it can be concluded that:

(1) From Table 2, the RMSE, MAPE and MAE of ARMA for DWP prediction are smaller than those of the SVM model, the R^2 value of ARMA for DWP prediction is larger than SVM. All the evaluation indices are in agreement that ARMA performs better than the SVM model.
(2) From Table 3, the RMSE, MAPE and MAE of ARMA for load prediction are smaller than those of the SVM model, the R^2 value of ARMA for load prediction is larger than that of SVM. All the evaluation indices agree that ARMA shows better prediction performance than SVM model and it is feasible and satisfactory for load prediction.
(3) The model evaluation indicates that prediction results of ARMA model is feasible for the next step of capacity planning for DWP.

4.2. Marginal Probability Density Function of Load and DWP

Based on the non-parametric estimation, the marginal probability density function of DWP and load can be obtained. The empirical distribution function is used as the standard for the actual distribution function and is used to determine the accuracy of the non-parametric estimation method.

Figures 4 and 5 show the comparison of marginal cumulative distribution by kernel distribution estimation with the corresponding empirical distribution function for DWP and load, respectively.

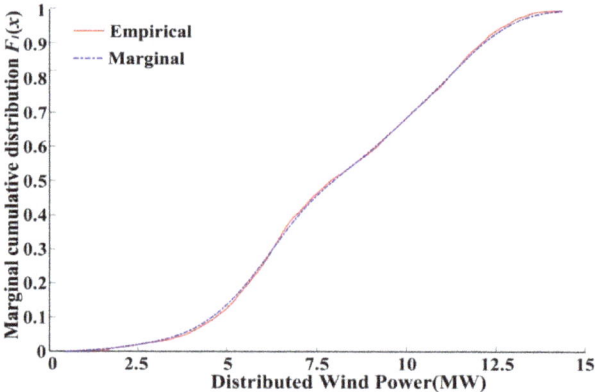

Figure 4. The comparison of marginal cumulative distribution and empirical distribution of DWP.

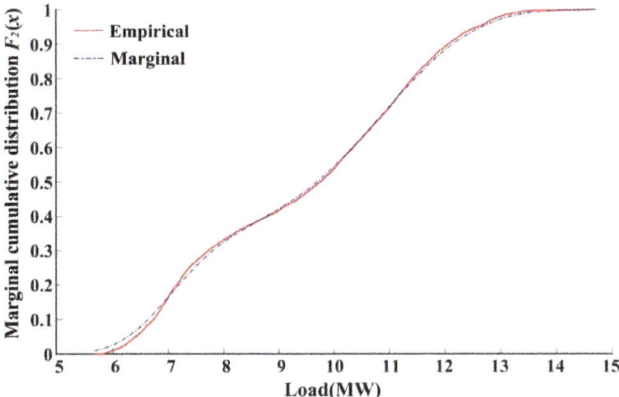

Figure 5. The comparison of marginal cumulative distribution and empirical distribution of load.

As shown in the figures, by comparing the gaps of the function graphically, the results of the non-parametric estimation are basically coincident with the empirical distribution, indicating a feasible estimation accuracy.

4.3. Parameter Estimation and Model Selection

With the time division strategy by month, the following is a detailed description of the phased copula selection based on DWP and load data in January, 2018 as an example. Based on the practical data of January, 2018, the binary frequency histogram of DWP and load is illustrated in Figure 6.

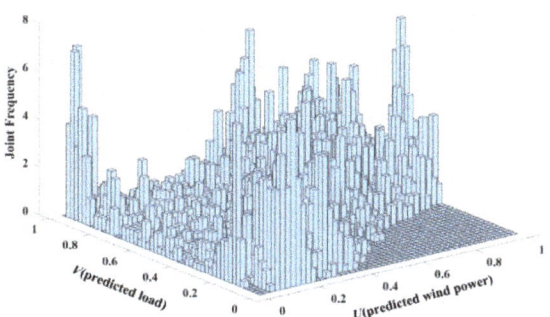

Figure 6. Binary frequency histogram of DWP and load in January 2018.

From Figure 6, the symmetric correlation of DWP and load is identified, and the joint distribution of the two variables are further examined by 5 copula functions. Figures 7–11 report the probability density and distribution function of each copula model of DWP and load in January, 2018 in a graphic view, and parameter estimation based on the MLE method for the copula models is shown in Table 4.

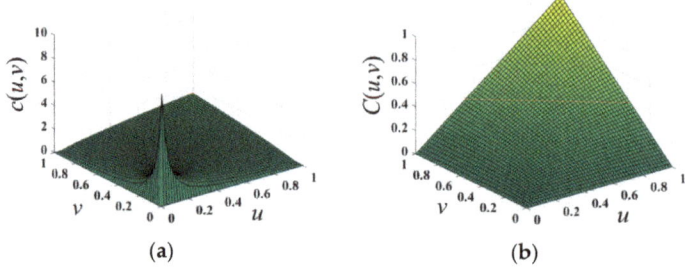

Figure 7. (**a**) Probability density function of Clayton copula; (**b**) distribution function of Clayton copula.

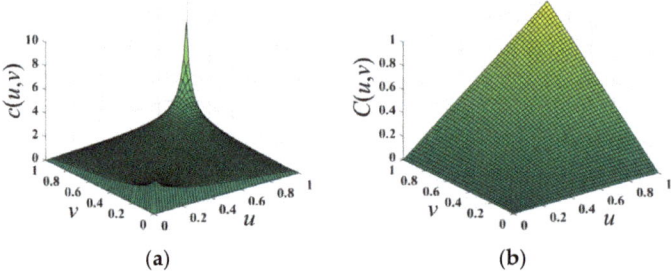

Figure 8. (**a**) Probability density function of Gumbel copula; (**b**) distribution function of Gumbel copula.

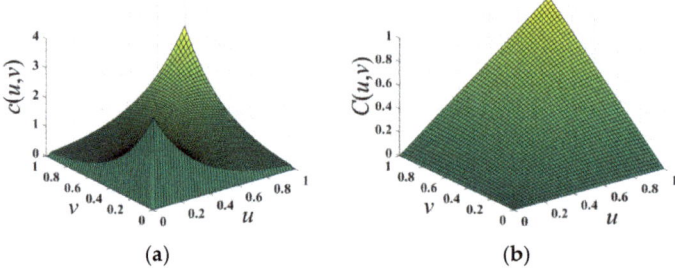

Figure 9. (**a**) Probability density function of Frank copula; (**b**) distribution function of Frank copula.

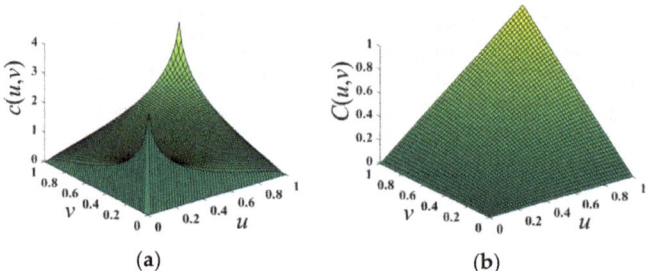

Figure 10. (**a**) Probability density function of Gaussian copula; (**b**) distribution function of Gaussian copula.

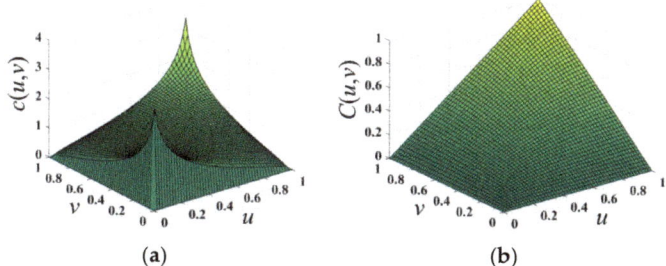

Figure 11. (a) Probability density function of t copula; (b) distribution function of t copula.

Table 4. Parameter estimation of copula models and evaluation indices.

Copula Type	Parameter Estimation	Evaluation Indices		
		Kendall	Spearman	Euclidean
Gaussian	0.3759	0.2453 ×	0.3611 ×	9.9062 ×
t	0.4870	0.3238 ○	0.4998 ○	6.8339 ○
Gumbel	0.3715	0.2709	0.3924	8.4066
Clayton	0.7454	0.2751	0.3953	7.4007
Frank	0.3373	0.3389 √	0.4920 √	2.1696 √
Empirical	0.4689	0.3377	0.4640	0.0000

Archimedean type copula has good properties including Clayton copula, Frank copula and Gumbel copula. Clayton copula excels at describing the asymmetric correlation and lower-tail characteristics of variables as shown in Figure 7.

From Figure 8, it can be found that the asymmetric correlation and upper-tail characteristics of variables are well depicted by the Gumbel copula.

The Frank copula can capture variables' negative and symmetric correlation. It can be found from Figure 9 that it can also indicate the progressive independence of both tails.

The ellipse type copula includes the Gaussian copula and t copula. From Figure 10, the asymmetric and progressive independence of tails are illustrated by Gaussian copula.

From t copula in Figure 11, the asymmetric tail characteristic of DWP and load is depicted.

Figure 12 draws the empirical copula distribution function.

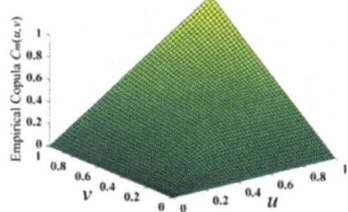

Figure 12. Empirical copula distribution function of DWP and load.

Different copulas show different characteristics of correlation and results based on different parameter estimation. To select a proper phased copula, the Kendall, Spearman and Euclidean distance indices of each copula are calculated and compared with those of the empirical copula in Table 4.

Based on the calculation of evaluation indices in Table 4, the two-stage filtering method [30] is carried out. When the type of copula model is inferior to other models under the evaluation criteria, it is marked by "×"; when this type of copula model is superior to other models, it is marked by "√"; when this type of copula model is closest among them apart from the optimal model, it is marked by "○". From Table 4, the Frank copula is determined to e the best fitting model by Kendall and Spearman correlation coefficients, and the Euclidean distance between the model and the empirical copula model is also the smallest. Since it receives the most "√", Frank copula function is selected as best fitting the correlation between DWP and load in January.

Similarly, by dividing the year into four quarters, the parameter estimation of each phased copula is obtained and the evaluation indices in each quarter of year are calculated in Table 5.

Table 5. The optimal Copula models and empirical Copula parameters for four quarters of the year.

	Copula Type	Parameter Estimation	Kendall	Spearman	Euclidean
1	Frank	1.1731	0.1286	0.1920	3.7201
	Empirical	0.1884	0.1262	0.1881	0.0000
2	t	0.2053	0.1316	0.1954	2.1972
	Empirical	0.1926	0.1328	0.1938	0.0000
3	t	0.4039	0.2647	0.3827	3.8865
	Empirical	0.3949	0.2793	0.3948	0.0000
4	Gumbel	1.5294	0.3461	0.4935	5.0976
	Empirical	0.4960	0.3579	0.4918	0.0000

In the comparison between the two time division strategies, the average value of the Euclidean distance between the best copula model and the empirical copula model in each month is 1.8864, whereas it is 3.7254 with quarter division. Therefore, the correlation between the DWP and load can be better fitted using the month division strategy.

4.4. Typical Scenario Generation

According to the variable-structure copula divided by month, the typical operation scenario of DWP and load is obtained by discretizing the continuous variable-structure copula function.

First, we discretize the phased copula model and generate a sample data of 96,000 × 2 dimensions. Next, set the number of typical scenarios to 6 and select the initial condensation point. Finally, use the K-means method to cluster the remaining discrete points and find the corresponding original quantiles. According to the steps in Section 2.3, Table 6 shows the generation results of typical scenarios.

Table 6. Quantile of DWP and load with probabilities in typical scenarios.

Scenario	Load (MW)	DWP (MW)	Probability P_L
1	8.3986	15.3398	0.183
2	6.2500	7.7391	0.160
3	8.2094	7.1449	0.137
4	5.6118	12.5456	0.180
5	4.8205	6.8032	0.182
6	7.7814	10.6478	0.158

It can be concluded from the results that each typical scenario has a similar incidence, which illustrates the rationality of dividing the initial points into six typical scenarios.

4.5. Capacity Planning of DWP

According to (7)–(9), the optimal solution of the objective function under the constraints of each scenario is obtained. Table 7 shows the results of capacity planning of DWP and the optimal value of the objective function without energy storage planning in each scenario.

Table 7. Optimal DWP output and objective function values for each scenario.

Scenario	Probability	DWP (p.u.)	F(×10⁴ RMB)
1	0.183	0.0883	1066.72
2	0.160	0.0648	889.63
3	0.137	0.0862	1015.56
4	0.180	0.0579	809.83
5	0.182	0.0495	686.76
6	0.158	0.0815	905.09

If capacity planning of DWP is conducted based on the minimum load from scenario 5 in Table 6, the planning result will be 4.95 MW, which is conservative. This will obviously cause a large amount of wind abandonment. The selection with the maximum capacity planning of DWP from scenario 1 will also lead to loss of economic profit. Taking into account the wind power consumption of the typical scenarios above and economic operation, the final planning capacity is the weighted sum with each scenario probability, that is:

$$P_{DWG*} = \sum_{i=1}^{k} P_{DWG(i)} P_{L(i)} = 0.0706 \text{ (p.u.)}, \tag{28}$$

where k is the number of scenarios, $P_{DWG}(i)$ is the planning capacity of DWP under scenario i, and $P_L(i)$ is the probability of scenario i. The final capacity planning of DWP is 7.06 MW.

4.6. Collaborative Capacity Planning of DWP and ESS

To maximize the consumption of wind power, it is necessary to employ the ESS so as to increase the planning capacity of DWP.

Based on the generation of typical scenarios, a typical schedule day is selected, and the collaborative capacity planning of DWP and ESS is examined based on the 24-h daily curve. In Figure 13, based on the ±10% fluctuation range of DWP and load in each typical scenario, several typical scenarios in Table 6 are included and marked in the typical daily curve.

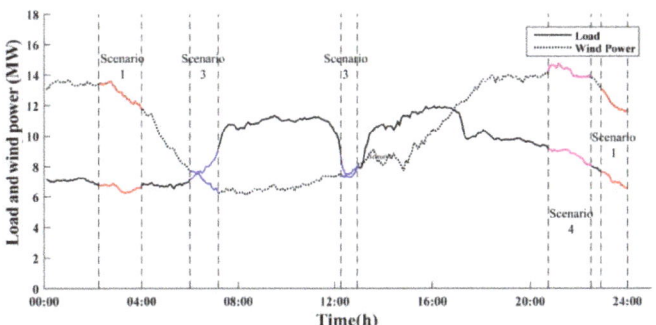

Figure 13. The typical daily curve with several typical scenarios.

Before optimization, the initial value of SOC is 0.6, and the initial state of ESS is discharge. Table 8 shows some specific parameters.

Table 8. Related parameters of DWP and ESS [33,35].

Parameter	Value
C_{wt} (RMB/kW)	4000
C_{sr} (RMB/kW)	3500
C_d (RMB/kWh)	0.68
C_p (RMB/kW)	4000
C_e (RMB/kVA)	3500
C_{yw} (RMB/kW)	20
r (%)	8
T (Year)	10
T' (Year)	15
Maximum charging/discharge power (MW)	150/150
Maximum/Minimum capacity (MWh)	600/10
Charging/discharging efficiency (%)	85/95

Based on the conditions above, fmincon optimization function in Matlab™ is employed to solve the proposed nonlinear constrained optimization problem.

Under the premise of allowing some wind abandonment, the optimal power output of DWP and the state of SOC in the typical day are obtained and shown in Figure 14. The corresponding charging and discharging power of ESS in the typical day is reported in Figure 15.

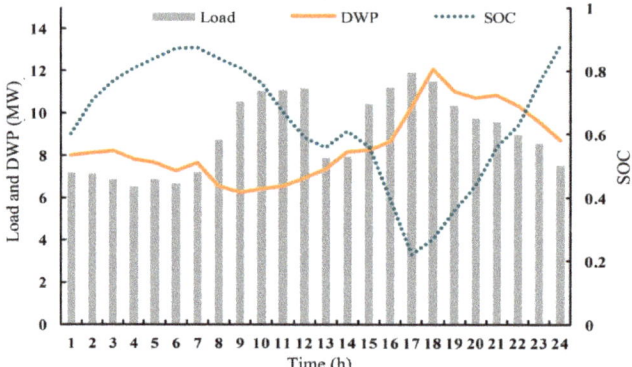

Figure 14. Optimal output of DWP and the state of SOC in the typical day.

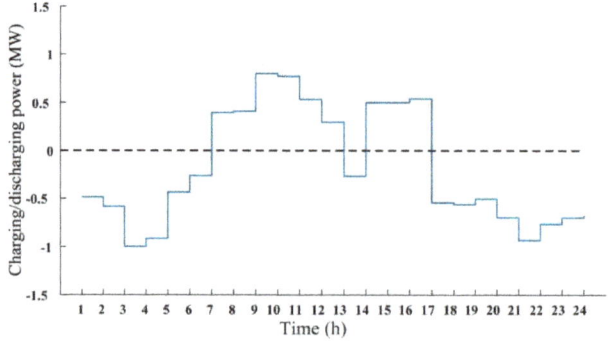

Figure 15. The charging and discharging power of ESS in the typical day.

Based on the optimal DWP planning and the states of ESS in the typical day, the final energy storage capacity planning is 4.63 MW, and the final DWP capacity planning is 12.07 MW. It can be concluded from this study that:

(1) The optimal planning of DWP and the SOC of ESS change with the fluctuation of load in the typical day. When load is smaller than actual wind power output, ESS charges and stores the extra wind power. When load is larger than actual wind power output, ESS discharges and supplies power to the load.
(2) The SOC of the ESS fluctuates within [0.1, 0.9], which meets the requirement of energy storage operation.
(3) Compared with the case without ESS, the DWP planning value increases from 7.06 MW to 12.07 MW, with the study results indicating that with the participation of energy storage, it is conducive to increasing the consumption of DWP.

Moreover, the active power loss of the distribution network, the annual cost of the system is further compared with that before the installation of ESS in Table 9.

Table 9. Capacity of DWP and economy before and after installing ESS.

	DWP (MW)	Network Loss Cost (10^4 RMB)	Average Annual Total Cost (10^4 RMB)
Without ESS	7.06	554.8	1066.7
With ESS	12.07	289.9	981.1

ESS can release power at peak load and it can absorb excess power when DWP output is higher than load. From Table 9, ESS can reduce the rate of wind abandonment, and the network loss cost is reduced with the collaborative planning of ESS, and the average annual cost is reduced with the benefited from time-of-day tariff in the distribution network.

The fluctuation of bus voltage is further studied and compared in the typical day. The voltage magnitude at some buses is lower than 0.95 in the case of without ESS installation, as shown in Figure 16, and the bus voltages all lie in the range of [0.95, 1.05] based on the collaborative planning of ESS as reported Figure 17.

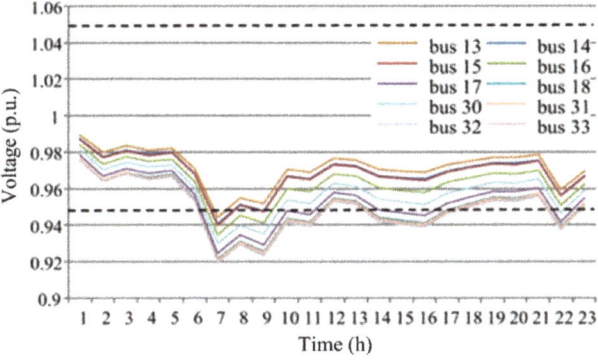

Figure 16. The fluctuation of bus voltage in the typical day before installing the ESS.

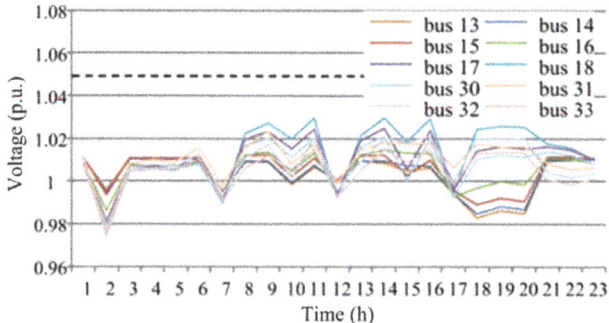

Figure 17. The fluctuation of bus voltage in the typical day after installing the ESS.

Based on the comparison of Figures 16 and 17, with the installation of the ESS, the voltage level is increased effectively at heavy load conditions. At the same time, the range of voltage fluctuation is reduced.

5. Conclusions

This paper proposes a DWP capacity planning method with participation of ESS optimization under multi-scenario conditions based on variable-structure copula and optimization theory. The conclusions can be reached as follows.

(1) The variable-structure copula models of DWP and load are established based on two different time segment strategies. The average Euclidean distance of the strategies by month is 1.8864, which is smaller than the strategies by quarter. It is concluded that the variable-structure copula model with time segment by month is better able to fit the changes of the correlation structure between DWP and load and can capture the correlation at each segment more accurately.

(2) The continuous correlation functions are discretized and typical scenarios of DWP and load are obtained by the clustering method. For each typical scenario, a feasible capacity planning model of DWP is established. In addition, the feasible optimal capacity of DWP is 7.06 MW, which is higher than the lowest active power demand of load. Therefore, the capacity planning model of DWP considering the correlation of DWP and load can effectively increase wind power consumption.

(3) Furthermore, a collaborative capacity planning model for DWP and ESS is proposed. Study results show that, compared with the case without participation of ESS, the capacity of DWP planning was increased to 12.07 MW, and the consumption of wind power was efficiently improved. For economy, collaborative capacity planning of DWP and ESS can reduce the cost of network loss of the distribution network and the annual cost of the system; moreover, for reliability, it can effectively satisfy the lower voltage limit and reduce the range of voltage fluctuation.

Future work includes the investigation of time-varying copula model and multi-point integration of wind and storage system for DWP planning.

Author Contributions: Conceptualization, Y.W. and R.Y.; methodology, Y.W.; software, Y.W. and R.Y.; validation, Y.W., R.Y. and Y.T.; formal analysis, Y.W. and R.Y.; investigation, R.Y.; resources, S.X.; data curation, S.X.; writing—original draft preparation, Y.W. and R.Y.; writing—review and editing, Y.W., R.Y., S.X. and Y.T.; visualization, R.Y.; supervision, Y.T.; project administration, Y.W.; funding acquisition, Y.W. All authors have read and agreed to the published version of the manuscript.

Funding: This research was funded by the National Natural Science Foundation of China, Grant Number 51507031.

Conflicts of Interest: The authors declare no conflict of interest.

Nomenclature

Symbols:

Y_t, Y_{t-i}	Time series of DWP, load at t, $t-i$
$\varepsilon_t, \varepsilon_{t-i}$	Prediction error term at t, $t-i$
α, β	Correlation coefficient of ARMA
p, q	Order of autoregressive, moving average process
σ^2	Variance of the predicted error term
n	Number of time series
$c(\cdot), C(\cdot)$	Probability density function, distribution functions of copula
$F(\cdot)$	Cumulative distribution function
$F_1(\cdot), F_2(\cdot)$	The marginal probability density functions of DWP, load
x, y	Samples of DWP, load (MW)
ρ_p, ρ_k, ρ_z	Pearson, Kendall, Spearman coefficient
u_i, v_i	Value of marginal probability density functions of DWP, load at i
$\bar{u}, \bar{v}$	Expected values of u_i, v_i (MW)
r	Number of the variables whose two attribute values have the same size relationship
d_i	Rank differences between u_i and v_i
$d_x{}^2$	Distance between basic copula and empirical copula model
$C_m(\cdot), C_x(\cdot)$	Distribution functions of empirical copula, basic copula
$F_m(\cdot), G_m(\cdot)$	Empirical distribution functions of DWP, load
$I(\cdot)$	Explanatory function
P_{DWG}	Planning capacity of DWP (MW)
C_{wt}	Annual initial investment cost (RMB/kW) of distributed wind farm
C_{sr}	Environmental income (RMB/kW)
r_0	Discount rate (%)
T, T'	Operating life of the distributed wind farm, ESS (year)
i, j	Bus number
N	Number of system buses
ΔU_{ij}	Voltage difference between bus i and bus j (kV)
Z_{ij}	Impedance of the branch i-j (Ω)
C_d	Electricity price (RMB/kWh)
φ	Power factor angle
P_{Gi}, Q_{Gi}	Active and reactive power from the reference bus (MW, Mvar)
P_{Di}, Q_{Di}	Active and reactive value of nodal load (MW, Mvar)
G_{ij}, B_{ij}	Conductance and susceptance of the branch i-j (S)
U_i, θ_i	Voltage magnitude (kV), phase angle of bus i (rad)
C_p	power cost (RMB/kW) of ESS
C_e	capacity cost (RMB/kVA) of ESS
C_{yw}	Annual operation and maintenance costs (RMB/kW)
P_e, Q_e, E_e	Active, reactive power, capacity of ESS (MW, Mvar, Mvar)
E_{BESS}	Energy absorbed by ESS (MVA)
$\sigma(t)$	Time-of-day tariff (RMB/kWh)
η	Charge and discharge efficiency

Acronyms:

DWP	Distributed wind power
ESS	Energy storage system
ARMA	Autoregressive moving average
AIC	Akaike Information Criterion
MLE	Maximum likelihood estimation
SVM	Support vector machine
RMSE	Root mean squared error
MAPE	Mean absolute percentage error
MAE	Mean absolute error
SOC	State of charging/discharging of ESS

Superscripts and subscripts:

min, max	The lower and upper limits of the variables
Subscript "−1"	The value of the previous moment

References

1. Shu, Z.; Jirutitijaroen, P. Optimal operation strategy of energy storage system for grid-connected wind power plants. *IEEE Trans. Sustain. Energy* **2014**, *5*, 190–199. [CrossRef]
2. Abdulgalil, M.A.; Khalid, M.; Alismail, F. Optimizing a distributed wind-storage system under critical uncertainties using benders decomposition. *IEEE Access* **2019**, *7*, 77951–77963. [CrossRef]
3. National Energy New Energy. *National Energy Administration Notice on Printing and Distributing the Interim Administrative Measures for the Development and Construction of Decentralized Wind Power Projects*; National Energy Administration: Beijing, China, 2018.
4. Huang, D.; Li, H.; Cai, G.; Huang, N.; Yu, N.; Huang, Z. An efficient probabilistic approach based on area grey incidence decision making for optimal distributed generation planning. *IEEE Access* **2019**, *7*, 93175–93186. [CrossRef]
5. De Oliveira, K.L.M.; Penido, D.R.R.; De Araujo, L.R. Planning of the wind farm distribution network using heuristic methods. *IEEE Latin Am. Trans.* **2018**, *16*, 2917–2924. [CrossRef]
6. Ehsan, A.; Yang, Q.; Cheng, M. A scenario-based robust investment planning model for multi-type distributed generation under uncertainties. *IET Gener. Transm. Distrib.* **2018**, *12*, 4426–4434. [CrossRef]
7. Zheng, Y.; Dong, Z.Y.; Meng, K.; Yang, H.; Lai, M.; Wong, K.P. Multi-objective distributed wind generation planning in an unbalanced distribution system. *CSEE J. Power Energy Syst.* **2017**, *3*, 186–195. [CrossRef]
8. Ouammi, A.; Dagdougui, H.; Sacile, R. Optimal planning with technology selection for wind power plants in power distribution networks. *IEEE Syst. J.* **2019**, *13*, 3059–3069. [CrossRef]
9. Wang, G.; Xin, H.; Wu, D.; Ju, P.; Jiang, X. Data-driven arbitrary polynomial chaos-based probabilistic load flow considering correlated uncertainties. *IEEE Trans. Power Syst.* **2019**, *34*, 3274–3276. [CrossRef]
10. Chen, J.; Yu, Q.; Li, Q.; Lin, Z.; Li, C. Probabilistic energy flow analysis of MCE system considering various coupling units and the uncertainty of distribution generators. *IEEE Access* **2016**, *7*, 100394–100405. [CrossRef]
11. Sun, G.; Guan, X.; Yi, X.; Zhao, J. Conflict evidence measurement based on the weighted separate union kernel correlation coefficient. *IEEE Access* **2018**, *6*, 30458–30472. [CrossRef]
12. Ouyang, J.; Li, M.; Zhang, Z.; Tang, T. Multi-timescale active and reactive power-coordinated control of large-scale wind integrated power system for severe wind speed fluctuation. *IEEE Access* **2019**, *7*, 51201–51210. [CrossRef]
13. Wu, Y.; Lee, C.; Chen, C.; Hsu, K.; Tseng, H. Optimization of the wind turbine layout and transmission system planning for a large-scale offshore wind farm by AI technology. *IEEE Trans. Ind. Appl.* **2014**, *50*, 2071–2080. [CrossRef]
14. Jiang, C.; Mao, Y.; Chai, Y.; Yu, M.; Tao, S. Scenario generation for wind power using improved generative adversarial networks. *IEEE Access* **2018**, *6*, 62193–62203. [CrossRef]
15. Ehsan, A.; Cheng, M.; Yang, Q. Scenario-based planning of active distribution systems under uncertainties of renewable generation and electricity demand. *CSEE J. Power Energy Syst.* **2019**, *5*, 56–62. [CrossRef]
16. Nelson, R.B. *An Introduction to Copulas*; Springer: New York, NY, USA, 2006; pp. 109–115.
17. Ehsan, A.; Yang, Q. Coordinated investment planning of distributed multi-type stochastic generation and battery storage in active distribution networks. *IEEE Trans. Sustain. Energy* **2019**, *10*, 1813–1822. [CrossRef]
18. Khalid, M.; Akram, U.; Shafiq, S. Optimal planning of multiple distributed generating units and storage in active distribution networks. *IEEE Access* **2018**, *6*, 55234–55244. [CrossRef]
19. Xia, S.W.; Bu, S.Q.; Wan, C.; Lu, X.; Chan, K.W.; Zhou, B. A fully distributed hierarchical control framework for coordinated operation of DERs in active distribution power networks. *IEEE Trans. Power Syst.* **2019**, *34*, 5184–5197. [CrossRef]
20. Liu, Y.; Xiao, L.Y.; Yao, G.D.; Bu, S.Q. Pricing-based demand response for a smart home with various types of household appliances considering customer satisfaction. *IEEE Access* **2019**, *7*. [CrossRef]
21. Sklar, A. *Fonctions de Répartition àn Dimensions et Leurs Marges*; University Paris Press: Paris, France, 1959; pp. 229–231.
22. Cui, M.; Krishnan, V.; Hodge, B.; Zhang, J. A copula-based conditional probabilistic forecast model for wind power ramps. *IEEE Trans. Smart Grid* **2019**, *10*, 3870–3882. [CrossRef]
23. Papaefthymiou, G.; Kurowicka, D. Using copulas for modeling stochastic dependence in power system uncertainty analysis. *IEEE Trans. Power Syst.* **2009**, *24*, 40–49. [CrossRef]

24. Ma, Z.; Chen, H.; Chai, Y. Analysis of voltage stability uncertainty using stochastic response surface method related to wind farm correlation. *Prot. Control Mod. Power Syst.* **2017**, *2*, 1–9. [CrossRef]
25. Wang, Y.R.; Luo, Y.N. Research of wind power correlation with three different data types based on mixed copula. *IEEE Access* **2018**, *6*, 77986–77995. [CrossRef]
26. Stephen, B.; Galloway, S.J.; McMillan, D.; Hill, D.C.; Infield, D.G. A copula model of wind turbine performance. *IEEE Trans. Power Syst.* **2011**, *26*, 965–966. [CrossRef]
27. Holeňa, M.; Bajer, L.; Ščavnický, M. Using copulas in data mining based on the observational calculus. *IEEE Trans. Knowl. Data Eng.* **2015**, *27*, 2851–2864. [CrossRef]
28. Hazarika, S.; Biswas, A.; Shen, H. Uncertainty visualization using copula-based analysis in mixed distribution models. *IEEE Trans. Vis. Comput. Graph.* **2018**, *24*, 934–943. [CrossRef]
29. Chen, Z.; Li, S. CDO pricing using Archimedean copula with dynamic structure. In Proceedings of the 2014 International Conference on Management Science & Engineering 21th Annual Conference Proceedings, Helsinki, Finland, 17–19 August 2014; pp. 1121–1126.
30. Wang, Y.; Luo, Y.; Dai, Z. Study and comparison of wind power correlation using two types of dataset. In Proceedings of the 2018 IEEE Power & Energy Society General Meeting (PESGM), Portland, OR, USA, 5–9 August 2018; pp. 1–5.
31. Eryilmaz, S. Modeling dependence between two multi-state components via copulas. *IEEE Trans. Reliab.* **2014**, *63*, 715–720. [CrossRef]
32. Sedghi, M.; Ahmadian, A.; Aliakbar-Golkar, M. Optimal storage planning in active distribution network considering uncertainty of wind power distributed generation. *IEEE Trans. Power Syst.* **2016**, *31*, 304–316. [CrossRef]
33. Ma, M.T.; Yuan, T.J.; Chen, G.Y. Analysis of comprehensive economic benefits of energy storage participating in wind power auxiliary service. *Power Syst. Technol.* **2016**, *40*, 3362–3367.
34. Jinlei, S.; Lei, P.; Ruihang, L.; Qian, M.; Chuanyu, T.; Tianru, W. Economic operation optimization for 2nd use batteries in battery energy storage systems. *IEEE Access* **2019**, *7*, 41852–41859. [CrossRef]
35. Jiang, P.; Yuan, T.J.; Che, Y.; Liu, B. Research on capacity planning of the integration of decentralized wind and storage system. *Power Capacit. React. Power Compens.* **2018**, *39*, 138–146.

© 2020 by the authors. Licensee MDPI, Basel, Switzerland. This article is an open access article distributed under the terms and conditions of the Creative Commons Attribution (CC BY) license (http://creativecommons.org/licenses/by/4.0/).

Article

Adaptive Multi-Model Switching Predictive Active Power Control Scheme for Wind Generator System

Hongwei Li [1,*], Kaide Ren [1], Shuaibing Li [2] and Haiying Dong [2,*]

[1] School of Automation and Electrical Engineering, Lanzhou Jiaotong University, Lanzhou 730070, China; renkaide@mail.lzjtu.cn
[2] School of New Energy and Power Engineering, Lanzhou Jiaotong University, Lanzhou 730070, China; shuaibingli@mail.lzjtu.cn
* Correspondence: lihongwei@mail.lzjtu.cn (H.L.); hydong@mail.lzjtu.cn (H.D.); Tel.: +86-0931-4938001 (H.L.); +86-0931-4938391 (H.D.)

Received: 16 February 2020; Accepted: 10 March 2020; Published: 12 March 2020

Abstract: To deal with the randomness and uncertainty of the wind power generation process, this paper proposes the use of the clustering method to complement the multi-model predictive control algorithm for active power control. Firstly, the fuzzy clustering algorithm is adopted to classify actual measured data; then, the forgetting factor recursive least square method is used to establish the multi-model of the system as the prediction model. Secondly, the model predictive controller is designed to use the measured wind speed as disturbance, the pitch angle as the control variable, and the active power as the output. Finally, the parameters and measured data of wind generators in operation in Western China are adopted for simulation and verification. Compared to the single model prediction control method, the adaptive multi-model predictive control method can yield a much higher prediction accuracy, which can significantly eliminate the instability in the process of wind power generation.

Keywords: wind power generation; multi-model predictive control; fuzzy clustering

1. Introduction

The uncertainty of wind speed makes the output power of the wind power generation system fluctuate greatly [1–3]. Frequent switching control will result in a transient overload of the transmission chain and an overshoot of output power. As a result, the system shows very strong nonlinearity. Since the multi-model predictive control (MMPC) method can effectively cope with complex nonlinear systems—i.e., wind generation systems—it has been widely used recently.

In [4], a model predictive control (MPC)-based optimal active power control scheme for a doubly-fed induction generator (DFIG) was proposed, which was applied to wind farms with a distributed energy storage system (ESS). For such a multi-input and multi-output (MIMO) wind generation system, the dynamic characteristics of the converter and wind turbines (WTs) were considered in MPC. By using MPC, the reference for active power between WTs and ESS was optimized according to the local wind conditions. The results showed that such a control scheme can greatly reduce the control error of active power for WTs. Similarly, an online model-based predictive control method was proposed in [5], which was used for the real-time optimal operation of a wind power integrated system including demand response (DR) and ESS. This method took into account all the interaction effects of the control facilities according to the estimated output of the future wind farms and realized the maximum utilization of wind power. The model has good universality and adaptability, and is suitable to resolve the high uncertainty of wind power generation and customer behavior. Compared with the online no-prediction method and offline prediction method, the daily wind energy utilization of this method is increased by 13.9% and 4.9%, respectively. In [6], an active power control architecture based on the combination of MPC and wind turbine state classification was proposed, where an equivalent model of an MPC controller was established. Compared with traditional methods, this method is more suitable for the active power control t the wind farm level since it ensures that the wind farm operates as a single controllable entity on the grid and acts like a traditional power plant.

Considering that wind farms always cooperate and interact with traditional thermal power plants, a comprehensive model of the hybrid system including both a wind farm and thermal power plant was established in [7] by combining the wind farm model with the well-known thermal power model. In [8], an improved load frequency control (LFC) method based on MPC was proposed, and the simulation results showed that this method could effectively improve the frequency response level of both the wind farm and thermal power plant, thus improving the frequency performance of the connected system.

To effectively coordinate the control of active power and reactive power, an MPC-based distributed information synchronization scheme and distributed coordination control scheme for wind farms was proposed and verified on a wind farm containing 10 WTs in [9]. For the distributed coordination control scheme, the pitch angle and generator torque of the WTs were optimized to reduce the fatigue load of the WTs and track the power reference of the wind farm. In [10], the MPC was utilized to replace the proportional integral (PI) controller for the speed control of a permanent magnet synchronous generator (PMSG). Similarly, the MPC was also applied to replace the traditional PI controller in [11,12]. The MPC avoids the disadvantages of the traditional PI controller; e.g., a slow dynamic response, coupling, complex structure, difficulty in determining the PI's parameters, etc. Compared with the PI control, the MPC has better performance in speed tracking, irrespective of whether the step speed is increased or step speed is decreased.

In [13], a variable-weight MPC strategy was introduced to optimize the coordination of mechanical load and the power of the wind power generation system. In the variable-weight MPC strategy, the pitch and torque are coordinated by Pareto analysis to optimize the output and load of the generator. Through the evaluation of the wind condition, the weight matrix of MPC can be updated adaptively. Finally, compared with the traditional gain scheduling PI control, the results showed that the strategy is effective. According to the operation areas of the wind turbine, three kinds of controllers—classic MPC and two kinds of economic model predictive control (EMPC)—were designed in [14]. The simulation results showed that the performance of the three predictive control strategies was better than that of the traditional LQR controller. EMPC can not only effectively reduce the fatigue load of all operation areas, but also increase the utilization of wind energy when the wind speed fluctuates near the rated wind speed. When the power grid operates under an unbalanced condition, the low-voltage ride-through capability becomes a difficulty for managers. Therefore, a finite set model predictive control strategy was proposed in [15], which could ensure the converter provides a balanced current and both active and reactive power support during grid connection. Compared with the traditional MPC, this method increases the control delay, and the waveform quality is much better.

Besides, the MPC method was also applied to a new type of converter control a the wind turbine with PMSG in [16], and the separation control of the machine and the grid side was realized by minimizing the value function. Further study on MMPC for PMSG control was also reported in [17]. The MMPC can ensure the smooth control of the active power output of WTs while ensuring high wind energy utilization efficiency. In the low wind speed condition, the fluctuation of active power can be reduced. The overshoot of active power can be reduced when WT operates near the rated wind speed. A smooth control can be realized, and the fluctuation of active power can also be significantly reduced when the WT operates above the rated wind speed.

From the above studies, it is apparent that the MPC method can solve the uncertainties and random changes of wind speed in the wind power generation system. However, there is little discussion about the short-term rapid change of wind speed in the above studies, especially for the case in which the wind speed varies through three different wind speed sections in a few seconds.

In view of this situation, this paper proposes an MMPC method for wind turbines under all wind conditions. The main contents of this paper include the following: (1) the classification of the collected

1000 groups of field-collected data using the fuzzy clustering method (FCM) and establishment of a multivariable prediction model using the forgetting factor recursive least square method (FFR-LSM), (2) the design of the multi-model predictive controller (MMPCR) with measurable wind speed as disturbance signal and pitch angle as the control variable, (3) the application of the designed MMPCR to the actual power control of wind turbine, and a comparison of the results with those of the single model predictive control method. The rest of this paper is organized as follows: Section 2 introduces the modeling of the wind turbine data from actual field using fuzzy clustering, Section 3 presents the design of the generalized predictive controller and Section 4 details the realization of multi-model switching control. Case study and simulation analysis are carried out in Section 5, while the conclusions are finally drawn in Section 6.

2. Fuzzy Clustering Modeling of the Field-Collected Wind Turbine Data

2.1. Mathematical Model of Variable Speed Variable Pitch Wind Turbine

Generally, a variable speed variable pitch wind turbine is mainly composed of a wind wheel, transmission part, and generator. The output power can be described as [18]

$$P = \begin{cases} \eta P_r, & v_{cutin} \leq v < v_N \\ P_N, & v_N \leq v \leq v_{cutout} \end{cases} \tag{1}$$

with

$$P_r = 0.5\rho v^3 S C_p(\lambda, \beta) \tag{2}$$

and

$$C_P = (0.44 - 0.0167\beta)\sin\left[\frac{\pi(\lambda-3)}{15-0.3\beta}\right] - 0.00184(\lambda-3)\beta \tag{3}$$

$$\lambda = \frac{2\pi nR}{v} = \frac{\omega R}{v} \tag{4}$$

where η is the conversion efficiency of wind energy, P is the mechanical power of WT, ρ represents air density, S stands for the swept area of the blade, λ is the tip speed ratio, β is the pitch angle, v is the wind speed, ω is the speed of the main shaft, and R is the diameter of the wind turbine.

2.2. Fuzzy Clustering of Data Sets

The measured data of WTs including the wind speed, pitch angle and active power of the generator are classified by the subtraction clustering algorithm–fuzzy c-means clustering, and then multiple models are generated by FFR-LSM. The fuzzy c-means clustering algorithm determines the category of each data point according to the membership degree. When the fuzzy objective function is the smallest, the data group is divided into C fuzzy categories. Details of the algorithm can be found in [19] and [20]. The Davies Bouldin index (DBI) is used to evaluate the cluster performance. The process of subtractive clustering is as follows [21,22]:

- Step 1: Determine the number of categories C and fuzzy weight index m, and initialize the clustering center V.
- Step 2: Calculate the fuzzy membership matrix U according to Equation (5).

$$u_{ij} = \begin{cases} \left[\sum_{k=1}^{C} \frac{\|x_i - V_j\|^{\frac{2}{m-1}}}{\|x_i - V_k\|^{\frac{2}{m-1}}}\right]^{-1} & \|x_i - V_k\| \neq 0 \\ 1 & \|x_i - V_k\| = 0 \text{ and } k = j \\ 0 & \|x_i - V_k\| = 0 \text{ and } k \neq j \end{cases} \tag{5}$$

where u_{ij} is the membership of x_i belonging to category j, and V_j or V_k is the clustering center of category j or k.

- Step 3: Calculate the category center V_j using Equation (6).

$$V_j = \frac{\sum_{i=1}^{n} u_{ij}^m x_i}{\sum_{i=1}^{n} u_{ij}^m} \quad (6)$$

- Step 4: Calculate the objective value J of cluster V_j using Equation (7) and judge whether target value is met or not. If it is met, the clustering will end; otherwise, return to step 2.

$$J = \sum_{i=1}^{n} \sum_{j=1}^{c} (u_{ij})^m \|x_i - V_j\| \quad (7)$$

In this paper, a total of 1000 groups of field measured data set M ($m1$, $m2$, $m3$) are classified, where $m1$ stands for the pitch angle, $m2$ represents wind speed, and $m3$ is the active power of the generator. The clustering result shows that a minimum DBI can be obtained when the number of clusters is equal to five. The clustering result is shown in Table 1.

Table 1. The clustering result of field-collected wind turbine data. DBI: Davies Bouldin index.

Number of Clustering	3	4	5	6	7
DBI	1.4473	0.6923	0.1978	1.1791	0.9733

When the number of clusters is five, all clustering center of the data can be summarized as

$$\text{Center} = \begin{bmatrix} 0.00 & 2.07 & 661.09 \\ 0.04 & 4.43 & 709.23 \\ 5.94 & 7.02 & 870.10 \\ 10.07 & 9.78 & 1300.79 \\ 13.08 & 11.06 & 1508.43 \end{bmatrix}$$

Considering that some of the data may overlap with each other—e.g., the wind speed can increase or decrease in a short time—the output power is consequently affected by the wind speed. Therefore, the wind generation system can be regarded as a system with a strong disturbance.

2.3. Least Square Modeling

Based on the results of the above clustering data, the wind speed and pitch angle are used as inputs, and the output power is taken as the output for the least squares modeling and to perform the index function determination. To overcome the shortcoming of the poor correction ability of the least square method (LSM), the forgetting factor recursive LSM (FFR-LSM) is adopted in this paper [22]. The performed index function is

$$J = \sum_{k=1}^{L} \lambda^{L-k} [y(k) - \varphi^T(k)\hat{\theta}(k)] \quad (8)$$

where $\hat{\theta}(k)$ is the parameter under-identification, $\varphi(k)$ is the observation matrix, λ is the forgetting factor, L is the times of observation, and $y(k)$ is the output of the system.

For the objective function given in Equation (8), Equation (9) can be used to express the recursive least square parameter estimation formula of the forgetting factor.

$$\begin{cases} \hat{\theta}(k) = \hat{\theta}(k-1) + K(k)\left[y(k) - \varphi^T(k)\hat{\theta}(k-1)\right] \\ K(k) = \frac{P(k-1)\varphi(k)}{\lambda + \varphi^T(k)P(k-1)\varphi(k)} \\ P(k) = \frac{1}{\lambda}\left[1 - K(k)\varphi^T(k)\right]P(k-1) \end{cases} \quad (9)$$

with $K(k)$ representing the gain matrix and $P(k)$ standing for the covariance matrix.

According to the measured data and the operating characteristics of the wind generator system, the cumulative abnormal return (CAR) model expressed in Equation (11) is used to identify the parameters. Combined with the classification results of the field-collected wind turbine data, five mathematical models are established.

$$Y(k+1) = \varphi^T(k)\theta \quad (10)$$

with

$$Y(k+1) = [y_1(k+1), y_2(k+1), y_3(k+1), y_4(k+1), y_5(k+1)]^T$$

$$\theta = \begin{bmatrix} a_{11} & a_{12} & a_{13} \\ a_{21} & a_{22} & a_{23} \\ a_{31} & a_{32} & a_{33} \\ a_{41} & a_{42} & a_{43} \\ a_{51} & a_{52} & a_{53} \end{bmatrix}$$

$$\varphi^T(k) = \begin{bmatrix} -y_1(k) & u_1(k) & \xi_1(k) \\ -y_1(k) & u_2(k) & \xi_2(k) \\ -y_1(k) & u_3(k) & \xi_3(k) \\ -y_1(k) & u_4(k) & \xi_4(k) \\ -y_1(k) & u_5(k) & \xi_5(k) \end{bmatrix}$$

Taking the initial value $\theta(0) = 0$, $P(0) = 10^5 I$, $\lambda = 0.95$, I is the unit matrix, and five mathematical models of wind generation system can be obtained, as shown in Equation (11).

$$\begin{cases} y_1(k+1) = 0.9374 y_1(k) + 0.3711 u_1(k) + v_1(k) \\ y_2(k+1) = 0.9695 y_1(k) + 0.3371 u_1(k) + v_2(k) \\ y_3(k+1) = 0.9217 y_1(k) + 0.2091 u_1(k) + v_3(k) \\ y_4(k+1) = 0.9241 y_1(k) + 0.3821 u_1(k) + v_4(k) \\ y_5(k+1) = 0.9801 y_1(k) + 0.2891 u_1(k) + v_5(k) \end{cases} \quad (11)$$

where $y_i(k)$ is the output power of the wind generator, $u_i(k)$ is the pitch angle, and $v_i(k)$ is the wind speed. Therefore, Equation (11) is taken as the prediction model for the system.

3. Design of the Generalized Predictive Controller

In this paper, the controlled auto-regressive integrated moving-average (CARIMA) model is used as the prediction model [21,22]. The discrete difference equation is shown in Equation (12).

$$A(z^{-1})y(k) = B(z^{-1})u(k-1) + C(z^{-1})\xi(k)/\Delta \quad (12)$$

where $y(k)$ and $u(k)$ are the output and input of the controlled object, $\Delta = 1 - k^{-1}$ is the difference operator, $A(z^{-1})$, $B(z^{-1})$, and $C(z^{-1})$ are polynomials of backward operators, and $\xi(k)$ is the white noise sequence with zero mean value. If the time delay of the system $d > 1$, then let the coefficients of the first $d - 1$ terms of $B(z^{-1})$ be zero. The polynomials also satisfy the following:

$$\begin{cases} A(z^{-1}) = 1 + a_{1,1}z^{-1} + a_{1,2}z^{-2} + \cdots + a_{1,n_a}z^{-n_a} \\ B(z^{-1}) = b_{1,0} + b_{1,1}z^{-1} + b_{1,2}z^{-2} + \cdots + b_{1,n_b}z^{-n_b}, \quad l_{1,0} \neq 0 \\ C(z^{-1}) = 1 + c_{1,1}z^{-1} + c_{1,2}z^{-2} + \cdots + c_{1,n_c}z^{-n_c} \end{cases}$$

3.1. Objective Function

In general, for the sake of the following smoothness, tracking control is carried out by following the reference track after softening. The reference trajectory is shown as follows:

$$w(k+j) = a^j y(k) + \left(1 - a^j\right) y_r(k) \tag{13}$$

where $y_r(k)$ stands for a set value for output, $w(k)$ represents the reference trajectory, and α is the softening coefficient with $0 < \alpha < 1$.

The performance function can be represented by using the following equation:

$$\min J(k) = \sum_{j=1}^{n}[y(k+j) - w(k+j)]^2 + \sum_{j=1}^{m} \lambda(j)[\Delta u(k+j-1)]^2 \tag{14}$$

where n is the maximum prediction length, m is the control time-domain ($m \leq n$), and $\lambda(j)$ is the control weighting coefficient. To prevent the variation caused by a drastic change of control increment, a certain value is set for $\lambda(j)$, while $\Delta u(k)$ is the control increment of the system. Therefore, the generalized predictive control method can be understood to find a reasonable control increment sequence $\Delta u(k)$, $\Delta u(k+1), \ldots, \Delta u(k+m-1)$ to make the objective function J have the minimum value.

3.2. Output Prediction

In order to predict the output ahead of step j, the Diophantine equation is introduced in the generalized predictive control method:

$$E_j(z^{-1})A(z^{-1})\Delta + z^{-j}F_j(z^{-1}) = 1 \tag{15}$$

where E_j and F_j are determined by prediction length and system parameter.

The optimal output prediction value can be obtained by combining Equations (13) and (15):

$$Y = G\Delta U + f \tag{16}$$

with

$$Y = [y(k+1), y(k+2), \cdots, y(k+n)]^T$$

$$\Delta U = [\Delta u(k), \Delta u(k+1), \cdots, \Delta u(k+m-1)]$$

$$f = H\Delta u(k) + Fy(k) = [f(k+1), f(k+2), \cdots, f(k+n)]^T$$

$$G = \begin{bmatrix} g_0 & & 0 \\ g_1 & g_0 & \\ \vdots & \vdots & \ddots \\ g_{n-1} & g_{n-2} & \cdots & g_0 \end{bmatrix} = \begin{bmatrix} g_{11}z^{-1} + g_{12}z^{-2} + \cdots \\ g_{22}z^{-1} + g_{23}z^{-2} + \cdots \\ \vdots \\ g_{nn}z^{-1} + g_{n(n+1)}z^{-2} + \cdots \end{bmatrix}$$

$$H = \begin{bmatrix} G_1 - G_0 \\ z(G_2 - z^{-1}g_1 - g_0) \\ \vdots \\ z^{n-1}(G_n - z^{-n+1}g_{n-1} - \cdots - z^{-1}g_1 + g_0) \end{bmatrix}$$

$$F = [F_1, F_2, \cdots, F_n]^T$$

where G is the control matrix. The parameters of E_j, F_j, G_j, and H_j are determined by the prediction length and system parameter, and satisfy the following:

$$\begin{cases} E_j(z^{-1}) = e_{j,0} + e_{j,1}z^{-1} + \ldots + e_{j,j-1}z^{-(j-1)} \\ F_j(z^{-1}) = f_{j,0} + f_{j,1}z^{-1} + \ldots + f_{j,j-1}z^{-(j-1)} \\ G_j(z^{-1}) = g_{j,0} + g_{j,1}z^{-1} + \ldots + g_{j,j-1}z^{-(j-1)} \\ H_j(z^{-1}) = h_{j,0} + h_{j,1}z^{-1} + \ldots + h_{j,j-1}z^{-(j-1)} \end{cases}$$

3.3. Determination of the Optimal Control Law

The hypothesis is that $W = [w(k+1), w(k+2), \cdots, w(k+n)]^T$; when $\frac{\partial J}{\partial \Delta u} = 0$, the optimal performance index function of Equation (17) can be obtained, as shown in Equation (18).

$$\Delta U = (G^T G + \lambda I)^{-1} G^T (W - f) \tag{17}$$

$$u(k) = u(k-1) + \Delta u \tag{18}$$

To make full use of remaining control information and prevent the predictive control increament of $\Delta u(k)$ from being reduced due to interference, a smooth filter is used as the input to carry out weighted control, as shown in Equation (19).

$$u(k) = \frac{\sum_{i=j}^{s} q(i) u(k|k-i+1)}{\sum_{i=1}^{s} q(i)} \tag{19}$$

where s stands for the control steps in the time domain and $q(i)$ represents the control weighting coefficient.

The generalized predictive control method can make the control output close to the reference trajectory or target curve as much as possible through rolling optimization and feedback correction. In this paper, the control input $u(k+1)$ optimizes and corrects the control input $u(k)$ at step k through the optimal prediction at step $k+1$, which can slow down the excessive increase or decrease of the control input increment in a certain trend and effectively reduce the occurrence probability of overshoot. The actual control variable is

$$u_s(k) = \frac{u(k) + u(k+1/k)}{2} = u(k) + \Delta u \frac{(k+1/k)}{2} \tag{20}$$

with

$$\Delta u\left(\frac{k+1}{k}\right) = \frac{\sum_{i=1}^{m} \beta(k) \Delta u(k+i/k)}{\sum_{i=1}^{m} \beta(k)} \tag{21}$$

where $\Delta u(k+i/k)$ is the predictive control increment at time $k+1$ and $\beta(k)$ is the weighting factor ($0 < \beta < 1$).

4. Multi-Model Switching Control

The structure diagram of the MMPC system of the wind generation system is given in Figure 1, where the stability of the multi-model switching control system has been demonstrated in Equation [18]. In the figure, $\hat{y}_i (i = 1, 2, \ldots, 5)$ is the output of $M_1, M_2, \ldots, M_5$, and y_r is the reference input. At each sampling time, the optimal sub-model will be selected for predictive control [19,21].

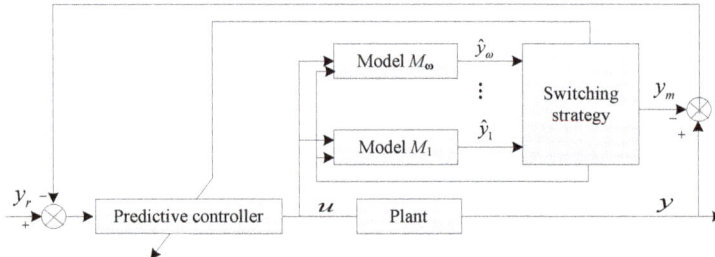

Figure 1. Structure diagram of the multi-model predictive control (MMPC) system.

Suppose at time k, $e_i(k) = y(k) - \hat{y}_i(k)$ represents the error between actual output and the output of sub-model i, define the switching performance index function:

$$J_i(k) = ae_i^2(k) + b\sum_{j=1}^{l} \rho^j e_i^2(k-j) \qquad (22)$$

where l is the time length, a and b ($a > 0$, $b > 0$) are the error weighting coefficients representing the matching degree of current and past time length l, respectively, while ρ is the forgetting factor. The smaller the $J_i(k)$, the higher the matching degree of model M_i.

5. Case Study and Simulation Analysis

In order to verify the effectiveness of the proposed method, the wind turbines with a doubly-fed induction generator (DFIG) in Guazhou Wind Farm of Western China are utilized for the case study. The parameters of wind turbine can refer to the Appendix A.

The wind generation system is given in Equation (11) and is adopted for simulation analysis (see in Figure 2), where the pitch angle is taken as the control variable and the wind speed is considered as a disturbance. The pitch angle is controlled according to the disturbance signal to ensure that the output of the wind turbine can follow the predetermined target. In this paper, the forgetting factor $\rho = 0.7$, the softening coefficient $\alpha = 0.35$, the unit value is adopted as the control length, and the value of control weighting coefficient $\lambda(j)$ is 0.9. The simulation results of the single MPC and the MMPC are given in the following figures.

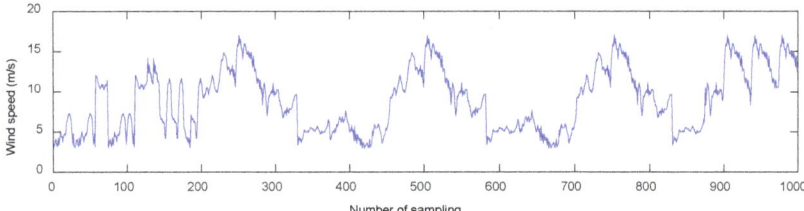

Figure 2. Field-measured wind speed data.

From Figure 3, it is apparent that the tracking effect of the MPC algorithm is satisfactory, and the maximum error of output power is 2.8 kW for the wind farm. As presented in Figures 3c and 4c (the local zoom of Figures 3a and 4a), it is apparent that the output active power from the results derived by the MMPC method is more smooth than that yield from the single MPC method. In comparison, the MMPC can derive a much smaller output error under the same model switch strategy shown in Figure 3e or Figure 4e. The maximum active output error using the single MPC is almost 3 kW (in Figure 3c), while that yield from MMPC is about 2kW (in Figure 4c). It is mainly because that the proposed adaptive multi-model switching predictive control method can switch to different models according to the wind speed variation, which can ensure the wind generation system operates in an optimal condition by selecting the most suitable model given in Figure 1. For the single MPC method, however, there is no additional choice but to adapt to the variation of the wind speed and tune the control parameters according to the predicted control error. Since the wind speed sometimes varies rapidly in a very short time, the performance of the MPC method becomes inferior when compared with the MMPC method.

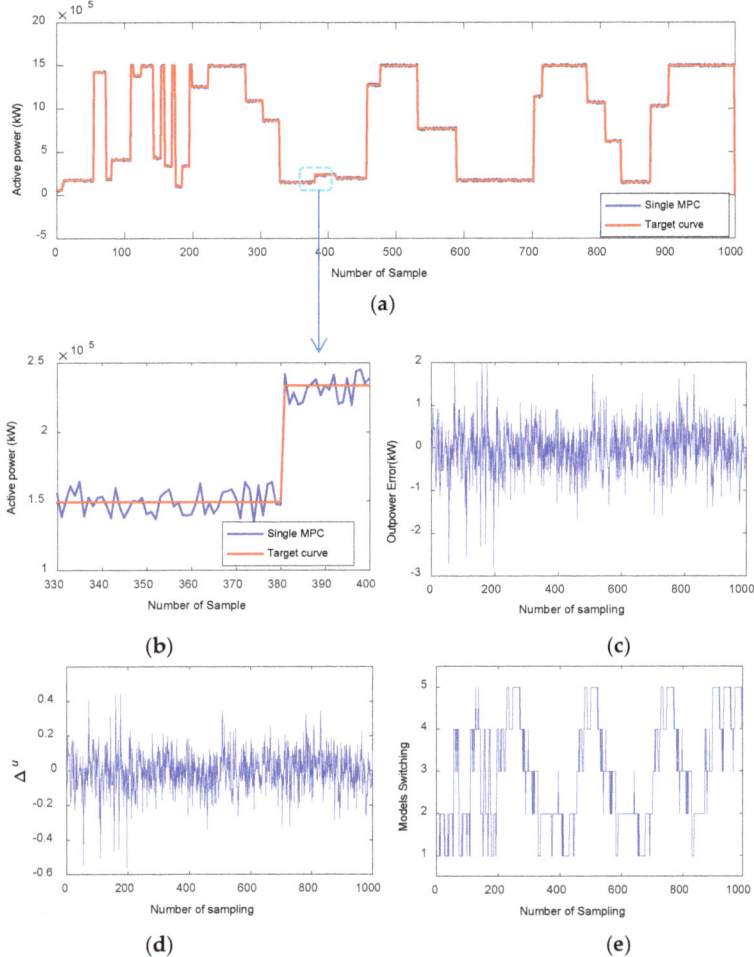

Figure 3. Simulation results using the single model predictive control (MPC). (**a**) The output power of wind farm using a single MPC; (**b**) Local zoom of (**a**), (**c**) output power error between target curve and control output, (**d**) control incremental of the single MPC, (**e**) model selection during the control process.

Additionally, as is reflected in Figures 3d and 4d, the switching control accuracy of MMPC is much higher than that of the single MPC. The results show that the MMPC strategy can reduce the tracking error in the dynamic process of the system and improve the convergence speed and tracking accuracy of the system. The simulation results proves that, with the fuzzy clustering method used, the wind speed data is pre-processed and classified to certain groups, and corresponding predictive control scheme with suitable model given in Figure 1 is assigned, which enables the MMPC outperformances the single MPC to a great extent.

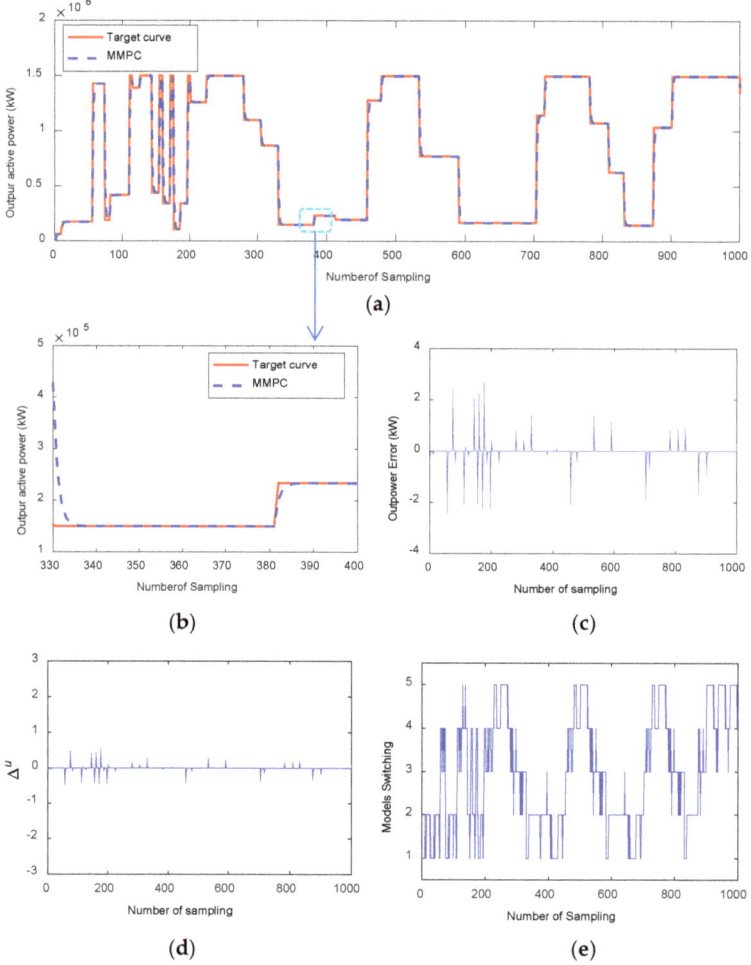

Figure 4. Simulation results using the MMPC. (**a**) The output power of wind farm using MMPC; (**b**) local zoom of (**a**), (**c**) output power error between the target curve and control output, (**d**) control incremental of the single MMPC, (**e**) model selection during the control process.

6. Conclusions

In this paper, up to 1000 groups of field-collected datasets are adopted for the case study. These data are firstly classified into five groups using the classical clustering algorithm–fuzzy c-means clustering. After this, the predictive controller is designed for multi-model switching control, and both the single MPC and MMPC are applied to the wind generation system for the active power control of a local wind farm in Gansu Province. For the case study and result analysis, it is proved that the proposed adaptive MMPC for the active power of a wind generator is effective and efficient, and the combination of the fuzzy c-means clustering and the MMPC solves the problem of randomness and uncertainty in the active power prediction of the wind generation system. By pre-clustering and multi-model switching, the impact of wind speed randomness and uncertainty on the output power is minimized, and the prediction accuracy is effectively improved. The proposed method is of guiding significance for the active power prediction in the system and the near future.

Author Contributions: Conceptualization, H.L. and H.D.; methodology, H.L. and K.R.; simulation analysis, S.L.; writing—original draft preparation, H.L. and S.L.; writing—review and editing, H.L. and H.D. All authors have read and agreed to the published version of the manuscript.

Funding: This research was funded by the Tianyou Youth Talent Lift Program, and in part by the Youth Science Foundation of Lanzhou Jiaotong University under Grant 2019029.

Acknowledgments: The authors would like to give their thanks to Dehai Wang for providing data support of this work.

Conflicts of Interest: The authors declare no conflict of interest.

Appendix A

Parameters of the Wind Turbine with DFIG Used for Case Study in Section 5	
Rated power	1.5 MW
Operation wind speed range	3 m/s~25 m/s
Rated wind speed	11 m/s
Semi-diameter of the rotor	31.4 m
Air density	1.225 kg/m^3

References

1. Gu, H.; Wang, J.; Lin, Q.; Gong, Q. Automatic Contour-Based Road Network Design for Optimized Wind Farm Micrositing. *IEEE Trans. Sustain. Energy* **2015**, *6*, 281–289. [CrossRef]
2. Chen, J.; Hu, W.; Cao, D.; Zhang, B.; Huang, Q.; Chen, Z.; Blaabjerg, F. An Imbalance Fault Detection Algorithm for Variable-Speed Wind Turbines: A Deep Learning Approach. *Energies* **2019**, *12*, 2764. [CrossRef]
3. Gao, X.; Meng, K.; Dong, Z. Cooperation-driven distributed control scheme for large-scale wind farm active power regulation. *IEEE Transactions on Energy Conversion* **2017**, *32*, 1240–1250. [CrossRef]
4. Huang, S.; Wu, Q.; Guo, Y.; Rong, F. Optimal active power control based on MPC for DFIG-based wind farm equipped with distributed energy storage systems. *Int. J. Electr. Power Energy Syst.* **2019**, *113*, 154–163. [CrossRef]
5. Arasteh, F.; Riahy, G.H. MPC-based approach for online demand side and storage system management in market based wind integrated power systems. *Int. J. Electr. Power Energy Syst.* **2019**, *106*, 124–137. [CrossRef]
6. Liu, M.; Zou, J.; Peng, C.; Xie, Y.; Li, M. Active power control for wind farms based on MPC combined with state classification. *IFAC-Pap. OnLine* **2017**, *50*, 2137–2144. [CrossRef]
7. Luo, Z.; Wei, S.; Chai, Y.; Liu, Y.; Sun, X. Simulation of wind farm scheduling algorithm based on predictive model control. In Proceedings of the 2017 Chinese Intelligent Systems Conference, Mudanjiang, China, 14–15 October 2017.
8. Liu, J.; Yao, Q.; Hu, Y. Model predictive control for load frequency of hybrid power system with wind power and thermal power. *Energy* **2019**, *172*, 555–565. [CrossRef]
9. Guo, Y.; Gao, H.; Wu, Q.; Østergaard, J.; Yu, D.; Shahidehpour, M. Distributed coordinated active and reactive power control of wind farms based on model predictive control. *Int. J. Electr. Power Energy Syst.* **2019**, *104*, 78–88. [CrossRef]
10. Mousa, H.H.H.; Youssef, A.R.; Mohamed, E.E.M. Model predictive speed control of five-phase permanent magnet synchronous generator-based wind generation system via wind-speed estimation. *Int. J. Electr. Power Energy Syst.* **2019**. [CrossRef]
11. Li, L.; Zhang, D. Model predictive control for wind farm integration through VSC-HVDC. In Proceedings of the 2018 13th IEEE Conference on Industrial Electronics and Applications, Wuhan, China, 1–2 June 2018.
12. Mallick, A.; Singh, S.N.; Mohapatra, A. Active power regulation by MPC based flywheel energy storage system. *Adv. Energy Power Syst.* **2018**, *508*, 57–71.
13. Lin, Z.; Chen, Z.; Liu, J.; Wu, Q. Coordinated mechanical loads and power optimization of wind energy conversion systems with variable-weight model predictive control strategy. *Appl. Energy* **2019**, *236*, 307–317. [CrossRef]
14. Cui, J.; Liu, X. Economic model predictive control of variable-speed wind energy conversation systems. *Control Eng. China* **2019**, *26*, 431–439. (In Chinese)

15. Ye, H.; Chen, C.; Li, S.; Ding, C. Model predictive current control of inverters to meet low-voltage ride-through requirements. *Control Eng. China* **2018**, *25*, 795–798.
16. Fan, X.; Lei, M. Research on three-level wind power generation system based on predictive control. *Electric Drive.* **2018**, *48*, 8–11+36.
17. Liu, X.; Wang, W.; Guo, J.; Guo, D. Research on predictive control of active power for direct-driven permanent magnet wind turbine generators. *Acta Energ. Sol. Si.* **2018**, *39*, 210–217. [CrossRef]
18. Wang, Y.; Yu, M.; Li, Y. Model predictive controller-based distributed control of wind turbine DC microgrid. *Trans. China Electrotech. Soc.* **2016**, *31*, 57–66. (In Chinese)
19. Zhou, K.; Yang, S.; Ding, S.; Luo, H. On cluster validation. *Syst. Eng. Theory Pract.* **2014**, *34*, 2417–2431.
20. Lu, X.; Wang, X.; Dong, H.; Ma, B. Research on Sliding Mode Predictive Control of Energy-saving Operation of High-speed Train. *Control Eng. China* **2016**, *23*, 389–393.
21. Guo, Z.; Song, A.; Mao, J. Nonlinear generalized predictive control based on least square support vector machine. *Control Decis.* **2009**, *24*, 520–525. [CrossRef]
22. Lu, X.; Dong, H. Application of multi-model active fault-tolerant sliding mode predictive control in solar thermal power generation system. *Acta Autom. Sin.* **2017**, *43*, 1241–1247.

 © 2020 by the authors. Licensee MDPI, Basel, Switzerland. This article is an open access article distributed under the terms and conditions of the Creative Commons Attribution (CC BY) license (http://creativecommons.org/licenses/by/4.0/).

Article

A Modified Reynolds-Averaged Navier–Stokes-Based Wind Turbine Wake Model Considering Correction Modules

Yuan Li [1,*], Zengjin Xu [2], Zuoxia Xing [3], Bowen Zhou [4,*], Haoqian Cui [5], Bowen Liu [1] and Bo Hu [6]

1. School of Science, Shenyang University of Technology, Shenyang 110870, China; sy_lbw@sina.com
2. School of Chemical Equipment, Shenyang University of Technology, Shenyang 110870, China; zengjin_xu@sut.edu.cn
3. School of Electrical Engineering, Shenyang University of Technology, Shenyang 110870, China; xingzuoxia@sut.edu.cn
4. College of Information Science and Engineering, Northeastern University, Shenyang 110819, China
5. Fuxin Electric Power Supply Company, State Grid Liaoning Electric Power Co. Ltd., Fuxin 123000, China; cpschq@sina.com
6. State Grid Liaoning Electric Power Co. Ltd., Shenyang 110004, China; dianlihubo@sina.com
* Correspondence: syliyuan@sut.edu.cn (Y.L.); zhoubowen@ise.neu.edu.cn (B.Z.); Tel.: +86-150-0401-9739 (B.Z.)

Received: 14 July 2020; Accepted: 25 August 2020; Published: 27 August 2020

Abstract: Increasing wind power generation has been introduced into power systems to meet the renewable energy targets in power generation. The output efficiency and output power stability are of great importance for wind turbines to be integrated into power systems. The wake effect influences the power generation efficiency and stability of wind turbines. However, few studies consider comprehensive corrections in an aerodynamic model and a turbulence model, which challenges the calculation accuracy of the velocity field and turbulence field in the wind turbine wake model, thus affecting wind power integration into power systems. To tackle this challenge, this paper proposes a modified Reynolds-averaged Navier–Stokes (MRANS)-based wind turbine wake model to simulate the wake effects. Our main aim is to add correction modules in a 3D aerodynamic model and a shear-stress transport (SST) k-ω turbulence model, which are converted into a volume source term and a Reynolds stress term for the MRANS-based wake model, respectively. A correction module including blade tip loss, hub loss, and attack angle deviation is considered in the 3D aerodynamic model, which is established by blade element momentum aerodynamic theory and an improved Cauchy fuzzy distribution. Meanwhile, another correction module, including a hold source term, regulating parameters and reducing the dissipation term, is added into the SST k-ω turbulence model. Furthermore, a structured hexahedron mesh with variable size is developed to significantly improve computational efficiency and make results smoother. Simulation results of the velocity field and turbulent field with the proposed approach are consistent with the data of real wind turbines, which verifies the effectiveness of the proposed approach. The variation law of the expansion effect and the double-hump effect are also given.

Keywords: Reynolds-averaged Navier–Stokes method; wind turbine wake model; 3D aerodynamic model; turbulence model; correction modules

1. Introduction

In the wake region of a wind turbine, the energy absorption rate of the wind turbine is decreased and the fatigue load is increased due to the decreasing wind speed and increasing intensity of turbulence [1–3]. The wind turbine wake model can provide a basic theory for micro-location, power prediction and the assessment of wind turbine performance in wind energy engineering [4]. However, in the existing wake model, corrections are not comprehensively considered in the aerodynamic model and turbulence model, which challenges its calculation accuracy [5,6].

The study of wind turbine wakes focuses on the velocity distribution, the wake zone width, and the details of the development and change of flow structure at different locations of the wake region. Figure 1 shows the ideal region division of wind turbine wakes, which is recognized as a hypothetical model in wake research. The area with a 3–5-time diameter behind the wind turbine is called the near wake region, and that with a 5-time diameter is called the far wake region [7,8].

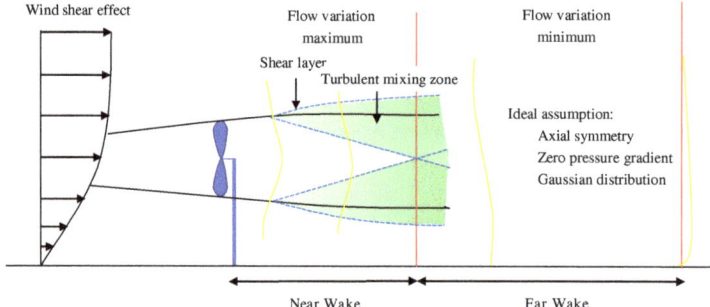

Figure 1. Ideal hypothesis and regional distribution of wind turbine wakes.

The existing wake models are divided into three categories: semi-empirical engineering models [9,10], vortex method models [11,12] and computational fluid dynamics (CFD) models [13]. The state-of-the-art CFD models can be divided into two classes: Reynolds-averaged Navier–Stokes (RANS) [14] and large-eddy simulation (LES) [15,16]. As the semi-empirical engineering model is one-dimensional, the computation is small, and the velocity distribution of the wake region can be located quickly. This model is often used in the software of engineering design, such as Windsim and Windfarmer. However, the calculation results are far from the measured data of wind farms [16]. The vortex method uses a lift line or lift face blade to simulate for the steady wind flow and simple unsteady conditions, especially for the aerodynamic characteristics and wake tip vortex of the wind turbine in the random wind velocity field with the wind shear effect [11]. The CFD method is a general method, which describes the flow field by using hydrodynamic control equations, and it obtains the numerical results of the flow field by using computer program design. The results are affected by the design of RANS and LES models in the CFD method, which needs to improve the accuracy of turbulence models. Moreover, computational cost is also an index to assess the CFD method [17–19]. The advantages and disadvantages of research on wind turbine wake models are shown in Table 1 [20–23].

Table 1. The advantages and disadvantages of research on wind turbine wake models.

Category	Models	Advantages	Disadvantages
Semi-empirical model	Jensen [20]	Widely used in industry	Suitable for the far wake region
	Fuga [21]	Widely used in industry with medium accuracy	Only uses for offshore wind farms
Simulation model	DNS [22]	Direct numerical simulation with high accuracy	Requires a lot of computing resources and simulation time
	LES [23]	Widely used in research with high accuracy	Appropriate assumptions are necessary

DNS (direct numerical simulation), LES (large-eddy simulation).

Concerning LES, a full-scale 3D wind turbine model by using sliding tetrahedron mesh is established to calculate the wakes in the rotor diameter range of 20 times [24]. The authors of [23] designs a non-full-scale model combined with the model of the wind turbine, and they discuss the function of each sub-grid model under the LES method. Concerning the RANS, a shear-stress transport (SST) k-ω turbulence model, a 2D high-precision computational grid, is used to calculate the wakes [15]. A tree crown model is applied in the turbulence model [13]. Its effect on the wind turbine in regard to incoming airflow is similar to that of tree crown on airflow. Based on the tree crown model, a modified calculation of the turbulence model is proposed by adding a source term [25]. Its calculation results are compared with the measured data of the Danish Nibe B wind turbine. The results show that the modified model is feasible in the near wake region. However, the wake velocity deficit in the far wake region is underestimated. Furthermore, Shen et al. obtain the surface pressure and friction of airfoil by X foil software, and distribute them into the 2D actuator surface model as volume force [26]. The blade surface lift force is calculated by utilizing the vortex method in [27]. In [28] and [29], the scholars develop an aeroelastic fatigue, aerodynamics, structures, and turbulence (FAST) model of a wind turbine and offshore wind farm application (OWFA) model combined with an actuation model considering the influence of blade deformation on wind turbine wakes. An improved k-ε turbulence model is presented for the numerical simulation of wind turbine wakes. However, in the vertical direction, the simulation results do not have good agreement with the experimental results [30].

In summary, the above approaches lack the consideration of comprehensive corrections in the aerodynamic model and turbulence model. Motivated by the above discussion, the velocity field and turbulence field of wind turbines are simulated by a modified RANS (MRANS) method in this paper. The main contributions of this paper are summarized as follows:

- This paper proposes a modified Reynolds-averaged Navier–Stokes (MRANS)-based wind turbine wake model to simulate the wake effects. Based on the correction module, the proposed blade element momentum (BEM) -fuzzy aerodynamic model can amend the inconsistent condition between the wake simulation and the experiment test. For the turbulence model, the turbulence attenuation is effectively avoided by adding the hold source term. The accuracy of the turbulence intensity distribution is improved by correcting the closure constant and the dissipation term.
- Simulation results of the velocity field and turbulent field with the proposed approach are consistent with the data of real wind turbines, which verifies the effectiveness of the proposed approach. Furthermore, the computation efficiency is significantly improved by the developed mesh partition method.

The rest of this paper is organized as follows: Section 2 describes the overall proposed wind turbine wake model. The modified wind turbine 3D model with BEM and fuzzy theory is studied in Section 3. The SST k-ω turbulence model is presented in Section 4. The variable size hexahedron mesh partition method is developed in Section 5. Numerical simulations are given in Section 6. Finally, conclusions are drawn in Section 7.

2. Overall Wind Turbine Wake Model

In this paper, the RANS model is adopted in the wind turbine wake model. Wind turbine wakes can be regarded as an incompressible flow field, and energy equations are not considered in this paper in accordance with the Reynolds-averaged equations [5]. The continuity equation is given by Equation (1).

$$\frac{\partial u_i}{\partial x_i} = 0, \qquad (1)$$

where u_i is the average velocity component in direction x_i, $i = 1, 2, 3 \ldots$

The momentum equations are given by Equation (2).

$$\frac{\partial(\rho u_i u_j)}{\partial x_j} = \frac{\partial p}{\partial x_i} + \frac{\partial(2\mu S_{ij} - \rho \overline{u'_i u'_j})}{\partial x_j} + f_i, \qquad (2)$$

where ρ is the air density, u_i is the i-th velocity component, p is positive stress, μ is the dynamic viscosity coefficient, $S_{ij} = \frac{1}{2}\left(\frac{\partial u_j}{\partial x_i} + \frac{\partial u_i}{\partial x_j}\right)$ is the average strain tensor, u'_i is the pulsation velocity component in direction x_i, and f_i is the volume source term.

In the process of Reynolds time averaging, an unknown tensor additional term $\rho \overline{u'_i u'_j}$ is derived, which is called the Reynolds stress tensor and represents energy transfer caused by turbulence pulsation. Due to this unknown additional term, the Reynolds time-averaged Equations (1) and (2) cannot be closed. The turbulence model can calculate the Reynolds stress tensor by the vorticity viscosity hypothesis. The Boussinesq equation is shown in Equation (3) [5].

$$\rho \overline{u'_i u'_j} = 2\mu_t S_{ij} - \frac{2}{3}\rho k \delta_{ij}, \qquad (3)$$

where the turbulent kinetic energy $k = 0.5\overline{u'_i u'_i}$, $\delta_{ij} = \begin{cases} i = j & 1 \\ i \neq j & 0 \end{cases}$, and μ_t is the eddy viscosity coefficient determined by the turbulence model.

To enhance the accuracy of the abovementioned Reynolds stress tensor and eddy viscosity coefficient, we develop a correction module for an SST k-ω turbulence model. Meanwhile, to further comprehensively complete the volume source term of the RANS model, another correction module is also introduced to the BEM-fuzzy 3D aerodynamic model. By combining the turbulence model and the aerodynamic model, we propose an M-RANS wake model; the overall modeling framework is shown in Figure 2.

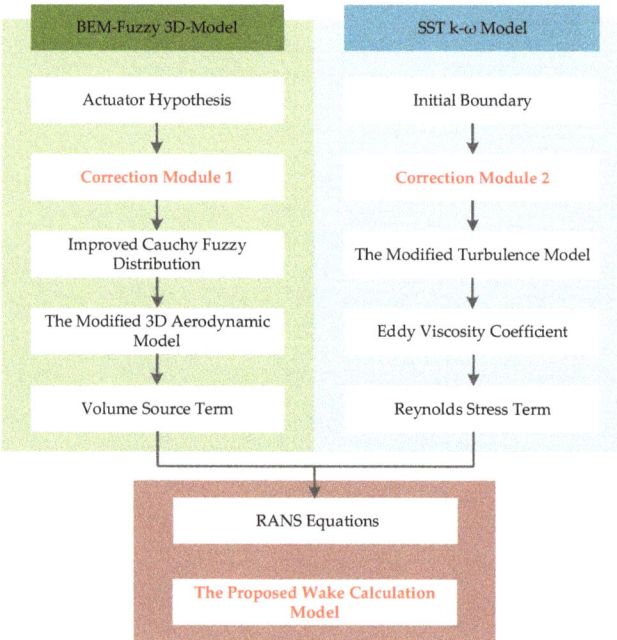

Figure 2. The overall framework of the modified Reynolds-averaged Navier–Stokes-based wind turbine wake model.

3. BEM-Fuzzy 3D Aerodynamic Model

As shown in Figure 2, the BEM-fuzzy model is established to obtain the volume source term. When compared to the traditional 2D BEM model of the actuator hypothesis, this model is modified by three corrections to the 2D model and by Cauchy fuzzy distribution of the 3D model.

Figure 3 illustrates the principle of the BEM theory calculation, which can effectively obtain the lift and resistance of wind turbines [31]. According to BEM theory, the element of local velocity is calculated by Equation (4).

$$V_r = \sqrt{[V_\infty(1-a)]^2 + [\Omega\, r(1+b)]^2}, \qquad (4)$$

where Ω is the angular velocity of wind wheel rotation, r is the span position of the blade element airfoil section, V_∞ is the axial atmospheric free flow velocity, and a and b are the axial induction factor and tangential induction factor, which are determined by iterative methods [32].

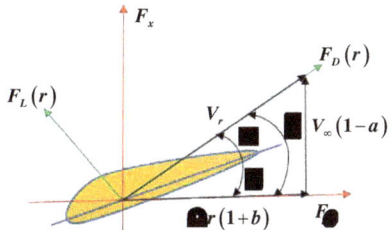

Figure 3. Blade element stress analysis diagram.

The inflow angle ϕ is calculated by Equation (5).

$$\phi = \tan^{-1}\left[\frac{V_\infty(1-a)}{\Omega\, \gamma(1+b)}\right], \qquad (5)$$

where $\alpha = \phi - \gamma$ is the attack angle, and γ is the local pitch angle.

The force vector of the blade element at the span position in the 2D plane is obtained by Equation (6).

$$F_{2D}(r) = \begin{bmatrix} F_L(r) \\ F_D(r) \end{bmatrix} = \frac{1}{2} n_b \rho V_r^2 c \begin{bmatrix} C_L e_L \\ C_D e_D \end{bmatrix}, \qquad (6)$$

where n_b is the number of blades, $C_L = C_L(a, R_e)$ and $C_D = C_D(a, R_e)$ (determined by the airfoil aerodynamic characteristic curve) are the lift coefficient and drag coefficient, R_e is the Reynolds number of the local chord length c, and e_L and e_D are the unit vectors for the directions of lift and drag, respectively. Due to the poor precision of the traditional 2D model, it is necessary to add corrections to improve accuracy. The traditional 2D model is modified from the following three aspects.

3.1. Tip Loss Correction

Due to the pressure difference between the lift surface and the pressure surface of the blade, the airflow at the tip and root of the blade will flow twice along the blade radial direction. The moment that acting on the blade is reduced, the blade element force at the tip has a great influence on the aerodynamic performance of the whole blade. Therefore, the tip loss correction is considered. The tip loss coefficient is defined by Equation (7) [33].

$$\eta_1 = \frac{2}{\pi}\cos^{-1}\left[\exp\left(-g\frac{n_b(D/2-r)}{2r\sin\phi}\right)\right], \qquad (7)$$

where the parameter g is obtained by Equation (8).

$$g = \exp[-c_1(n_b \Omega D/2V_\infty - c_2)] + c_3. \tag{8}$$

The recommended values of the empirical parameters c_1, c_2, c_3 are 0.125, 21, and 0.1, respectively.

3.2. Hub Loss Correction

The hub aerodynamic performance is affected by the separation of airflow to the root of the blade. Similar to the blade tip loss, the hub loss is considered. The hub loss coefficient is defined by Equation (9) [32].

$$\eta_2 = \frac{2}{\pi} \cos^{-1}\left[\exp\left(-g\frac{n_b(r - R_{hub})}{2R_{hub}\sin\phi}\right)\right], \tag{9}$$

where R_{hub} is the hub radius, and the correction factor η is $\eta = \eta_1 \cdot \eta_2$.

3.3. Attack Angle Correction

The blade has a certain thickness and width, especially at the root of the blade, which makes the direction of the airflow change greatly. In the front and rear edge of the airfoil, the circumferential velocity of the airflow increases. At the same time, the cross-section area of airflow decreases, and the axial velocity of airflow increases due to the thickness of airfoil. The thickness and width of the blade affect the attack angle. Attack angle changes are given by Equation (10) [34].

$$\begin{cases} \Delta\alpha_1 = \frac{1}{4}\left[\tan^{-1}\left(\frac{V_\infty(1-a)}{\Omega r(1+2b)}\right) - \tan^{-1}\left(\frac{V_\infty(1-a)}{\Omega r}\right)\right] \\ \Delta\alpha_2 = \frac{16}{15}\frac{\frac{V_\infty gc}{\Omega D^2}}{\left(\frac{2V_\infty}{D}\right)+\left(\frac{2r}{D}\right)}\frac{t_{max}}{c} \end{cases}, \tag{10}$$

where t_{max} is the maximum thickness of blade element airfoil, $\Delta\alpha_1$ is the attack angle change caused by the influence of blade width on the direction of airflow, and $\Delta\alpha_2$ is the attack angle change caused by the influence of blade thickness on airflow direction.

The attack angle is revised by Equation (11).

$$\alpha = \phi - \beta - \Delta\alpha_1 - \Delta\alpha_2 \tag{11}$$

The modified blade element aerodynamic force in the 2D plane is expressed by Equation (12).

$$F_{2D}(r) = \begin{bmatrix} F_L(r) \\ F_D(r) \end{bmatrix} = \eta\frac{1}{2}n_b\rho V_r^2 c\begin{bmatrix} C_L'e_L \\ C_D'e_D \end{bmatrix}. \tag{12}$$

Here, we have obtained the 2D model, and it is necessary to establish the 3D model via the improved Cauchy fuzzy distribution.

The thickness of wind turbine volume force distribution in the axis is uncertain. To ensure the stability of numerical simulation and increase the convergence rate of calculation, it is necessary to smooth the volume force to both sides of the neutral layer. Take an observation point of blade airfoil as an example; the volume force of wind turbine firstly remains unchanged in a small range of the axial direction. Subsequently, both ends of the axis are rapidly attenuated to zero. This is essentially a fuzzy problem, as shown in Figure 4. Therefore, the mapping can be obtained by Equation (13).

$$\begin{cases} A : U \to (0,1], A \in F(U) \\ l \to A(l) \end{cases}, \tag{13}$$

where the domain U is the volume force at a distance l from the origin, the fuzzy rule A is the stable condition of the domain element, and $A(l)$ is the membership function of domain U.

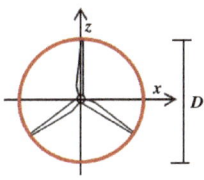 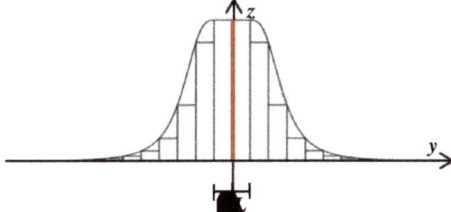

(a) Actuator hypothesis model   (b) Improved Cauchy fuzzy distribution

Figure 4. Schematic diagram of the 3D fuzzy model.

The membership function $A(l)$ can be determined by the improved Cauchy fuzzy distribution as follows.

$$A(l) = \begin{cases} \frac{1}{1+\alpha(\Delta L - l)^\beta} & if\, l \leq -\Delta L \\ 1 & if\, -\Delta L < l \leq \Delta L \\ \frac{1}{1+\alpha(l-\Delta L)^\beta} & if\, l > \Delta L \end{cases}, \tag{14}$$

where $\Delta L = \frac{\int_0^{\frac{D}{2}} c(r) dr}{D} \sin\phi$ is the interval element as shown in Figure 4b; the parameters α and β are 1.5 and 3.

The cut set $A_\lambda = \{l | l \in U, A(l) \geq \lambda\}$ under the accuracy λ level determines the calculation area. The axial distribution is treated as square intervals ΔL in Figure 4 to reduce the computation time.

The calculated thickness L under accuracy λ is expressed by Equation (15).

$$\begin{cases} L = 2N\Delta L \\ N = [\max(A_\lambda(l))/\Delta L] \end{cases}, \tag{15}$$

where [ ] is the downward rounding operation.

The attenuation coefficient at any location l within the calculated thickness is obtained by Equation (16).

$$A_n(l) = \{A(n\Delta L) | n = [l/\Delta L], l \in R\}. \tag{16}$$

According to the theory of actuator hypothesis [35], the average lift and drag distribution of the wind turbine in the 2D plane is calculated by Equation (17).

$$f_{2D}(r) = \begin{bmatrix} f_L(r) \\ f_D(r) \end{bmatrix} = n_b \rho V_r^2 c / (4\pi r) \begin{bmatrix} C_L e_L \\ C_D e_D \end{bmatrix}. \tag{17}$$

The volume force of the discrete fuzzy distribution along the normal direction of the wind turbine in 3D space is obtained by Equation (18).

$$f_{3D}(r,l) = \begin{bmatrix} f_L(r,l) \\ f_D(r,l) \end{bmatrix} = \eta n_b \rho V_r^2 c A_n(l) / (4\pi r \Delta L) \begin{bmatrix} C_L' e_L \\ C_D' e_D \end{bmatrix}. \tag{18}$$

The volume forces at three different dimensions are calculated by Equation (19).

$$f_{3D}(x,y,z) = \begin{bmatrix} f_x \\ f_y \\ f_z \end{bmatrix} = X_2 X_1 f_{3D}(r,l), \tag{19}$$

where the original point of the right angle coordinate system is the model center, x-direction is the axial direction, z-direction is the height direction, $X_2 = \begin{bmatrix} \cos\phi & \sin\phi \\ \sin\phi & -\cos\phi \end{bmatrix}$ is the transformation matrix for a direct coordinate system, $X_1 = \begin{bmatrix} 1 & 0 \\ 0 & \cos\psi \\ 0 & \sin\psi \end{bmatrix}$ is the transformation matrix for the normal force as well as tangential force, and Ψ is the polar angle in a polar coordinate system.

4. SST k-ω Turbulence Model

As shown in Figure 2, the modified SST k-ω model is established to obtain the Reynold stress term of the global RANS model.

The SST k-ω model includes a k-ω model in the outer layer (near the wall) and a k-ε model in the inner layer (the outer edge of the boundary layer, the free shear layer and the fully developed region of turbulence) [36]. This method combines the k-ω model and the k-ε model by the mixed function of weighted average. The k-ε model has less dependence on the far field condition, and the k-ω model has a high accuracy in the near wall simulation. To improve the simulation accuracy in the strong adverse pressure gradient and separation flow, a modified vortex viscosity coefficient is proposed. The SST k-ω two-equation model is expressed by Equation (20).

$$\begin{cases} \frac{\partial(\rho k)}{\partial t} + \frac{\partial(\rho u_i k)}{\partial x_i} = \underbrace{\min\left(\mu_t \frac{\partial u_i}{\partial x_j}\left(\frac{\partial u_i}{\partial x_j} + \frac{\partial u_j}{\partial x_i}\right), 10\beta^*\rho k\omega\right)}_{\text{turbulent kinetic energy generation term}} - \underbrace{\beta^*\rho k\omega}_{\text{dissipative term}} + \underbrace{\frac{\partial}{\partial x_i}\left[(\mu + \sigma_k\mu_t)\frac{\partial k}{\partial x_i}\right]}_{\text{diffusion term}} \\ \frac{\partial(\rho\omega)}{\partial t} + \frac{\partial(\rho u_i\omega)}{\partial x_i} = \underbrace{\alpha'\rho S^2}_{\text{turbulent kinetic energy generation term}} - \underbrace{\beta'\rho\omega^2}_{\text{dissipative term}} + \underbrace{\frac{\partial}{\partial x_i}\left[(\mu + \sigma_\omega\mu_t)\frac{\partial \omega}{\partial x_i}\right]}_{\text{diffusion term}} + \underbrace{2(1-F_1)\rho\sigma_{\omega 2}\frac{1}{\omega}\frac{\partial k}{\partial x_i}\frac{\partial \omega}{\partial x_i}}_{\text{cross diffusion term}} \end{cases}, \quad (20)$$

where

$$\mu_t = \frac{a_1 \rho k}{\max\left(a_1\omega, \sqrt{2S_{ij}S_{ij}}\tanh\left[\left(\max\left(\frac{2\sqrt{k}}{\beta^*\omega y}, \frac{500\mu}{\rho\omega y^2}\right)\right)^2\right]\right)},$$

$$\xi = F_1\xi_1 + (1-F_1)\xi_2 \ (\xi = \{\alpha', \beta', \sigma_k, \sigma_\omega\}), \quad (21)$$

$$F_1 = \tanh\left(\min\left(\max\left(\frac{\sqrt{k}}{\beta^*\omega y}, \frac{500\mu}{\rho\omega y^2}\right), \frac{4\rho\sigma_{\omega 2}k}{\max\left(2\sigma_{\omega 2}\rho\frac{1}{\omega}\frac{\partial k}{\partial x_j}\frac{\partial \omega}{\partial x_j}, 10^{-20}\right)}\right)\right)^4,$$

where y is the distance from the nearest wall to the calculation point, and the closed constants are set as:

$$\beta^* = 0.09, a_1 = 0.31$$
$$\alpha'_1 = 5/9, \beta'_1 = 3/40, \sigma_{k1} = 0.085, \sigma_{\omega 1} = 0.5 \ \text{in layer}$$
$$\alpha'_2 = 0.44, \beta'_2 = 0.0828, \sigma_{k2} = 1, \sigma_{\omega 2} = 0.856 \ \text{out layer}$$

If the two-equation model is used to simulate the free stream, the initial value of the inlet boundary will gradually decrease with the downstream flow. When the fluid approaches to the wind turbine, the local turbulent variable value is no longer the initial value of the inlet boundary, which may cause turbulence attenuation.

The turbulent kinetic energy generation term of the equations is zero due to no velocity gradient in the inlet free flow. The diffusion term and cross diffusion term can be ignored due to no gradient of turbulence variables. Solving Equation (20), we can obtain the result in the X direction as follows:

$$\omega = \omega_I\left(1 + \frac{\omega_I\beta'x}{u}\right)^{-1}, k = k_I\left(1 + \frac{\omega_I\beta'x}{u}\right)^{-\frac{\beta^*}{\beta'}} \quad (22)$$

where the index I is the initial inlet boundary, x is the downstream distance, and u is local wind velocity.

4.1. Hold Source Term

It can be seen from Equation (22) that the turbulent kinetic energy and the specific dissipation rate will be attenuated from the initial stage of the inlet boundary to the outlet of the computational domain. Therefore, the turbulence attenuation effect must be corrected in the wake calculation. To reduce the turbulence attenuation, the dissipative term caused by turbulence attenuation can be offset by adding the hold source term in the turbulence model. The turbulent equations of the inlet free flow can be obtained by Equation (23).

$$\begin{cases} \frac{\partial(\rho k)}{\partial t} + \frac{\partial(\rho u_i k)}{\partial x_i} = -\beta^* \rho k \omega + \beta^* \rho k_R \omega_R \\ \frac{\partial(\rho \omega)}{\partial t} + \frac{\partial(\rho u_i \omega)}{\partial x_i} = -\beta' \rho \omega^2 + \beta' \rho \omega_R^2 \end{cases}, \quad (23)$$

where k_R and ω_R are the atmospheric real environmental turbulence values calculated by Equation (24).

$$\begin{cases} k_R = \frac{3}{2}(U_0.I_0)^2 \\ \omega_R = \frac{\rho k_R}{\mu}\left(\frac{\mu_t}{\mu}\right)^{-1} \end{cases}, \quad (24)$$

where U_0 is the average velocity of incoming flow, I_0 is the atmospheric turbulence intensity, and $\frac{\mu_t}{\mu}$ is the vortex–viscosity ratio.

The new source term is smaller than the dissipative term of the original equation for the wind turbine wakes region due to the introduced turbulence attenuation. The addition of the hold source term has little effect on the original SST k-ω model. Meanwhile, to avoid the numerical oscillation caused by the addition and switch of the hold source term in the free flow region and the non-free flow region, the hold source term is adopted in the whole computational domain.

The overestimation of the turbulent dissipation rate will result in the slow recovery rate of the wakes predicted by the RANS model [35]. The higher turbulent kinetic energy can promote the convection–diffusion between the wake region and the surrounding free fluid, and accelerate the recovery of wake velocity, which can be obtained by reducing the dissipation term. We reduce the turbulent dissipation by closure constant correction and correction factor addition, as discussed in the following subsection.

4.2. Closure Constant Correction

Based on the equilibrium flow theory, the surface friction velocity u^* is introduced. The turbulent kinetic energy k is calculated by Equation (25).

$$k = \frac{u^{*2}}{\sqrt{\beta^*}}, \quad (25)$$

where u^{*2}/k belongs to [0.17, 0.18], and the turbulent attenuation ratio β^*/β' is 1.2 [37]. The constant values of the modified closure are obtained by Equation (26).

$$\begin{cases} \beta^* = 0.033 \to 0.090 \\ \beta'_1 = 0.025 \to 0.075 \end{cases}, \quad (26)$$

4.3. Correction Factor Addition

The purpose of the dissipative correction is to modify the dissipation in the near wake region, but the previously modified equation is not universal in the whole. In this paper, a dissipative correction adaptive factor with the full computational domain is proposed. This method can highlight the effect of the correction term in the near wakes and reduce the correction in other regions. Dissipative item revision for the k-ω equation is expressed by Equation (27).

$$-\left(1+\eta_3 \cdot \frac{\omega_I}{\omega}\right)\beta'\rho\omega^2, \tag{27}$$

where $\eta_3 = 1 + \exp\left(-\left(1-\frac{u}{u_0}\right)^{-1}\right)$.

5. Mesh Partition Method

The Semi-implicit method for pressure-linked equations (SIMPLE) solution is used under Fluent 6.0 solver considering two initial conditions. One is the neutral atmospheric condition, the other is the wind shear effect. The source term and turbulence are set by the user-defined function. The shear effect is represented by the logarithmic function.

$$U(z) = \frac{u^*}{K}\ln\left(\frac{z}{z_0}\right) \tag{28}$$

where $u^* = \sqrt{\tau_\omega/\rho}$ is the surface friction velocity, τ_ω is the shear stress, K is the Karman constant and the value is 0.41, z is the altitude from the ground, and z_0 is the surface roughness.

The 3D wind turbine model of the computational domain is completed by SolidWorks. The model is imported to the Ansys ICEM program to achieve mesh partition as shown in Figure 5. With the variable size regular hexahedron method, three different mesh average sizes (1/3D, 1/10D and 1/40D) are applied. The cartesian coordinate system is used, the center of the wind turbine is the coordinate origin, and the number of meshes is about 0.5 million. Compared with the general hexahedron sweep method, the developed method saves three-quarters of the mesh number.

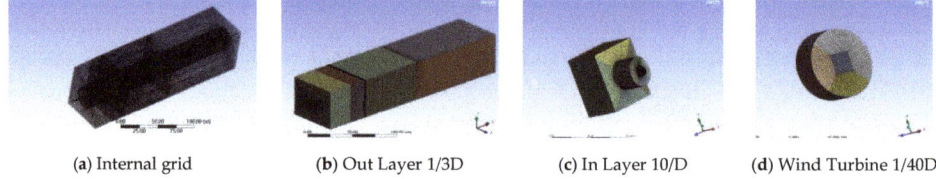

(a) Internal grid   (b) Out Layer 1/3D   (c) In Layer 10/D   (d) Wind Turbine 1/40D

Figure 5. Mesh design scheme.

Figure 6 shows the calculation domain dimension design. The computational domain is extended 25D in the streamwise direction (x-direction), 6D in the lateral direction (y-direction) and 6D in the vertical direction (z-direction). The wind turbine (mesh size is 1/40D) is located at 0D in the x-direction. The inner layer encryption region (mesh size is 1/10D) is close to the wind turbine. In the front view, the inlet of airflow direction is on the left end, and the exit is on the right.

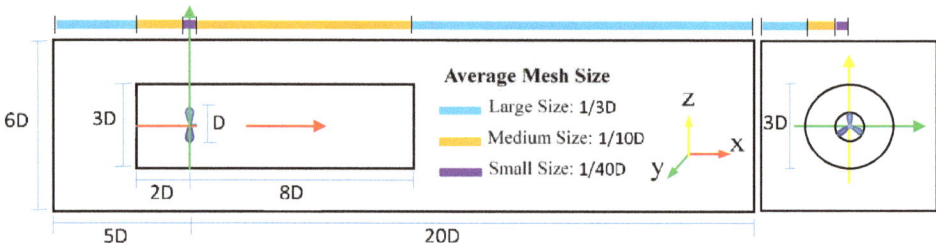

Figure 6. Calculation domain dimension design.

6. Simulation Results

6.1. Simulation Setup

To verify the effectiveness of the proposed method, experimental data are acquired from two wind turbines at Nibe, in Northern Denmark. The experiment data are tested by the actual measurement of the wind farm [38]. As shown in Figure 7, four meteorological masts (MMs) are placed in a line at downstream distances, 2.5D, 4D, 6D and 7.5D, concerning the Nibe B wind turbine. The data consist of average values every 1 min for two years. The specific operating parameters of the wind turbines are shown in Table 2. The neutral atmospheric condition with wind shear effect is used as the boundary condition in this paper.

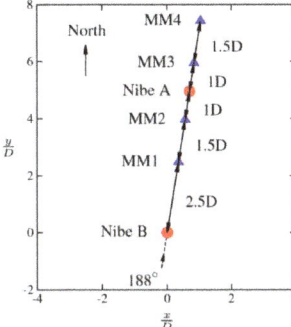

Figure 7. Measurement schematics of meteorological masts (MMs) at Nibe.

Table 2. Parameters of the wind turbines.

Parameters	Nibe Measurement
Wind turbine model	Nibe B
Wind speed (U_∞)	8.5 m/s
Turbulence intensity (I_∞)	10.1%
Boundary condition	Uniform
The wind turbine rotor diameter (D)	40 m
The wind turbine rotor rotating speed (Ω)	34 rpm
The coefficient of axial thrust (C_T)	0.89
The wind turbine hub height (Z_H)	45 m

Six modified items in two correction modules are proposed to improve the accuracy of the wake calculation, which are tip loss correction (Equation (7)), hub loss correction (Equation (9)), attack angle correction (Equation (10)), turbulence attenuation correction (Equation (23)), closure constant correction (Equation (26)) and dissipative item correction (Equation (27)). Thus, six models are designed and compared as shown in Table 3.

Table 3. Six models for comparison.

Correction Item	Model 1	Model 2	Model 3	Model 4	Model 5	Model 6
Tip loss correction	√	×	×	√	√	√
Hub loss correction	√	×	×	√	√	√
Attack angle correction	√	√	×	√	√	√
Turbulence attenuation correction	√	√	√	×	√	√
Closure constant correction	√	√	√	×	×	√
Dissipative item correction	√	√	√	×	×	×

6.2. Velocity Field under the Uniform Inflow Condition

6.2.1. Comparison and Analysis of Velocity Field Results at the Center Axis

Figure 8 shows the influence of each correction item on the change in wake velocity. The comparison between model 1 and exe1 (measured at Nibe) data shows that all the six correction items can ensure the agreement between the calculated results and the experimental values. The overall velocity changes to the lowest value of 0.48 at the position of 2.5D behind the wind turbine, increases rapidly in the near wake region, and reaches 0.64 and 0.75 at 4D and 6D, respectively. The overall velocity slowly increases in the far wake region and reaches 0.85 and 0.97 at 7.5D and 20D, respectively.

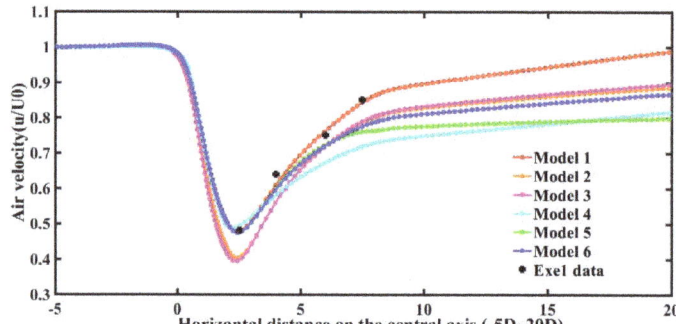

Figure 8. Velocity field at the center axis.

The comparison between models 2 and 3 and model 1 shows the influence of three correction items on the position and amount of wake velocity valley. Models 2 and 3 overestimate the wake drop velocity at the valley by 0.41 and 0.38, respectively. The results show that the correction of tip loss and hub loss reduces the overestimated volume source term by a large margin. The correction of the attack angle can also adjust the valley value.

The comparison between models 4–6 and models 1–3 shows the effects of the modified turbulence equation on the wake velocity recovery, especially in the far wake region. At 20D, the velocity recovery of models 4–6 is slower than that of model 1 in the far wake region with 0.79, 0.81, 0.87, respectively, while models 2 and 3 can still show a faster recovery velocity in the far wake region with 0.88, 0.89, respectively, which indicates that the three turbulence correction terms could affect the far wakes. The recovery of far wakes is significantly higher in model 1 than models 5 and 6. This shows that increasing the sealing parameter and reducing the dissipative term affect the increase in velocity recovery.

6.2.2. Correction Velocity Results at the Different Cross-Sections of Axial Direction

Figure 9a shows that the correction of the wind turbine model wake field is obvious in the near wake region. The comparison between model 2 and model 3 shows that the modification of tip loss and hub loss can effectively narrow the profile velocity change curve and reduce the expansion range of the wake flow field. At the same time, the velocity valley value is increased from 0.38 to 0.42 due to the decreasing of the volume source term by the blade tip and hub modification. Meanwhile, the valley difference between model 1 and model 4 is only 2%. The whole curvature is nearly closed to model 1, which indicates that the correction of the turbulence equation in the near wakes has less effect.

Figure 9b indicates that the effect of turbulence correction increases gradually with the increase in horizontal distance, while the effect of wind turbine correction gradually decreases. The comparison between model 1 and model 4 indicates that the wake velocity recovery without turbulence correction is not satisfactory, as the velocity changes from higher than 2% at 2.5D to lower than 5% at 4D.

Figure 9c shows that the correction of the turbulence equation in the far wakes of the wind turbine is obvious. Compared with models 4 and 5, the correction effect of the decreasing dissipation term in model 6 is obvious in the range of −1.5D to 1.5D in the y-direction, which causes the wind velocity curve to reach to the maximum. On the whole, comparing the 7.5D section with the 2.5D section, the velocity from the 0.5D to 1D in the y-direction gradually decreases in the axial direction.

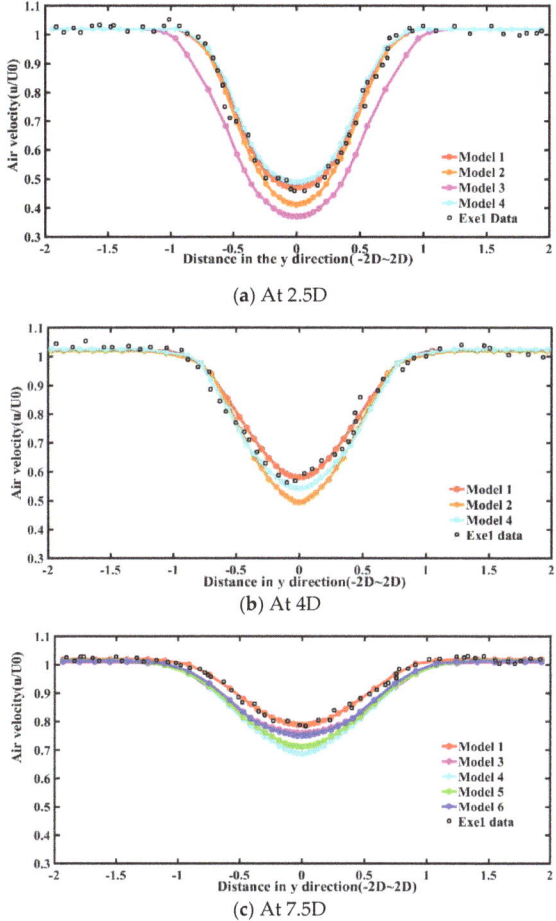

Figure 9. Correction velocity at the different cross-sections of axial direction.

6.2.3. Velocity at Different Altitudes

Figure 10 demonstrates that the valley position moves forward slightly from 2.5D to the horizontal distance with the increasing height. There is no significant change in the near and far wake regions. The wake drop slows down after 0.3D and increases from 3% to 10–20% per 0.1D in the direction of height.

Figure 11 indicates that the expansion effect is obvious in the 10D range behind the wake, and shows a nonlinear boundary curve. The expansion range decreases gradually, and the nonlinearity of the expansion boundary curve decreases with the increasing elevation. When compared with Figure 11a,b, the velocity valley decreases from 0.48 at 0D to 0.58 at 0.3D, and the velocity gradient decreases gradually. The range of wake expansion at 0D is a reference for the wake velocity model, which is used in the microscopic site selection of a flat surface.

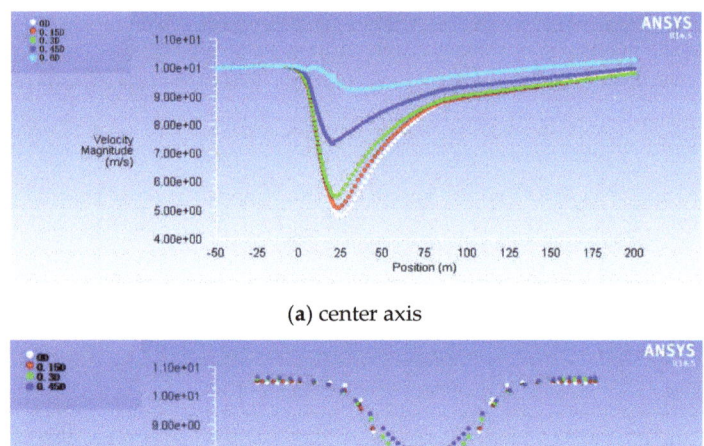

(**a**) center axis

(**b**) the cross-section at 2.5D

Figure 10. Velocity comparison at 0D, 0.15D, 0.3D and 0.45D heights.

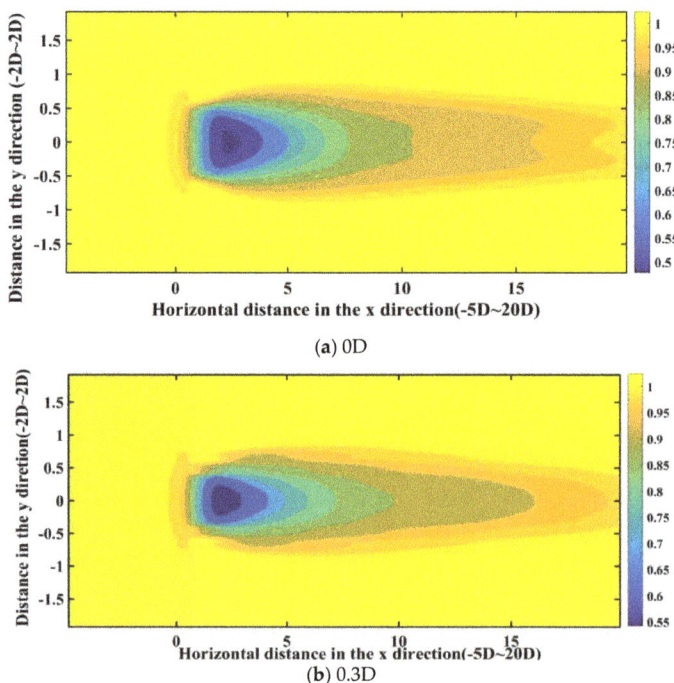

(**a**) 0D

(**b**) 0.3D

Figure 11. Wake velocity clouds on the neutral layer of different heights.

6.3. Turbulence Field Result under the Uniform Inflow Condition

It can be seen from Figure 12 that the initial turbulence intensity of models 1, 5 and 6 is maintained in the free flow region. The result of model 4 shows that the turbulence attenuation is not corrected. In the near wake region of the wind turbine, the turbulence intensity simulated by models 1, 5 and 6 is higher than that of model 4. Notably, the result of model 5 is about 30% higher than that of model 4. With the increasing position, the wake effect starts to weaken, and the turbulence intensity decreases correspondingly. The performance of model 1 is the best, as the result of the modified model is also extremely close to the experimental value with 16% at the peak. As the dissipative term of model 1 is about twice that of the original one, the specific dissipation rate is lower, and the turbulence intensity is larger.

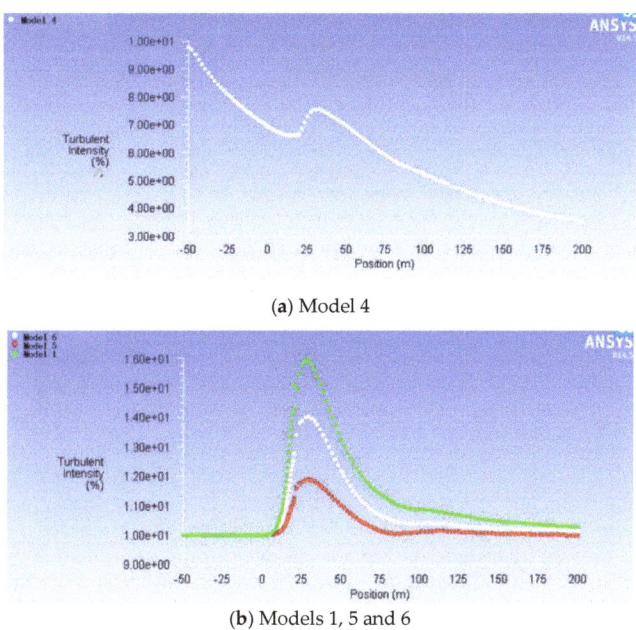

(a) Model 4

(b) Models 1, 5 and 6

Figure 12. The turbulence of different models at the central axis.

As shown in Figure 13, the result of the 2.5D position shows that the standard SST k-ω model underestimates the magnitude of turbulence intensity qualitatively, but the model can well predict the "double peak" effect of turbulence intensity. The "double peak" effect of turbulent intensity weakens gradually with the development and fragmentation of the tip vortex convection–diffusion of wakes, and the shear mixing layer continues to expand, which can be seen from the distribution of turbulence intensity at 4D. Through the comparison of models 1 and 3–6, it can be seen that the modified wind turbine model has no obvious effect on the turbulence field. At the same time, the effect of each turbulence correction term is obvious. The proposed model is more effective than model 5.

The turbulence intensity above the $Z = 0D$ plane is shown in Figure 14. The maximum turbulence intensity is 21.8%. The whole variation of the "double peak" effect can be seen. In the y-direction, the turbulent peak occurs mainly at ±0.5D (the blade tip), and the diffusion range is from −1D to 1D. In the x-direction, a rapid decay state that approaches the inlet turbulence value after 2.5 D is observed.

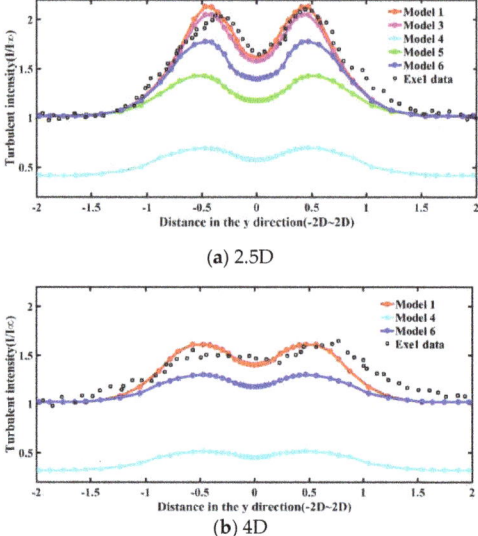

(a) 2.5D

(b) 4D

Figure 13. Turbulence intensity at the different cross-sections of the axial direction.

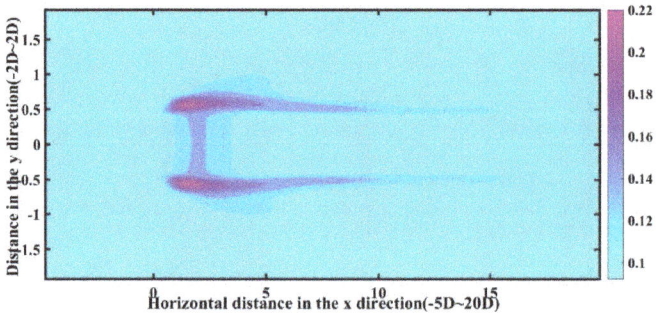

Figure 14. Distribution of turbulence intensity on the z = 0 plane.

6.4. Computation Evaluation

Table 4 shows the computational cost of different schemes.

Table 4. Mesh number for three items.

Average Size	Proposed Scheme (Thousand)	Scheme A [32] (Thousand)	Scheme B [22] (Thousand)
1/3D	30	20	150
1/10D	500	400	900
1/40D	2000	1500	3800

Compared with Denmark's 2D mesh scheme (Scheme A) [25], the mesh amount of our scheme is slightly larger than scheme A. However, it should be noted that scheme A sacrifices the third dimension of mesh modeling. Compared with Xu's half-axis hypothesis 3D scheme (Scheme B) [34], the mesh amount of our scheme is substantially less than scheme B under the same average size, although scheme B uses axisymmetric assumptions to build only a half 3D model.

7. Conclusions

This paper proposes a modified Reynolds-averaged Navier–Stokes (MRANS)-based wind turbine wake model to simulate wake effects. Based on the correction module, the proposed BEM-fuzzy aerodynamic model can amend the inconsistent condition between wake simulation and experiment tests, which affects the simulation of the near wake. For the turbulence model, the turbulence attenuation is effectively avoided by adding the hold source term to ensure the global correctness of boundary conditions. The accuracy of the turbulence intensity distribution is improved by correcting the closure constant and the dissipation term. The turbulence model affects the whole wakes, especially the far wakes.

The simulation results of the velocity field and turbulent field with the proposed approach are consistent with the data of real wind turbines, which verifies the effectiveness of the proposed approach. Furthermore, the computation efficiency is significantly improved by the developed mesh partition method.

Author Contributions: All the authors conceived and designed the study. B.L. and Z.X. (Zengjin Xu) performed the simulation and wrote the manuscript with guidance from Y.L.; Z.X. (Zuoxia Xing) reviewed and revised the manuscript; B.Z., H.C. and B.H. collaborated in the experiment verification and the analysis of results. All authors have read and agreed to the published version of the manuscript.

Funding: This research was funded by the National Natural Science Foundation of China (61703081), the Fundamental Research Funds for the Central Universities (N2004030), Liaoning Provincial Education Department Key Project (LZGD2017039), the Liaoning Revitalization Talents Program (XLYC1801005), and the State Key Laboratory of Alternate Electrical Power System with Renewable Energy Sources (LAPS19005).

Conflicts of Interest: The authors declare no conflict of interest.

References

1. Barthelmie, R.J.; Frandsen, S.T.; Nielsen, M.N.; Pryor, S.C.; Rethore, P.E.; Jørgensen, H.E. Modelling and measurements of power losses and turbulence intensity in wind turbine wakes at Middelgrunden offshore wind farm. *Wind Energy* **2007**, *10*, 517–528. [CrossRef]
2. Shao, Z.; Wu, Y.; Li, L.; Han, S.; Liu, Y. Multiple wind turbine wakes modeling considering the faster wake recovery in overlapped wakes. *Energies* **2019**, *12*, 680. [CrossRef]
3. Tao, S.; Kuenzel, S.; Xu, Q.; Chen, Z. Optimal micro-siting of wind turbines in an offshore wind farm using Frandsen-Gaussian wake model. *IEEE Trans. Power Syst.* **2019**, *34*, 4944–4954. [CrossRef]
4. Wang, H.; Yang, J.; Chen, Z.; Ge, W.; Hu, S.; Ma, Y.; Li, Y.; Zhang, G.; Yang, L. Gain scheduled torque compensation of pmsg-based wind turbine for frequency regulation in an isolated grid. *Energies* **2018**, *11*, 1623. [CrossRef]
5. Hand, B.; Cashman, A.; Kelly, G. A low-order model for offshore floating vertical axis wind turbine aerodynamics. *IEEE Trans. Ind. Appl.* **2017**, *53*, 512–520. [CrossRef]
6. Wang, X.; Gao, W.; Scholbrock, A.; Muljadi, E.; Gevorgian, V.; Wang, J.; Yan, W.; Zhang, H. Evaluation of different inertial control methods for variable-speed wind turbines simulated by fatigue, aerodynamic, structures and turbulence (FAST). *IET Renew. Power Gener.* **2017**, *11*, 1534–1544. [CrossRef]
7. Behrouzifar, A.; Darbandi, M. An improved actuator disc model for the numerical prediction of the far-wake region of a horizontal axis wind turbine and its performance. *Energy Convers. Manag.* **2019**, *18*, 482–495. [CrossRef]
8. Castellani, F.; Astolfi, D.; Mana, M.; Piccioni, E. Investigation of terrain and wake effects on the performance of wind farms in complex terrain using numerical and experimental data. *Wind Energy* **2017**, *20*, 1277–1289.
9. Sedaghatizadeh, N.; Arjomandi, M.; Kelso, R.; Cazzolato, B.; Ghayesh, M.H. The effect of the boundary layer on the wake of a horizontal axis wind turbine. *Energy* **2019**, *18*, 1202–1221. [CrossRef]
10. Reda, S.; Teng, W. A semi-empirical model for mean wind velocity profile of landfalling hurricane boundary layers. *J. Wind Eng. Ind. Aerodyn.* **2018**, *180*, 249–261.
11. Zhang, Y.; Ye, Z.; Li, C. Wind turbine aerodynamic performance simulation based on improved lifting surface free wake method. *Acta Energ. Sol. Sin.* **2017**, *38*, 1316–1323.

12. Zhang, Y.; Fernandez-Rodriguez, E.; Zheng, J.; Zheng, Y.; Zhang, J.; Gu, H.; Zang, W.; Lin, X. A review on numerical development of tidal stream turbine performance and wake prediction. *IEEE Access* **2020**, *8*, 79325–79337. [CrossRef]
13. Laan, M.P.; Sørensen, N.N.; Réthoré, P. An improved k-ε model applied to a wind turbine wake in atmospheric turbulence. *Wind Energy* **2015**, *18*, 889–907. [CrossRef]
14. Almohammadi, K.M.; Ingham, D.B.; Ma, L.; Pourkashanian, M. 2-D-CFD analysis of the effect of trailing edge shape on the performance of a straight-blade vertical axis wind turbine. *IEEE Trans. Sustain. Energy* **2015**, *6*, 228–235. [CrossRef]
15. Yin, X.; Zhao, X. Deep neural learning based distributed predictive control for offshore wind farm using high fidelity LES data. *IEEE Trans. Ind. Electron.* **2020**. [CrossRef]
16. Seim, F.; Gravdahl, A.R.; Adaramola, M.S. Validation of kinematic wind turbine wake models in complex terrain using actual windfarm production data. *Energy* **2017**, *123*, 742–753. [CrossRef]
17. Kleusberg, E.; Mikkelsen, R.F.; Schlatter, P.; Ivanell, S. High-order numerical simulations of wind turbine wakes. *J. Phys. Conf.* **2017**, *854*, 012025. [CrossRef]
18. Barrett, R.; Ning, A. Comparison of airfoil precomputational analysis methods for optimization of wind turbine blades. *IEEE Trans. Sustain. Energy* **2016**, *7*, 1081–1088. [CrossRef]
19. Arteaga-López, E.; Ángeles-Camacho, C.; Bañuelos-Ruedas, F. Advanced methodology for feasibility studies on building-mounted wind turbines installation in urban environment: Applying CFD analysis. *Energy* **2019**, *16*, 181–188. [CrossRef]
20. Liang, H.; Zuo, L.; Li, J.; Li, B.C.; He, Y.; Huang, Q. A wind turbine control method based on Jensen model. In Proceedings of the 2016 International Conference on Smart Grid and Clean Energy Technologies (ICSGCE), Chengdu, China, 19–22 October 2016; IEEE: Piscataway, NJ, USA, 2016.
21. Sørensen, O.; Jacob, B. *Linearised CFD Models for Wakes*; Risø National Laboratory: Roskilde, Denmark, 2011.
22. Johansen, J.; Sørensen, N.N.; Michelsen, J.A. Detached-eddy simulation of flow around the NREL Phase VI blade. *Wind Energy* **2002**, *5*, 185–197. [CrossRef]
23. Sarlak, H.; Meneveau, C.; Sorensen, J.N. Role of subgrid-scale modeling in large eddy simulation of wind turbine wake interactions. *Renew. Energy* **2015**, *77*, 386–399. [CrossRef]
24. Mo, J.O.; Choudhry, A.; Arjomandi, M.; Lee, Y. *Adelaide Research and Scholarship: Large Eddy Simulation of the Wind Turbine Wake Characteristics in the Numerical Wind Tunnel Model*; Elsevier Science Bv: Amsterdam, The Netherlands, 2013.
25. Zhang, B.; Soltani, M.; Hu, W.; Hou, P.; Huang, Q.; Chen, Z. Optimized power dispatch in wind farms for power maximizing considering fatigue loads. *IEEE Trans. Sustain. Energy* **2018**, *9*, 862–871. [CrossRef]
26. Farhan, A.; Hassanpour, A.; Burns, A.; Motlagh, Y.G. Numerical study of effect of winglet planform and airfoil on a horizontal axis wind turbine performance. *Renew. Energy* **2019**, *13*, 1255–1273. [CrossRef]
27. Lee, H.; Lee, D.J. Numerical investigation of the aerodynamics and wake structures of horizontal axis wind turbines by using nonlinear vortex lattice method. *Renew. Energy* **2019**, *13*, 1121–1133. [CrossRef]
28. Tang, D.; Xu, M.; Mao, J.; Zhu, H. Unsteady performances of a parked large-scale wind turbine in the typhoon activity zones. *Renew. Energy* **2020**, *14*, 617–630. [CrossRef]
29. Roggenburg, M.; Esquivel-Puentes, H.A.; Vacca, A.; Evans, H.B.; Garcia-Bravo, J.M.; Warsinger, D.M.; Ivantysynova, M.; Castillo, L. Techno-economic analysis of a hydraulic transmission for floating offshore wind turbines. *Renew. Energy* **2020**, *15*, 1194–1204. [CrossRef]
30. El-Askary, W.A.; Sakr, I.M.; Abdelsalam, A.M.; Abuhegazy, M.R. Modeling of wind turbine wakes under thermally-stratified atmospheric boundary layer. *J. Wind Eng. Ind. Aerodyn.* **2017**, *160*, 1–15. [CrossRef]
31. Richmond-Navarro, G.; Calderón-Muñoz, W.R.; LeBoeuf, R.; Castillo, P. A magnus wind turbine power model based on direct solutions using the blade element momentum theory and symbolic regression. *IEEE Trans. Sustain. Energy* **2017**, *8*, 425–430. [CrossRef]
32. Hansen, M. *Aerodynamics of Wind Turbines*, 3rd ed.; Routledge: Abingdon, UK, 2015.
33. Shen, W.Z.; Zhang, J.H. The actuator surface model: A new Navier–Stokes based model for rotor computations. *J. Sol. Energy Eng.* **2009**, *131*, 284–289. [CrossRef]
34. Parker, M.A.; Soraghan, C.; Giles, A. Comparison of power electronics lifetime between vertical- and horizontal-axis wind turbines. *IET Renew. Power Gener.* **2016**, *10*, 679–686. [CrossRef]
35. Vermeer, L.J.; Sørensen, J.N.; Crespo, A.S. Wind turbine wake aerodynamics. *Prog. Aerosp. Ences* **2003**, *39*, 467–510. [CrossRef]

36. Regodeseves, P.G.; Morros, C.S. Unsteady numerical investigation of the full geometry of a horizontal axis wind turbine: Flow through the rotor and wake. *Energy* **2020**, *202*, 117674. [CrossRef]
37. Boris, C. Wind Resource Accessment in Complex Terrain by Wind Tunnel Modelling. Ph.D. Thesis, Université d'Orléans, Orléans, France, 2012.
38. Taylor, G.J. *Wake Measurements on the Nibe Wind Turbines in Denmark*; National Power-Technology and Environment Centre: Leatherhead, UK, 1990.

© 2020 by the authors. Licensee MDPI, Basel, Switzerland. This article is an open access article distributed under the terms and conditions of the Creative Commons Attribution (CC BY) license (http://creativecommons.org/licenses/by/4.0/).

MDPI
St. Alban-Anlage 66
4052 Basel
Switzerland
Tel. +41 61 683 77 34
Fax +41 61 302 89 18
www.mdpi.com

Energies Editorial Office
E-mail: energies@mdpi.com
www.mdpi.com/journal/energies

www.ingramcontent.com/pod-product-compliance
Lightning Source LLC
LaVergne TN
LVHW070458100526
838202LV00014B/1750